Smart Grid Planning and Implementation

Smart Grid Planning and Implementation

Clark W. Gellings, P.E.

River Publishers

Routledge
Taylor & Francis Group
LONDON AND NEW YORK

Published 2020 by River Publishers
River Publishers
Alsbjergvej 10, 9260 Gistrup, Denmark
www.riverpublishers.com

Distributed exclusively by Routledge
4 Park Square, Milton Park, Abingdon, Oxon OX14 4RN
605 Third Avenue, New York, NY 10158

First published in paperback 2024

Library of Congress Cataloging-in-Publication Data

Gellings, Clark W.
Smart Grid Planning and Implementation/ Clark W. Gellings.
pages cm
Includes bibliographical references and index.
ISBN 0-88173-750-X (alk. paper) -- ISBN 978-1-4987-4781-3 (Taylor & Francis distribution : alk. paper) -- ISBN 978-8-7702-2327-0 (electronic) -- 1. Smart power grids. I. Title.

TK3105.G45 2015
333.793'20684--dc23

2015019323

Smart Grid Planning and Implementation / Gellings, Clark W.
First published by Fairmont Press in 2015.

Routledge is an imprint of the Taylor & Francis Group, an informa business

Publisher's Note
The publisher has gone to great lengths to ensure the quality of this reprint but points out that some imperfections in the original copies may be apparent.

ISBN: 978-0-88173-750-9 (The Fairmont Press, Inc.)
ISBN: 978-8-7702-2327-0 (online)

ISBN: 978-1-4987-4781-3 (hbk)
ISBN: 978-87-7004-483-7 (pbk)
ISBN: 978-0-13-873957-7 (ebk)
ISBN: 978-1-003-15196-8 (eBook+)

Table of Contents

Chapter 1

What is the Smart Grid?

WHAT IS THE SMART GRID?

The Smart Grid as defined here is based upon the descriptions found in the Energy Independence and Security Act of 2007. The term "Smart Grid" refers to a modernization of the electricity delivery system so it monitors, protects and automatically optimizes the operation of its interconnected elements—from the central and distributed generator through the high-voltage network and distribution system, to industrial users and building automation systems, to energy storage installations and to end-use consumers and their thermostats, electric vehicles, appliances and other household devices.

The Smart Grid will be characterized by a two-way flow of electricity and information to create an automated, widely distributed energy delivery network. It incorporates into the grid the benefits of distributed computing and communications to deliver real-time information and enable the near-instantaneous balance of supply and demand at the device level.

SMART GRID CHARACTERISTICS: DRIVERS AND OPPORTUNITIES

The definition of the smart grid builds on the work done by Electric Power Research Institute (EPRI), in EPRI's Intelligrid Program, (intelligrid.epri.com), in the Modern Grid Initiative (MGI) (NETI 2007), and in the GridWise Architectural Council (GWAC) (gridwise.org). These considerable efforts have developed and articulated the vision statements, architectural principles, barriers, benefits, technologies and applications, policies, and frameworks that help define what the Smart Grid is.

Smart Grid Benefits

Smart Grid benefits can be categorized into 5 types:

- **Power reliability and power quality**. The Smart Grid provides a reliable power supply with fewer and briefer outages, "cleaner" power, and self-healing power systems, through the use of digital information, automated control, and autonomous systems.
- **Safety and cyber security benefits**. The Smart Grid continuously monitors itself to detect unsafe or insecure situations that could detract from its high reliability and safe operation. Higher cyber security is built in to all systems and operations including physical plant monitoring, cyber security, and privacy protection of all users and customers.
- **Energy efficiency benefits**. The Smart Grid is more efficient, providing reduced total energy use, reduced peak demand, reduced energy losses, and the ability to induce end-user use reduction instead of new generation in power system operations.
- **Environmental and conservation benefits**. The Smart Grid is "green." It helps reduce greenhouse gases (GHG) and other pollutants by reducing generation from inefficient energy sources, supports renewable energy sources, and enables the replacement of gasoline-powered vehicles with plug-in electric vehicles.
- **Direct financial benefits**. The Smart Grid offers direct economic benefits. Operations costs are reduced or avoided. Customers have pricing choices and access to energy information. Entrepreneurs accelerate technology introduction into the generation, distribution, storage, and coordination of energy.

Stakeholder Benefits

The benefits from the Smart Grid can be categorized by the three primary stakeholder groups:

- **Consumers**. Consumers can balance their energy consumption with the real time supply of energy. Variable pricing will provide consumer incentives to install their own infrastructure that supports the Smart Grid. Smart grid information infrastructure will support additional services not available today.
- **Utilities**. Utilities can provide more reliable energy, particularly

during challenging emergency conditions, while managing their costs more effectively through efficiency and information.

- **Society**. Society benefits from more reliable power for governmental services, businesses, and consumers sensitive to power outage. Renewable energy, increased efficiencies, and PHEV support will reduce environmental costs, including carbon footprint.

A benefit to any one of these stakeholders can in turn benefit the others. Those benefits that reduce costs for utilities lower prices, or prevent price increases, to customers. Lower costs and decreased infrastructure requirements ameliorate social justice concerns around energy to society. Reduced costs increase economic activity which benefits society. Societal benefits of the Smart Grid can be indirect and hard to quantify, but cannot be overlooked.

Other stakeholders also benefit from the Smart Grid. Regulators can benefit from the transparency and audit-ability of Smart Grid information. Vendors and integrators benefit from business and product opportunities around Smart Grid components and systems.

Modern Grid Initiative (MGI) Smart Grid Characteristics

Smart Grid characteristics are prominent attributes, behaviors, or features that help distinguish the grid as "smart." The MGI developed a list of seven behaviors that define the Smart Grid. Those working in each area of the Smart Grid can evaluate their work by reference to these behaviors. These behaviors match those defined by similar initiatives and workgroups.

The behaviors of the Smart Grid as defined by MGI are:

- **Enable Active Participation by Consumers**. The Smart Grid motivates and includes customers, who are an integral part of the electric power system. The smart grid consumer is informed, modifying the way they use and purchase electricity. They have choices, incentives, and disincentives to modify their purchasing patterns and behavior. These choices help drive new technologies and markets.

- **Accommodate All Generation and Storage Options**. The Smart Grid accommodates all generation and storage options. It supports large, centralized power plants as well as Distributed Energy Re-

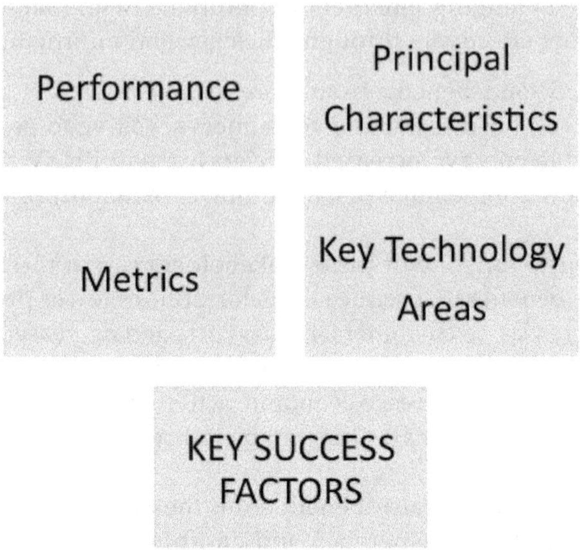

Figure 1-1. MGI's Principle Characteristics are Part of Their Smart Grid System Vision for Measuring Success (Source: EPRI Report to NIST, 2009)

sources (DER). DER may include system aggregators with an array of generation systems or a farmer with a windmill and some solar panels. The Smart Grid supports all generation options. The same is true of storage, and as storage technologies mature, they will be an integral part of the overall Smart Grid solution set.

- **Enable New Products, Services, and Markets**. The Smart Grid enables a market system that provides cost-benefit tradeoffs to consumers by creating opportunities to bid for competing services. As much as possible, regulators, aggregators and operators, and consumers can modify the rules of business to create opportunity against market conditions. A flexible, rugged market infrastructure exists to ensure continuous electric service and reliability, while also providing profit or cost reduction opportunities for market participants. Innovative products and services provide 3rd party vendors opportunities to create market penetration opportunities and consumers with choices and clever tools for managing their electricity costs and usage.
- **Provide Power Quality for the Digital Economy**. The Smart Grid provides reliable power that is relatively interruption-free. The

power is "clean" and disturbances are minimal. Our global competitiveness demands relatively fault-free operation of the digital devices that power the productivity of our 21st century economy.

- **Optimize Asset Utilization and Operate Efficiently**. The Smart Grid optimizes assets and operates efficiently. It applies current technologies to ensure the best use of assets. Assets operate and integrate well with other assets to maximize operational efficiency and reduce costs. Routine maintenance and self-health regulating abilities allow assets to operate longer with less human interaction.
- **Anticipate and Respond to System Disturbances (Self-heal)**. The Smart Grid independently identifies and reacts to system disturbances and performs mitigation efforts to correct them. It incorporates an engineering design that enables problems to be isolated, analyzed, and restored with little or no human interaction. It performs continuous predictive analysis to detect existing and future problems and initiate corrective actions. It will react quickly to electricity losses and optimize restoration exercises.
- **Operate Resiliently to Attack and Natural Disaster**. The Smart Grid resists attacks on both the physical infrastructure (substations, poles, transformers, etc.) and the cyber-structure (markets, systems, software, communications). Sensors, cameras, automated switches, and intelligence are built into the infrastructure to observe, react, and alert when threats are recognized within the system. The system is resilient and incorporates self-healing technologies to resist and react to natural disasters. Constant monitoring and self-testing are conducted against the system to mitigate malware and hackers.

SMART GRID CHALLENGES

The Smart Grid poses many procedural and technical challenges as we migrate from the current grid with its one-way power flows from central generation to dispersed loads, toward a new grid with two-way power flows, two-way and peer to peer customer interactions, and distributed generation. These challenges cannot be taken lightly—the Smart Grid will entail a fundamentally different paradigm for energy generation, delivery, and use.

Procedural Challenges

The procedural challenges to the migration to a smart grid are enormous, and all need to be met as the Smart Grid evolves:

- **Broad Set of Stakeholders**. The Smart Grid will affect every person and every business in the United States. Although not every person will participate directly in the development of the Smart Grid, the need to understand and address the requirements of all these stakeholders will require significant efforts.

- **Complexity of the Smart Grid**. The Smart Grid is a vastly complex machine, with some parts racing at the speed of light. Some aspects of the Smart Grid will be sensitive to human response and interaction, while others need instantaneous, automated responses. The smart grid will be driven by forces ranging from financial pressures to environmental requirements.

- **Transition to Smart Grid**. The transition to the Smart Grid will be lengthy. It is impossible (and unwise) to advocate that all the existing equipment and systems to be ripped out and replaced at once. The smart grid supports gradual transition and long coexistence of diverse technologies, not only as we transition from the legacy systems and equipment of today, but as we move to those of tomorrow. We must design to avoid unnecessary expenses and unwarranted decreases in reliability, safety, or cyber security.

- **Ensuring Cyber Security of Systems**. Every aspect of the Smart Grid must be secure. Cyber security technologies are not enough to achieve secure operations without policies, on-going risk assessment, and training. The development of these human-focused procedures takes time—and needs to take time—to ensure that they are done correctly.

- **Consensus on Standards**. Standards are built on the consensus of many stakeholders over time; mandating technologies can appear to be an adequate short cut. Consensus-based standards deliver better results over.

- **Development and Support of Standards**. The open process of developing a standard benefits from the expertise and insights of a broad constituency. The work is challenging and time consuming but yields results more reflective of a broad group of stakeholders, rather than the narrow interests of a particular stakeholder group.

Ongoing engagement by user groups and other organizations enables standards to meet broader evolving needs beyond those of industry stakeholders. Both activities are essential to the development of strong standards.

- **Research and Development**. The smart grid is an evolving goal; we cannot know all that the Smart Grid is or can do. The smart grid will demand continuing R&D to assess the evolving benefits and costs, and to anticipate the evolving requirements.

Technical Challenges to Achieving the Smart Grid

Technical challenges include the following:

- **Smart equipment**. Smart equipment refers to all field equipment which is computer-based or microprocessor-based, including controllers, remote terminal units (RTUs), intelligent electronic devices (IEDs). It includes the actual power equipment, such as switches, capacitor banks, or breakers. It also refers to the equipment inside homes, buildings and industrial facilities. This embedded computing equipment must be robust to handle future applications for many years without being replaced.

- **Communication systems**. Communication systems refer to the media and to the developing communication protocols. These technologies are in various stages of maturity. The smart grid must be robust enough to accommodate new media as they emerge from the communications industries and while preserving interoperable, secured systems.

- **Data management**. Data management refers to all aspects of collecting, analyzing, storing, and providing data to users and applications, including the issues of data identification, validation, accuracy, updating, time-tagging, consistency across databases, etc. Data management methods which work well for small amounts of data often fail or become too burdensome for large amounts of data—and distribution automation and customer information generate lots of data. Data management is among the most time-consuming and difficult task in many of the functions and must be addressed in a way that will scale to immense size.

- **Cyber Security**. Cyber security addresses the prevention of damage to, unauthorized use of, exploitation of, and, if needed, the restoration of electronic information and communications systems and services (and the information contained therein) to ensure confidentiality, integrity, and availability.

- **Information/data privacy**. The protection and stewardship of privacy is a significant concern in a widely interconnected system of systems that is represented by the Smart Grid. Additionally, care must be taken to ensure that access to information is not an all or nothing at all choice since various stakeholders will have differing rights to information from the Smart Grid.
- **Software applications**. Software applications refer to programs, algorithms, calculations, and data analysis. Applications range from low level control algorithms to massive transaction processing. Application requirements are becoming more sophisticated to solve increasingly complex problems, are demanding ever more accurate and timely data, and must deliver results more quickly and accurately. Software engineering at this scale and rigor is still emerging as a discipline. Software applications are at the core of every function and node of the Smart Grid.

THE SMART GRID CONCEPTUAL MODEL

The Smart Grid Conceptual Model is a diagram and description that are the basis for discussing the characteristics, uses, behavior, interfaces, requirements and standards of the Smart Grid. This does not represent the final architecture of the Smart Grid; rather it is a tool for describing, discussing, and developing that architecture. The conceptual model provides a context for analysis of interoperation and standards, both for the rest of this document, and for the development of the architectures of the Smart Grid. The top level of the conceptual model is shown in Figure 1-2.

The conceptual model consists of several *domains*, each of which contains many *applications* and *actors* that are connected by *associations*, which have *interfaces* at each end:

- **Actors** may be devices, computer systems or software programs and/or the organizations that own them. Actors have the capability to make decisions and exchange information with other actors through interfaces.

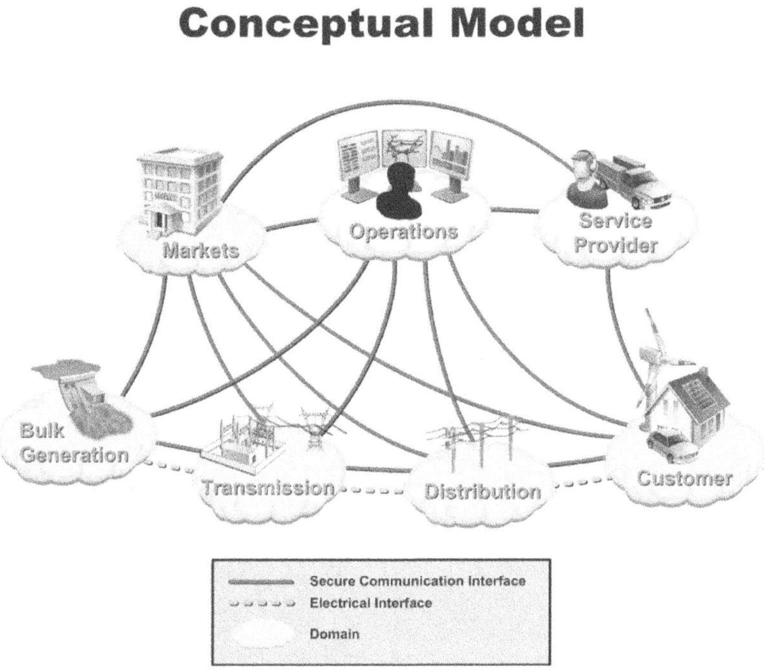

Figure 1-2. Smart Grid Conceptual Model—Top Level (Source: EPRI Report to NIST, 2009)

- **Applications** are the tasks performed by the actors within the domains. Some applications are performed by a single actor, others by several actors working together.
- **Domains** group actors to discover the commonalities that will define the interfaces. In general, actors in the same domain have similar objectives. Communications within the same domain may have similar characteristics and requirements. Domains may contain other domains.
- **Associations** are logical connections between actors that establish bilateral relationships. At each end of an association is an interface to an actor.
- **Interfaces** show either electrical connections or communications connections. In, the electrical interfaces are shown as yellow lines and the communications interfaces are shown in blue. Each of these

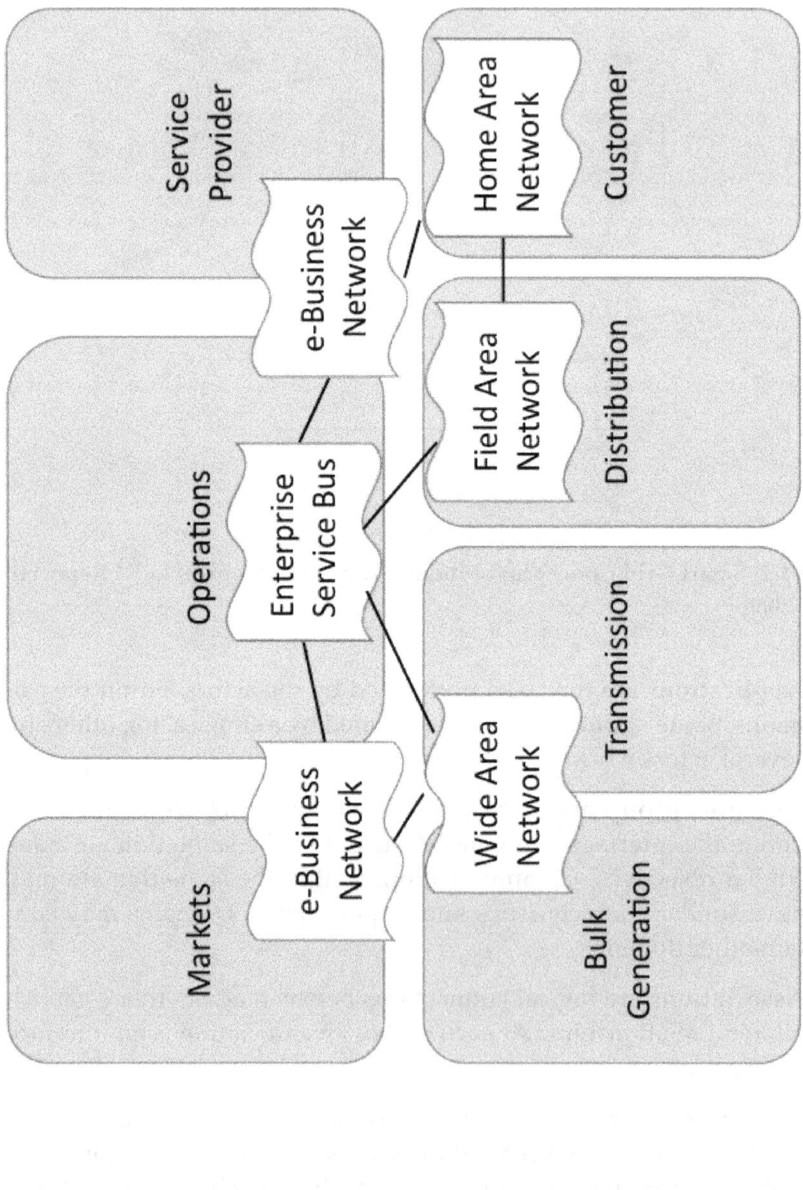

Figure 1-3. Smart Grid Generation and Transmission Domain (EPRI)

interfaces may be bi-directional. Communications interfaces represent an information exchange between two domains and the actors within; they do not represent physical connections. They represent logical connections in the smart grid information network interconnecting various domains (as shown in Table 1-1). The domains of the Smart Grid are listed briefly in Table 1-1. In Figure 1-6, domains are shown as clouds.

Table 1-1. Domains in the Smart Grid Conceptual Model

Domain	Actors in the Domain
Customers	The end users of electricity. May also generate, store, and manage the use of energy. Traditionally, three customer types are discussed, each with its own domain: home, commercial/building, and industrial.
Markets	The operators and participants in electricity markets
Service Providers	The organizations providing services to electrical customers and utilities
Operations	The managers of the movement of electricity
Bulk Generation	The generators of electricity in bulk quantities. May also store energy for later distribution.
Transmission	The carriers of bulk electricity over long distances. May also store and generate electricity.
Distribution	The distributors of electricity to and from customers. May also store and generate electricity.

It is important to note that domains are NOT organizations. For instance, an ISO or RTO may have actors in both the Markets and Operations domains. Similarly, a distribution utility is not entirely contained within the Distribution domain—it is likely to also contain actors in the Operations domain, such as a Distribution Management System, and in the Customer domain, such as meters

ADDITIONAL CHALLENGES

Work Force Issues

The utility work force is undergoing a revolution. One-half of the 500,000 to 600,000 utility workers will be eligible to retire in the next five years. They need to be replaced with an appropriately trained and motivated work force. Introducing Smart Grid technologies requires employees with greater skills to support the implementation, maintenance, and operation of the systems with digital components. Accomplishing this when it is already difficult to get highly skilled employees with technical experience will be challenging. In addition, these staffing requirements must be met at the same time as staffs are being reduced to hold the line on costs (Lave, 2007)

Outage management systems can be used to save costs associated with incorrect outage reports by verifying power outages at customer facilities. (PECO estimates that they avoided 7,500 crew dispatches in 2005 because they were able to see that the customer-reported outage was inaccurate.)

Regulatory Challenges to Achieving the Smart Grid

Smart Grid technologies offer unprecedented challenges to regulators in encouraging and adjudicating decisions regarding Smart Grid investments. Power systems have largely operated without "smart" technology for decades. As long as reliability levels have been maintained (the lights were still on) and costs were low (rates have been essentially flat for decades), it is conceptually difficult to understand how a basket of digital-based technologies can improve the current value of today's power system.

Among the regulatory hurdles regulated utilities face are the potential to lose sales and revenue from improved end-use energy efficiency. In some cases, they may be faced with reducing rates as part of seeking approval for Smart Grid investments and, subsequently, losing revenue. Often regulated utilities are reluctant to embrace energy-efficiency measures on the distribution system or those involving end-use customers for fear of the negative business impact in the regulated environment.

An additional regulatory challenge is to understand the incremental value of Smart Grid investments. Often, for logical reasons, Smart Grid technologies are implemented in stages, with each stage requiring a business plan for regulators to approve. However, the benefits of many of the Smart Grid efforts come from the synergistic applications of a portfolio of Smart Grid technologies.
This issue was summarized nicely by the Illinois Commerce Commission in a 2010 report (ISSGC, 2010):

> "The issue of smart grid cost recovery has been a matter of controversy and litigation for several years. Disagreements exist about whether recov-

> ery of a utility's smart grid costs should be restricted to the 'traditional' rate-base method, or whether a 'non-traditional' method (e.g., 'rider' recovery) should be used.
>
> Some stakeholders are concerned that utility proposals for cost recovery of smart grid investments would lead to significantly higher monthly bills and a shift in the risk of investment from utilities to ratepayers. Others believe that non-traditional cost recovery would be essential to accelerate deployment of smart grid technologies."

Finally, regulators and utilities are accustomed to utility asset life in the range of 30 to 50 years and business plans and rate cases for regulated utilities are typically based on that supposition. When digital equipment is involved—particularly at the early stages of evolution—the depreciation rates for a significant amount of capital investment may be 5 to 15 years. Appropriate depreciation rates must be allowed in order to pay for asset renewal without increasing costs more than necessary. In short, a forward view of technology must be embraced by utilities and regulators in order for the Smart Grid to be successful.

Drivers Stimulating Smart Grid Investment

A wide variety of policies and economic trends have begun to stimulate and drive U.S. investment in Smart Grid technology, including:

- **The Energy Independence and Security Act (EISA) of 2007** established national policy for grid modernization, created new federal committees, defined their roles and responsibilities, and provided incentives for investment.
- **The American Recovery and Reinvestment Act of 2009** provided more than $3.4 billion in stimulus funding for Smart Grid technology development and demonstration, plus $615 for Smart Grid storage. In October, 2009, 100 Smart Grid Investment Grants were awarded. These were 50/50 matching grants over a three-year period leading to an infusion of $7-8 billion in Smart Grid investments that are specifically targeted at projects that can be emulated by others.
- **Renewable portfolio standards** have been established in 30 states plus the District of Columbia stimulating rapid expansion of renewable technology and accelerating the need for Smart Grid tech-

nology for grid integration. EPRI's Prism analysis anticipates 135 GW of renewables by 2030 (EPRI 1020389). A number of states have also enacted policies to address specific environmental concerns.

- **Smart Grid interoperability standards,** called for by the EISA, moved forward with NIST's release in September, 2009, of a roadmap for interoperability standards. The starting point for NIST was EPRI's draft interim roadmap report released in June, 2009. While not a driver of Smart Grid investment of itself, these standards facilitate Smart Grid deployment.

- **Critical Infrastructure Protection (CIP) Cyber Security Standards,** maintained by the North American Electric Reliability Corporation (NERC) and approved by the Federal Energy Regulatory Commission (FERC) in 2006, are intended to ensure the protection of the critical cyber assets that control or effect the reliability of North America's bulk electric systems. The CIP Cyber Security Standards are mandatory and enforceable across all users, owners, and operators of the bulk power system (LogRhythm, 2009).

- **Demand response programs** have accelerated. The Federal Energy Regulatory Commission's (FERC's) 2008 Demand Response (DR) and Automated Metering Infrastructure (AMI) Survey indicated that advanced metering penetration reached 4.7% of total meters, up from less than 1% in 2006, and that 8% of U.S. customers were currently involved in DR programs. DR continues to gain ground through state legislative initiatives and utility regulation. FERC's "A National Assessment of Demand Response Potential—Staff Report" in June 2009 and FERC's staff report "National Action Plan on Demand Response" in June 2010 highlight the potential (FERC, 2009 and 2010).

- **Market demand** for Smart Grid technologies has drawn the interest of every major information technology company, from Cisco and Intel, to Google, IBM, and Microsoft who want to participate in one of the most attractive business opportunities of the future. The Smart Grid is viewed as the market equivalent of the Internet in terms of its trillion-dollar potential worldwide. It is the enabling infrastructure for the accelerated deployment of electric vehicles and plug-in hybrid electric vehicles.

- **Venture capital** is also entering the Smart Grid domain in a large way, promising to bring faster and more concentrated technical innovation in the areas of Automated Metering Infrastructure (AMI), communications and network technologies. More than $1 billion in VC funding has been extended to key startups, typified by GridPoint and Silver Spring Networks (green tech media, 2010).
- **Smart Grid roadmaps** are being developed by many electric utilities to optimize their investment strategy going forward. Objectives and starting points can be quite different from company to company, and the optimal pathway difficult to assess. EPRI has been working with SCE, FirstEnergy, and SRP, among others, to create roadmaps for Smart Grid investment. What is critical today is reliable data on benefits from Smart Grid demonstrations. Every effort must be made to measure the actual benefits realized through Smart Grid demonstrations as projects are deployed.
- **National transmission corridors** have been identified. The Energy Policy Act of 2005 authorized the DOE to conduct national electric transmission congestion studies and to designate National Corridors if appropriate. In 2007, DOE designated a Mid-Atlantic National Corridor and a Southwest Area National Corridor (DOE, 2007).
- **Outage prevention** becomes increasingly important in an information-service-based economy. There have been five major blackouts in the last 40 years, three of which occurred in the last decade. The Northeast blackout of 2003 resulted in an estimated $7 to $10 billion in losses to the region. Less disruptive but more pervasive power quality problems are estimated to now cost the U.S. $119 to $188 billion per year (EPRI 1006274).

A BRIGHT FUTURE?

There is a growing sentiment among industry observers in the western world that the electric utility as we know it as an institution is nearing its end of life. In particular, investor owned utilities are most at risk. Their business model allowed them to earn a fair, albeit regulated, return on assets they invested in order to provide electricity. For many decades demand grew and investors eagerly stepped forward to supply the needed capital to build power plants, transmission systems and distribution infrastructure.

Now that the demand for electricity is declining in many places and consumers are beginning to generate some of their own power from such as photovoltaic power systems there is some relative uncertainty that investment dollars will manifest themselves. As recently noted by Peter Kelley-Detwiler in Forbes (7/19/2013) "While not a big deal today…it will become noticeable in five years."

WILL ELECTRIC SERVICE GO THE WAY OF THE TELECOM INDUSTRY?

The telecommunications industry transformed from a capital-intensive, technology-focused model to a user-centric, service-delivery model. This transformation was largely enabled by Internet Protocol- (IP-) based technology and the introduction of new competitors which allowed telecom services to be provided without use of the legacy network owned by the original telephone providers. The Telecommunications Act of 1996 let any communications business compete in any market against any other. It was a combination of the Telecommunications Act and the evolution of digital technology which enabled the transformation of telecommunication industry. In today's market, a consumer can buy telecom services from traditional landline providers, a number of different cell providers, or cable TV companies. In some cases, these companies sell flat-rate access and do not charge for the volume of transmission by metering minutes or bytes.

Some have begun to question whether this is the end of paying for electricity by the kilowatt-hour. Urban Keussen, Senior Vice President of Technology and Innovation at E.ON suggested in a recent forum that "we'll see something similar to telecom." Instead of paying per minute or byte, many phone plans have moved to flat fees for services. (Greentech Media 2013)

A telecom-like transition cannot happen to the power sector for several reasons:

1. There is an absence of enabling technology which could parallel the grid. The telecom transition resulted in large part to the establishment of a completely separate system—cell networks which could displace land lines while providing the added feature of portability. There are cases where a local generation system may

be economical, but overall from a societal perspective, an interconnected system is the lowest cost electric system.

2. The telecom system is also an electrical system—however, the energy levels used in sending and receiving bytes is nearly insignificant compared to the massive levels in an electrical system. In addition, messages and digital files can be broken into groups, called packets. Electrical energy is transmitted not by the movement of electrons, but by electromagnetic vibrations. These vibrations require wires—substantial wires—to act as pathways. Electrical energy cannot be broken into packets and directed to take different pathways.
3. Substantially reducing power sector CO_2 emissions requires harvesting bulk wind and solar power from regions where there is a limited population and low electrical demand, and transmitting it to regions where electricity is needed. A grid is needed to provide this transmission.
4. While the cost and overall impact of communicating a packet of digits is minimal, the equivalent delivery of kilowatt hours has substantial cost and associated environmental impact. Each kilowatt hour is substantially more valuable than a packet and must be metered so as to equitably charge for its production. In addition, studies have shown that when electricity remains unmetered (such as in some multi-family apartment buildings), consumers tend to be somewhat wasteful.

An example of a partial transformation in one industry which was successful is the adoption of the ATM (automated teller machine) network. Today, ATM machines have made banking completely distributed because of a secure, intelligent and robust infrastructure than seamlessly connects ATM terminals with central banks.

Much like the ATM "industry," a robust and intelligent grid—A Smart Grid—can enable the most effective, low-cost use of resources.

FUTURE OPPORTUNITIES WHICH THE SMART GRID ENABLES

Developments in electric power systems are stimulating the integration of a communications system throughout much of the power system. Once that system is present, connectivity to the ultimate consumers will be enhanced with communications. This enhancement will allow enhanced electricity services (e.g. added billing information or real-time pricing); and services related to electricity (e.g., home security or appliance monitoring).

The integration of the electric and communications infrastructures with consumer facilities and equipment opens a wide variety of opportunities to enhance consumer energy services, to provide the consumer with access to a variety of new value-added services while enhancing the performance and operation of the power systems. Communications with consumer buildings and equipment has the potential to profoundly impact the future of the nation's energy systems as well as to improve energy efficiency.

A number of technological trends are evolving to support this:

- Society is becoming fully digital using all manner of networked electric/electronic devices, ultimately each individual may be functioning within an invisible, portable "information field." (Galvin 2005) Continuous and ubiquitous computing will become universally accessible throughout the U.S. Wireless internet access, available almost everywhere, will combine with the proliferation of inexpensive wireless computing platforms and specialized appliances to dramatically transform work and entertainment for many Americans.
- As a result of the previous trend, the number and diversity of portable electronic devices carried by the average individual will increase.
- Advances in genetics, pharmacology, and biotechnology will continue to expand the possibilities for medical treatment and human performance enhancement. The always-on, always-connected digital environment will enable continuous remote health monitoring and diagnostics.
- Role of nanotechnology: Will nanotech-based health monitoring and diagnostics become available (and widely accepted) within the next twenty years?
- Robotics and healthcare: Will robotics-based assistive technologies for the aging provide more cost-effective options than pharmacology? How will the demand for, and cost of, new therapies further exacerbate social tensions over healthcare costs?

- Telecommuting: Will universal broadband access and public policies that encourage telecommuting enable and encourage U.S. businesses increasingly to outsource knowledge-based work to domestic "contractors" without creating adverse, globally competitive consequences?
- Smart vehicles: Will "smart" vehicles and smart roads provide solutions for urban congestion and commuting?
- Alternatives to travel: Will advances in "tele-immersion," enabling a more robust simulation of an individual's presence at a distance than video conferencing, emerge as a practical alternative to business travel?
- Decentralized fabrication: Will nanotechnology give rise to decentralized fabrication, where modestly sized devices can produce a wide range of physical artifacts to spec and on demand?
- Bio-engineering: Will bioengineering, including the ability to synthesize functionally specific genomes, enable new methods of high-efficiency matter/energy conversion? Will bio-engineered organisms improve renewable energy production and utilization? Will genetically engineered organisms transform pollutants and waste products into commercially viable raw materials?
- Local Energy Resources: Will be integrated into residential, commercial and public structures, reducing the external energy demands of these structures with communications interfaces.

In 1997 the author asked the following questions at an Edison Electric Institute (EEI) Conference: "Is your Refrigerator on the Internet" (Gellings 1997). While this seems still to be largely a rhetorical question, it is possible that the near future, Internet communications will be built into many everyday appliances. Cellular phones now have built-in Web browsing capabilities. Security companies leverage the Internet (ADT-pulse)* as do garage door openers (Lift Master)† and smart thermostats (NEST)¶. Several companies are considering providing customers with

*ADT Pulse is a WiFi enabled security system
†LiftMaster is a WiFi enabled garage door opener
¶NEST is the Nest Learning Thermostat Introduced by NEST + Lab is sensor-driven, Wi-Fi enabled programmable thermostat.

remote diagnostic services for appliances, saving customers time and money or repair personnel dispatches. Appliances with built-in Internet communications would be able to provide their owners with health assessments via Internet Appliances; in effect, a dishwasher could tell its owner it's not feeling well. By delivering energy prices directly to customers' appliances, an energy company could also help customers save money by using appliances at the most cost-effective time of day. Merchandising tie-ins will be possible where a service provider will be able to inform mass market customers of new technology availability, price, delivery schedules, estimated annual savings, and probable appliance payback period.

The Smart Grid will enable consumer portal (CP) which represents the technology, data standards, and protocols that enable the two-way communication of electronic messages which consumer-owned networks and intelligent equipment and provides a pathway to the full development and implementation of a variety of energy-related services. Value can be added by leveraging the ability to communicate with consumer owned energy management systems and intelligent energy consuming equipment.

The consumer portal concept can be thought of as part "portal" part "gateway" and part "smart electric meter." The consumer portal concept provides a view into consumer facilities and carries the definition further to include communications with energy management systems and end-use subsystems and equipment. The CP briefly described, is a device or set of devices that enable intelligent equipment within consumer facilities to communicate two ways with remote systems over wide-area network. The CP can perform the functions of a communications gateway that physically and logically inter-connects a wide-area access network with a consumer's local network. The CP has locally available computing resources and can support local monitoring, data processing, management, and storage.

"Internet-ready" appliances have the potential to radically change the way power is sold to the mass market. When low cost communications are widely available, retailers such as Circuit City, or even appliance manufacturers such as Whirlpool, will be able to bundle energy with each appliance they sell, effectively aggregating a refrigerator, hot water heater or air conditioner load on a nationwide basis.

HAN EXPANSION IN CALIFORNIA

The expansion of HAN in California is an example of the interest in devices which could use a consumer portal.

The California market for Home Areas Network (HAN) devices and system is currently addressed mainly by 3rd party market players ranging from solar installers, cable companies, to consumer devices manufacturers. The market trajectory is being driven by California Public Utilities Commission (CPUC) regulation to create competitive, third-party retail market that allows consumers to choose and deploy their own HAN devices independent of the utilities. IN parallel, the CPUC is driving implementation of the US Federal Green Button Initiative to make AMI date to approved 3rd parties to accelerate a third-party market for customer energy services independent of the utilities.

California is going the route of allowing 3rd partied to provide energy services using AMI data and HAN systems. The main driver for Home Energy Management Systems is consume references for mobile apps and control. Low cost devices are transforming it into a consumer market. CA utilities are primarily interested in HAN from the viewpoint of using market available technologies for low cost demand response.

Home Area Network (HAN) systems have transitioned in the last five years from bleeding edge technology to commercially viable products. In the process, the business model has evolved from being predominantly early stage companies working through utilities to consumer products provided by larger, well established companies.

PRESSING ISSUES AND OPPORTUNITIES

As described earlier in this chapter, there are a number of pressing issues and timely opportunities currently facing the electricity sector. Several of these are highlighted in the remaining pages of this chapter.

The Rise of Distributed Resources

As described in outlining the need for a smart grid, the electric power system is changing rapidly with the rise of distributed energy resources (DER), particularly photovoltaic power generation (PV) on rooftops and in larger arrays connected to the distribution system. DER already has noticeable impact on electricity sales and, in some cases, the operation of the power system. Figure 1-4 illustrates the expected growth of rooftop PV installations in Southern California Edison's (SCE) Service territory. By 2018, SCE predicts that is will have more than 1600 MW of rooftop capacity on its system.

Customer-owned generation options—specifically PV and combined-heat and power (CHP)—are increasingly expected to meet future demand for electricity. Penetration levels could well exceed those of today. Currently, the trend is from conventional to renewable resources, which also can be seen as moving from dispatchable fixed-output technologies to less predictable variable-output or customer-controlled technologies. Through a combination of technological improvements, policy interventions, and consumer choices in technology and service, the role of DER is likely to become more important in the future. The obstacles to this expansion are the potential for instability and congestion on the grid, as well as the resultant increasingly volatile markets.

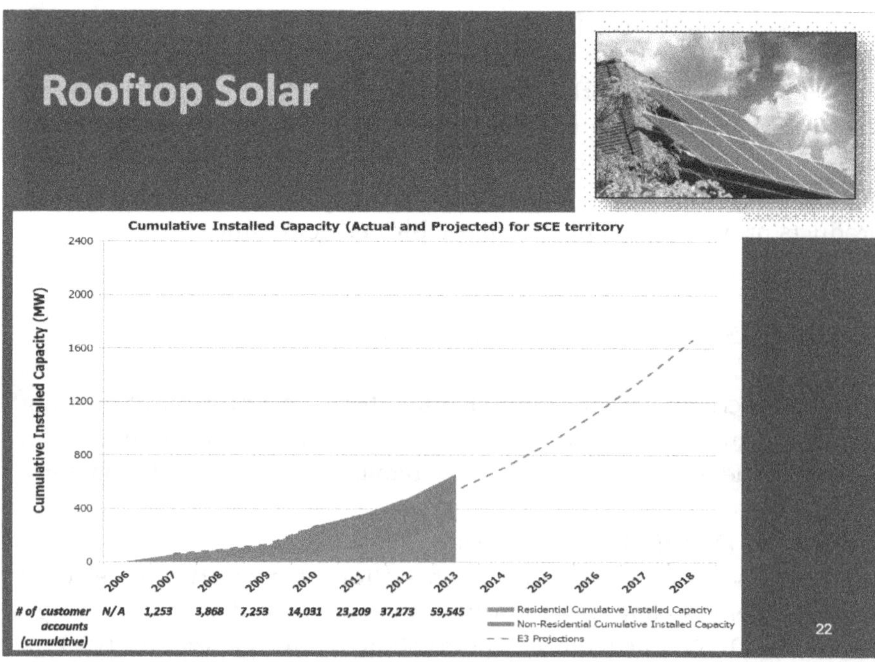

Figure 1-4. The actual and projected installation of rooftop solar in Southern California Edison's territory. (SCE as published in EPRI 3002004689)

The Scarcity of Capacity

For proper operation, power systems depend on an optimum mix of generation and other resources to provide both energy and capacity. The total resource requirement is based on an estimation of the generation required to meet demand for energy, plus needs for ramping, load following, voltage regulation, and frequency response. This mix is comprised of conventional synchronous generation coupled with increasing amounts of non-synchronous or variable generation. Flexible operation of conventional generation, augmented by demand response and energy storage, is used to supplement the newer resources.

Changing generation characteristics, coupled with changes in both the pattern and amount of consumer consumption, are resulting in an energy surplus in many markets, which is resulting in lower wholesale energy prices. Low prices provide an inadequate market signal for the deployment of new generation capacity.

In addition, wind and solar energy output is variable and difficult to predict accurately. It dispatches when available, forcing coal, nuclear, natural gas generation, and hydropower plants, to adjust their output in order to maintain reliability. System planners need to credit wind and solar systems with some portion of their nameplate capacity when estimating the need for generation capacity.

Figure 1-5 illustrates the situation in Germany during 2013. During that year, wind generation varied between 126 MW and 25,741 MW—from less than 1% to a maximum of 78% of installed capacity. Solar generation varied between near zero at night up to a high of 23,952 MW—reaching only 72% of installed capacity. With these fluctuations, Germany's power system operators cannot count on these renewable resources over all hours of the year. On October 17, 2013, the combined "in feed" from about 70,000 MW of wind and solar capacity met only 0.4% of total needs during a number of hours.

The Lack of Sufficient Synchronous Generation

The ability of a grid to serve load and meet quality and reliability standards—is influenced by all of the electrical devices connected to it and by medium and large rotating turbine-generator systems capable of synchronous operation. The rotating mass of the turbine-generator has inertia (potential energy), which is transformed into electrical inertia. The inertia afforded by synchronous generators provides the ability to ride through disturbances, such as the loss of a line, tree contact, or a

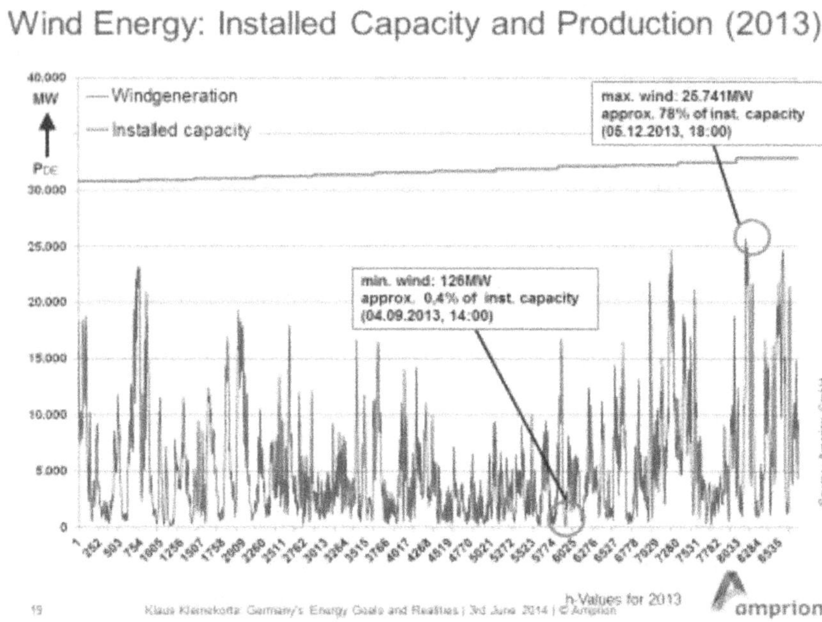

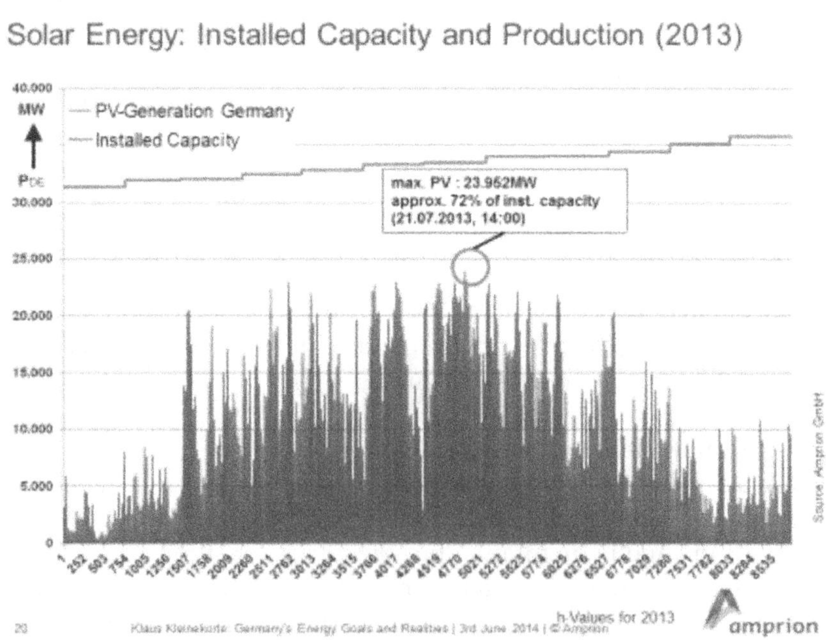

Figure 1-5. Wind energy and solar energy installed capacity and production in Germany in 2013. (Amprion as published in EPRI 3002004689)

sudden loss of a unit on the system. It also provides frequency control, volt/VAR control, inertial response, and other ancillary services.

Very few wind and PV installations have the necessary power electronics and electrical storage to provide the same electrical functionalities as do synchronous generators. As wind and solar displace conventional generation, their highly variable nature creates the need for additional reserve capacity.

As the penetration of variable generation increases, so too does the need for synchronous units and resources to provide balancing, protection, and regulation services. Figure 1-6 displays renewable energy targets for Europe, illustrating future pressures on the role of synchronous generation in meeting energy and capacity needs. Figure 1-7 illustrates where these trends can lead for a region with high levels of wind deployment: In Ireland, non-synchronous penetration cannot exceed 50% without increasing levels of wind curtailment or higher levels of system services.

Market Signals

In many countries, wholesale energy markets have been established for some years now and can be considered rather mature. Capacity markets were not initially considered necessary because capacity margins were adequate, and there were far fewer policies and incentives directed toward renewable generation. Market-based frameworks are now emerging to provide market signals to new types of resources so that the power system will have adequate capacity and other ancillary services, in addition to adequate energy.

Ireland has had an all-island, one-zone, single electricity market (SEM) since 2007. As a result of the country's renewable energy mandate, the SEM includes an energy market that has facilitated growth of renewable generation from approximately 8% to 20% of total energy production in 2012. New wind installations are increasing by 350 MW per year, but expansion will be hampered since non-synchronous penetration cannot exceed 50% without increasing curtailment.

SEM's energy markets have the consequence of reducing running hours for thermal generators and currently do not compensate them for the flexibility they provide to the system. Without a capacity payment mechanism, thermal generation will exit the market. To establish a more equitable market, Ireland is revising its SEM to include incentives related to the character of generation resources, referred to as grid services or

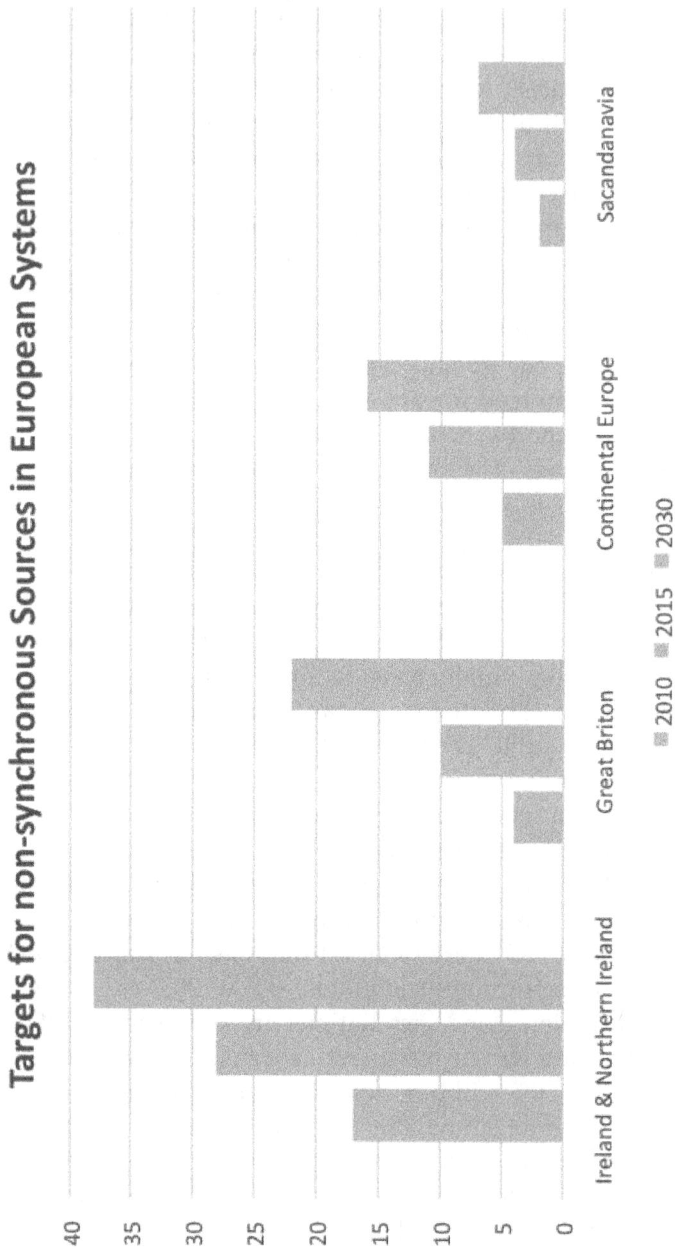

Figure 1-6. illustrates the possible evolution of non-synchronous generation vs. synchronous generation in European power systems. (Electricity Research Centre—University City Dublin as published in EPRI 3002004689)

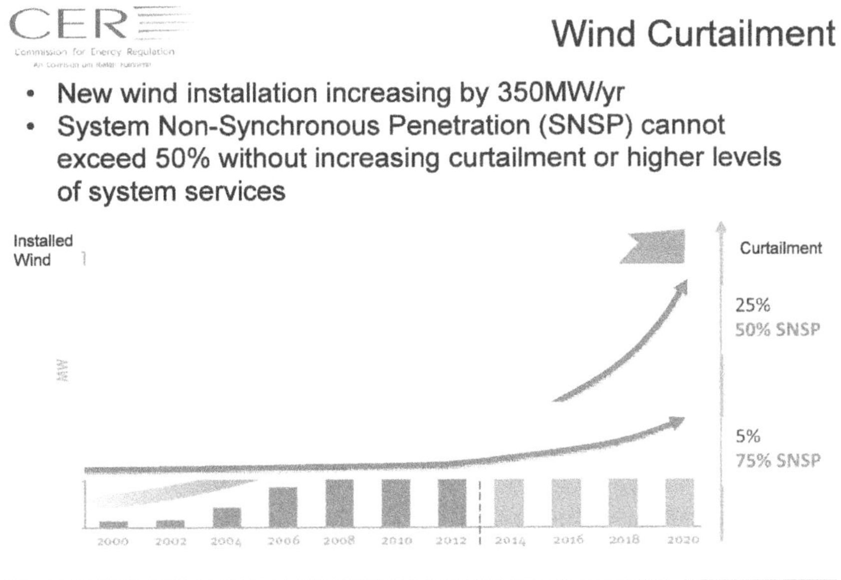

Figure 1-7. Wind production will be limited by the lack of thermal generation. (Commission for Energy Regulation as published in EPRI 3002004689)

"System Services."

In certain U.S. regions, demand response (DR)—in the form of price-induced programs or direct load control—can bid into markets. A recent U.S. Federal Energy Regulatory Commission (FERC) Order requires that DR must be compensated equally with generation resources.

In the U.S. Northeast, capacity markets managed by independent system operators (ISOs) are continuing to evolve. Rule changes are focusing on creating an equal playing field for all resources. PJM is enabling DR to be available not only during summer but throughout the year, and it is shortening the time for a DR aggregator to respond from 2 hours to 30 minutes.

The California ISO (CAISO) and PacifiCorp have proposed that the Western Electricity Coordinating Council create an Energy Imbalance Market (EIM) focused on enhancing reliability in the face of potential capacity scarcity. The reliability benefits of the proposed EIM include better management of imbalances, enhanced ability to manage flows within system operating limits, and enhanced opportunities to deliver energy from a diverse set of conventional and emerging technologies, including Demand Response.

Integration Standards

In Europe, the European Network of Transmission System Operators—Electric (ENTSO-E) is responsible for overall system harmonization among the Transmission System Operators (TSOs). This is especially critical given that over 80% of the European system plans over the next 10 years are driven in one way or another by renewables, with social acceptance of new infrastructure presenting a significant challenge. To facilitate this evolution, ENTSO-E is establishing network codes to set up common requirements and rules for the European internal electricity market as detailed in Table 1-2.

In Europe, frequency control already is required of generators of all sizes.

In the United States and other countries outside of Europe, the revised standard developed by the Institute of Electrical and Electronics Engineers (IEEE) referred to as 1547 is expected to provide the framework to allow for appropriate grid integration standards for inverters. Some U.S. states are developing their own standards, including in California where an ongoing regulatory process is expected to result in standards that may contain seven elements:

- Anti-islanding protection
- Low- and high-voltage ride-through
- Low- and high-frequency ride-through
- Dynamic volt-VAR optimization

Table 1-2. ENTSO-E Proposed Network Codes (ENTSO-E as published in EPRI 3002004689)

Connection Network Codes	Market Network Codes	Operational Network Codes
Set requirements for	***Set market rules for***	***Set common rules for***
Generators	Day-Ahead & Intra-Day Capacity Calculations	Assessing Adequacy
Demand-Side Resources	The engagement of Demand-Side Resources for Long-Term Timeframes	Planning Outages
High-Voltage DC Connections	System Balancing	Ensuring Security
		Managing Emergencies

- Ramp rates
- Fixed power factor
- Soft-start reconnection

Integration Technologies

Advanced or smart inverters are expected to allow the system to accommodate lower electrical inertia at high non-synchronous penetration and to facilitate wind and solar curtailment. Table 1-3 summarizes the existing services inverters can provide and the new ones advanced inverters are being designed to provide.

In the last few months, there has been a virtual explosion of "passive systems," also known as home energy management (HEM) systems or energy management systems (EMS)*. While not 100% passive, these "connected home" technologies allow grid-interactive building technologies to be operated based on consumer preference, local intelligence,

Table 1-3. Grid Support Services Categories. Traditional Services are those which today's inverters are capable of supplying. New services are those which advanced inverters could additionally provide. (SMA as published in EPRI 3002004689)

Category	Frequency support	Voltage support	System restoration
Traditional services	Frequency containment reserve (<5, 10 or 30 seconds)	Normal operation: control of power factor; reactive power; or voltage	Black start
Traditional services	Frequency restoration reserve (< 15 minutes)		
Traditional services	Replacement reserve (15 minutes to hours)		
New services	Fast frequency response – synthetic inertia (<2 seconds)	Fast reactive current injection	Islanding
New services	Ramping margin (1,3, 8 hours ahead)		

*The term Energy Management Systems (EMS) is also used to describe the grid operating system which is used to control areas of the grid.

and supervisory control from the utility system. They may simply involve a smart thermostat or an advanced PV inverter or a HEM or building energy management systems. Figure 1-8 illustrates one such system developed by the vendor Passivsystems. Their model incorporates a grid management feature that turns on-demand loads, such as hot water or space conditioning, into dispatchable ones. On-board microprocessor communications integrated into consumer devices make them an active part of the power system.

THE NEED FOR ELECTRIFICATION

Some power system experts are beginning to recognize that EVs and other electrification technologies, such as heat pumps, to be of equal importance to distributed generation in shaping the industry's future through 2030. Electricity is a refined energy form can be generated from diverse sources, and it offers unmatched precision and quality at the point of end use. Through different applications, it provides light, heat,

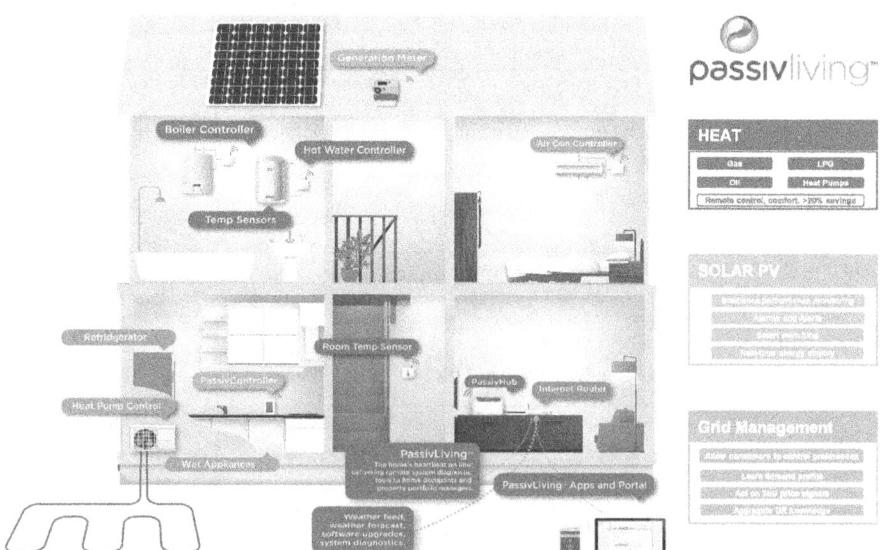

Figure 1-8. An example of a smart home with sensors, communications, local intelligence and grid connectivity. (Passivsystems as published in EPRI 3002004689)

comfort, and mechanical work. Electrification is the key toward reduction in direct fossil fuel consumption in buildings and industry.

Changes in Consumer Expectations

The electricity consumer continues to value high levels of power quality and reliability as well as affordable bills. Increasingly, they are demanding clean energy with efficiency and first-class service quality often extending to community engagement for prosumers. Most practitioners agree that as the electricity industry becomes more dynamic, consumer expectations are growing. Those expectations are shaped by other customer experiences, where choice and empowerment are part of every business, and new needs and wants can emerge from diverse directions.

A key challenge for today's electricity industry is to identify where future value creation will develop. There is an increasing need to monitor trends in behavior and technology in order for suppliers and providers to stay in tune. The advent of potentially disruptive innovations and new services in communications, information technology, self-generation, hyper-efficient appliances, DR, home automation, and other areas is fueling change, with their common element—connectivity—becoming increasingly important for tomorrow's electricity landscape.

Expanding Connectivity

Following from the disruptive transformation that began just over 2 decades ago with mass-market access to the Internet, the tools people use to access, process, communicate, and apply information have changed dramatically in the last few years. The iPhone's introduction in 2007 and the iPad's launch 3 years later have motivated an expanding and diversifying class of commercial analogues and spin-off developments. Digital communications have become a pervasive part of everyday life and integral to modern lifestyles, including world-changing mediums such as Facebook and Twitter. Consumers are seemingly always tethered to the on-line world—a trend even more pronounced among younger generations who have grown up with on-line ubiquity and the expectation of information and opportunity for action on demand.

Mobile devices and social networking are beginning to impart a transformational change in how consumers access and use energy-related information and grid-connected devices. Smart phones and tablets alone provide numerous energy-related opportunities for consumer engagement, energy storage, workforce productivity, and simple load control while eliminating small frictions that can radically alter decisions, perhaps most importantly willingness to participate. This presents an opportunity for the electricity industry to reconnect with consumers in a new way and enhance operational efficiency.

Novel services are being tested around the world, building on consumer-controlled examples such as the Nest (www.nest.com) thermostat that offer set-and-forget thermostat and simplicity in load control, plus the opportunity for on-demand intervention from any location. New services could also be provided to enterprises, which rely on digital communications and information analytics to run their systems, manage their facilities, and drive energy-related planning functions and business decisions. Individually and collectively, abilities for independent and coordinated action are expanding.

In tomorrow's world, mass collaboration will involve machine-to-machine, person-to-machine, and person-to-person interactions. Cisco Systems refers to this as the "Internet of Everything"—10 billion connected devices today, doubling by 2020 to 20 billion, multiplied by 50 billion additional devices and sensors within homes, buildings, etc. The resultant network will span 1 trillion "energy harvesters" connecting everything: people, process, data, and things. People will connect in more relevant, valuable ways. Process will enable the delivery of the right information to the right person (or machine) at the right time. Data will be leveraged into more useful information. And things—the physical devices and objects connected to the Internet and each other—will facilitate intelligent decision-making.

Cisco believes this will allow expanding customer engagement with new technology- and process-enabled services so as to create new business lines, particularly from the expansion in customer adoption of DER. In addition, the "Internet of Everything" will integrate workforce mobility and automation to enhance operations by enabling connectivity and shared intelligence for new business opportunities. Finally, it will underlie a truly integrated grid capability of accommodating new technologies, greater variability in electricity generation, expanded flexibility at all scales, and prosumers delivering and demanding novel services.

Consumer Services

As market rules and regulations advance to value flexibility, turn demand into dispatchable load, empower DER to provide grid support services, and require all variable-output generation to interconnect using smart inverters, business opportunities will be created. However, traditional utilities are not seen as agents of change and instead are viewed by the financial community as too slow and too rigid.

References

- "Energy Independence and Security Act of 2007 http://www.eia.doe.gov/oiaf/aeo/otheranalysis/aeo_2008analysis papers/eisa.html
- "The Aging Workforce: Electricity Industry Challenges and Solutions," L.B. Lave, M. Ashworth and C.W. Gellings, Electricity Journal, January 2007.
- "The Power to Reduce CO_2 Emissions: The Full Portfolio: 2009 Technical Report," EPRI, Palo Alto, CA: 2009. 1020389
- "The Modern Grid Initiative Version 2.0." conducted by the National Energy Technology Laboratory for the U.S. Department of Energy Office of Electricity Delivery and Energy Reliability, January 2007, http://www.netl.doe.gov/moderngrid/resources.html
- "Report to NIST on the Smart Grid Interoperability Standards Roadmap," (Contract No. SB1341-09-CN-0031), D. Von Dollen, EPRI, June 17, 2009
- LogRhythm and NERC CIP Compliance, www.logrhythm.com, 2009.
- "The Cost of Power Disturbances to Industrial and Digital Economy Companies," EPRI, Palo Alto, CA: 2009. 1006274
- "A National Assessment of Demand Response Potential," Federal Energy Regulatory Commission, June 2009
- "National Action Plan on Demand Response, " Federal Energy Regulatory Commission, June 2010
- "Smart Grid VC Investment Surge." greentechgrid, January 21, 2010, www.greentechmedia.com
- A Consumer Portal at the Junction of Electricity, Communications, and Consumer Energy Services, Clark Gellings. The Electricity Journal, November 2004
- The Path to Perfect Power: New Business Opportunity for A Customer-Demand Driven Electricity World, GF Energy, LLC, Galvin Electricity Initiative, 2006
- EPRI 2014 International Technology Innovation Summit, EPRI 3002004689, November 2014.
- Is your Refrigerator on the Internet? Presentation by C.W. Gellings, Edison Electric Institute (EEI) Annual Conference: Critical Issues Forum, Philadelphia, PA, June 10, 1997

Chapter 2

Smart Grid Technologies

There are a variety of Technologies which may be part of the smart grid. These are usually grouped onto three categories: Transmission Systems and Substations; Distribution Systems and Customer Systems.

TRANSMISSION SYSTEMS AND SUBSTATIONS

The Transmission Systems and Substations Category typically includes transmission and substation equipment including, overhead and underground transmission lines and substations along with appurtenances over 69kv. Distribution technologies typically includes all technologies from the medium voltage 4 to 69kV in the substation through to the customer's meter. This includes the communications infrastructure essential to the smart grid as well as controllers used to automate the distribution systems and the metering infrastructure. Customer technologies include all of those identified as "behind the meter."

The high-voltage transmission system is the "backbone" of the power delivery system. It transmits very large amounts of electric energy between regions and sub-regions. Transmission system equipment which fails and causes power outages much less frequently than distribution equipment. But when transmission equipment fails, many more customers are affected, and outage costs can be much higher, compared to the impact of a distribution equipment-related outage. This fact, combined with the high cost per mile or per piece of transmission equipment, has historically led to greater attention to transmission system reliability.

In general, monitoring of transmission assets is more cost-effective and beneficial than any other asset class (EPRI 1016055). Although transmission lines are one of the critical core backbone elements of the power grid, thousands of miles are unattended and not monitored in any way. Transmission lines have seasonal ratings that need to be considered by

Table 2 1. Transmission Line Miles (EPRI 1022519)

Voltage (kV)	Miles
230 AC	85,048
345 AC	59,767
500 AC	32,870
765 AC	4,715
250-500 DC	3,307
Total Miles	**184,707**

operations and planning. For the most part, there is little if any real-time monitoring so far, other than at substations that provide operators with loading information.

Transmission investment trended downward for more than two decades, declining from $4.8 billion in 1975 to $2.25 billion in 1997, then leveled off before beginning to climb again. It reached roughly $5 billion in 2000 and is $11 billion in 2010. Smart Grid functionality should help to increase the value of future transmission investment over and the expansion needed to meet load growth.

Dynamic Thermal Circuit Rating (DTCR)

Dynamic rating including dynamic thermal circuit rating (DTCR) and real-time monitoring of transmission lines are becoming important tools to maintain system reliability while optimizing power flows. Dynamic ratings can be considered a low-cost alternative for increased transmission capacity. Dynamic ratings are typically 5 to 15% higher than conventional static ratings. Application of dynamic ratings can benefit system operation in several ways, in particular by increasing power flow through the existing transmission corridors with minimal investments. (EPRI 1022519)

Dynamic rating increases the functionality of the Smart Grid because it involves the monitoring of real-time system data that can be used in various applications:

- Real-time monitors yield a continuous flow of data to system operations—line sag, tension or both, wind speed, conductor temperature, etc.—traditionally not available to operators.

- Monitored data can be processed to spot trends and patterns.
- Real-time monitored data may be turned into useful operator predictive intelligence (e.g., critical temperature and percent load reduction needed in real time).

For example, The New York Power Authority (NYPA) is evaluating the instrumentation and dynamic thermal ratings for overhead transmission lines. An area of possible application for DTCR is the growing penetration of wind generation; when turbines are operating, one expects higher dynamic ratings because of increased wind speed. NYPA uses DTCR software, which uses real-time or historical weather and electrical load data to calculate dynamic ratings for overhead lines in real time based on actual load and weather conditions that generally are accessed through the utility's SCADA/EMS system.

AC transmission lines rated 115 kV to 230 kV in the U.S. are most susceptible to being thermally limited. Lines rated 345 kV and above are more likely voltage-limited. Albeit, there may be a few lines rated at these higher voltages that are thermally limited—they would be an exception and not the rule. Furthermore, of the lines which are potentially thermally limited, only 50% actually are. There are 85,048 miles of lines at 230 kV. To dynamically rate all 85,048 miles would require one unit per 7.5 miles or 11,340 units. The cost of DTCR deployed in quantity is estimated to be $20,000 initially, declining to $10,000 over 10 years.

Sensors

The Smart Grid will require a more diverse and wider array of sensors throughout the power system to monitor conditions in real time. In particular, sensors in transmission corridors and in substations can address multiple applications:

- Safety: The application of sensors for transmission line or substation components will allow for the monitoring and communication of equipment conditions. Information that a transmission line or substation component is in imminent risk of failure will enable actions to be taken to address the safety of utility personnel.
- Workforce Deployment: If the condition of a component or system is known to be at risk, personnel can be deployed to prevent an outage.

- Condition-Based Maintenance: Knowledge of component condition enables maintenance actions to be initiated at appropriate times rather than relying on interval-based maintenance.
- Asset Management: Improved knowledge of the condition of equipment and stresses that they have been subjected to will allow managers to better manage the assets. Sensor data used together with historic performance information, failure databases and operational data allows better allocation of resources.
- Increased Asset Utilization: The rating of transmission components is influenced by a range of factors such as ambient weather conditions, loading history and component configuration. In order to address this complexity, static ratings are usually based on conservative assumptions of these factors. Higher dynamic ratings can be achieved with more precise, real-time knowledge of the asset's condition.
- Forensic and Diagnostic Analysis: After an event occurs, there is limited information to understand the root cause. Sensors allow the capture of pertinent information in real time for a more rigorous analysis.
- Probabilistic Risk Assessment: Increased utilization of the grid is possible if contingency analyses are performed using a probabilistic, rather than deterministic, methods. To use probabilistic methods, knowledge of the condition of components and the risks they pose are needed

The transmission system of the future will utilize a synergistic concept for the instrumentation of electric power utility towers with sensor technology designed to increase the efficiency, reliability, safety, and security of electric power transmission. The system concept is fueled by a list of sensing needs illustrated in Figure 2-1 (EPRI 1016921). This illustration is limited to transmission line applications (i.e., 69 kV and above), not distribution, with the focus on steel lattice and pole structures, not wooden.

Figures 2-1 depicts some of the high-level concepts that are listed below:

- Sensors distributed on transmission structures and/or conductors.

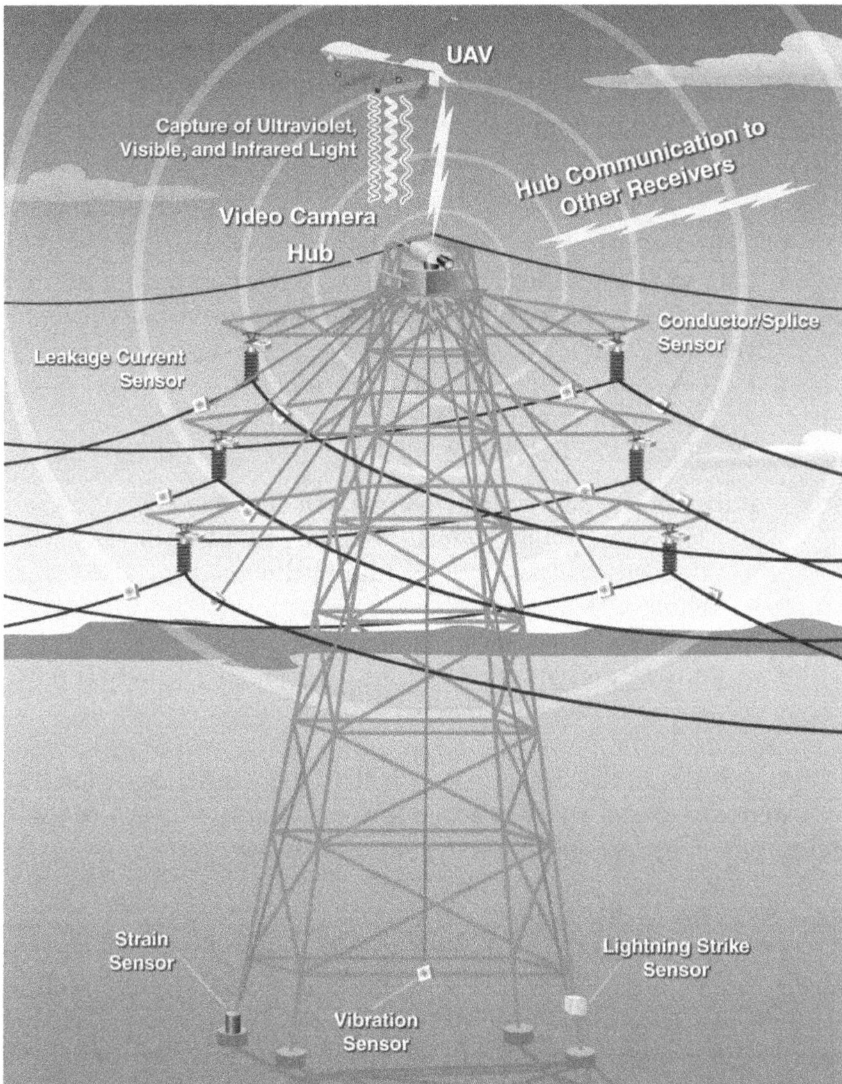

Figure 2 1. Image Showing a Single Structure Illustrating Some Sensor Concepts (EPRI 1016921)

- Sensors that may or may not communicate with "hub" installed on the structure— either wireless or wired.
- Sensor information is collected, stored and analyzed in a "central

database" which is part of the utility's current data management system. The data are collected/communicated from the sensors/" hubs" to the "central database" using one of the following methods:

— Wirelessly back to the "central database" from the individual structure "hub," e.g. Radio Frequency (RF) directly, via satellite or cell phone network.

— Collected using a vehicle traveling the length of the line. The data from the collection vehicle is transferred during or after the inspection. The following is a list of possible data collection vehicles:
  - Unmanned Airborne Vehicle (UAV)
  - Manned Aerial Vehicle
  - "Line Crawler" Robot

— If the vehicle data collection approach is utilized:
  - The vehicle may collect the data wirelessly directly from the sensors (possibly excluding the need for a structure "hub").
  - The vehicle may also have sensors aboard recording data during the collection process, (e.g. video, Ultraviolet (UV), Infrared (IR), still images)

These concepts are discussed in detail in the following sections.

There are a number of possible sensors to address each of these applications. Table 2-2 elucidates the range of sensor needs.

Sensor Systems Architecture

The sensor system architecture is comprised of a) sensors that acquire diagnostic data from components of interest and b) communications hubs that collect the sensor data and relay it to a central repository. Sensors may be directly attached to the item being monitored, or may be remotely located such as in the case of a camera. Communications hubs may be mounted on or near towers or may be located on a wide variety of mobile platforms, such as manned airplanes or unmanned line crawlers or UAVs. Sensors and hubs may operate and be polled periodically (e.g., at intervals of minutes, hours, days) or continuously monitored (e.g., a real time alarm) depending on the application. In any case, sensors communicate their results via hubs to a central repository. Figure 2-2 illustrates the architecture and flow of data.

Table 2-2. Sensor Needs (EPRI 1016921) (*Cont'd*)

	Item	Cause	Result	Update Interval	Probability	Consequence	Sensing Technologies
1	System Tampering	Terrorism	Tower/line down	Real-time	Low	High	Vibration, Acoustic, E-Field, Optical
2	System Encroachment	Man-made	Safety hazard, Less reliable	3-12 mo	High	Med	Optical, Satellite, Proximity, Vibration, E-Field
3	System Encroachment	Vegetation	Flashover, Fire	3 mo	High	High	Optical, Satellite, LIDAR, Line-of-Sight, Proximity
4	System Encroachment	Avian Nesting, Waste	Flashover	6-12 mo	High	High	Optical, Vibration, Leakage Current, Proximity, E-Field
5	Shield Wire	Corrosion	Flashover, Outage	3-6 years	Med	High	Optical, IR Spectroscopy, Eddy Current, MSS
6	Shield Wire	Lightning	Flashover, Outage	1 year	Med	High	Optical, IR Spectroscopy, Eddy Current, MSS, Lightning Detection, Vibration
7	Insulator (Polymer)	Age, Material Failure	Outage	6 years	Med	High	Optical, Vibration, RFI, UV, IR
8	Insulator (Ceramic)	Age, Material Failure	Outage	12 years	Low	High	Optical, Vibration, RFI, UV, IR
9	Insulator	Contamination	Flashover	3 mo	Med	Med	Optical, RFI, UV, IR, Leakage Current
10	Insulator	Gun Shot	Outage	Real-time, 3 mo	Med	High	Optical, Vibration, RFI, UV, IR, Acoustic
11	Phase Conductor	External strands broke	Line Down, Fire	1 year	Low	High	Optical, Vibration, RFI, UV, IR

Table 2-2. Sensor Needs (EPRI 1016921) (*Cont'd*)

	Item	Cause	Result	Update Interval	Probability	Consequence	Sensing Technologies
12	Phase Conductor	Internal strands broke	Line Down, Fire	1 year	Low	High	E-MAT, MSS, Electromagnetic
13	Phase Conductor	Corrosion of steel core	Line Down, Fire	1 year	Low	High	E-MAT, MSS, Electromagnetic, IR Spectroscopy, Optical
14	Connector Splice	Workmanship, thermal cycling, age	Line Down, Fire	1 year	Med	High	Direct Contact Temperature, IR Temperature, Ohmmeter, RFI, E-MAT, MSS
15	Hardware	Age	Line Down, Fire	6 years	Low	High	Optical, IR Spectroscopy
16	Phase Spacer	Age, galloping event	Line Down, Fire	6 years	Low	Med	Optical, UV, RFI
17	Aerial Marker Ball	Vibration Damage, Age	Safety concerns	1 year	Low	Med	Optical, UV, RFI
18	Structure (Steel Lattice)	Corrosion	Reliability Concerns	10 years	Med	Med	Optical, IR Spectroscopy
19	Structure (Steel Lattice)	Bent, damaged members	Reliability Concerns	1 year	Med	Med	Optical, Strain, Position, Tilt
20	Structure (Steel Pole)	Corrosion, age	Reliability Concerns	10 years	Med	Med	Optical, IR Spectroscopy
21	Structure (Steel Pole)	Internal Deterioration	Reliability Concerns	1 year	Med	Med	Optical, MSS, Ultrasonics
22	Foundation (Grillage)	Age, corrosion	Reliability Concerns	10 years	High	High	Excavation, MSS, Radar, GPR Imaging, Half Cell, Voltage Potential
23	Foundation (Anchor Bolt)	Age, corrosion	Reliability Concerns	10 years	Low	High	Optical, Ultrasonics, E-MAT, Vibration

Table 2-2. Sensor Needs (EPRI 1016921) (*Concluded*)

	Item	Cause	Result	Update Interval	Probability	Consequence	Sensing Technologies
24	Foundation (Preform)	Age, corrosion	Reliability Concerns	10 years	Med	High	Optical, Ultrasonics, E-MAT, Vibration
25	Foundation (Stub Angles)	Age. Corrosion	Reliability Concerns	10 years	Low	High	Optical, Ultrasonics, E-MAT, Vibration
26	Foundation (Direct Embedment)	Age, corrosion	Reliability Concerns	10 years	High	High	Excavation, MSS, Half Cell, Voltage Potential
27	Foundation (Anchor Rods, Screw-In)	Age, corrosion	Reliability Concerns	10 years	High	High	Excavation, MSS, Half Cell, Voltage Potential, Ultrasonics
28	Grounding	Age, corrosion, tampering	Reliability, Lightning, Safety concerns	6 years	Med	Med	AC impedance, DC resistance, Impulse
29	TLSA (Transmission Line Surge Arrestor)	Lightning Strikes, age	Reliability, Lightning Concerns	1 year	Med	Med	Optical, IR, Leakage Current, Lightning Strike Counter

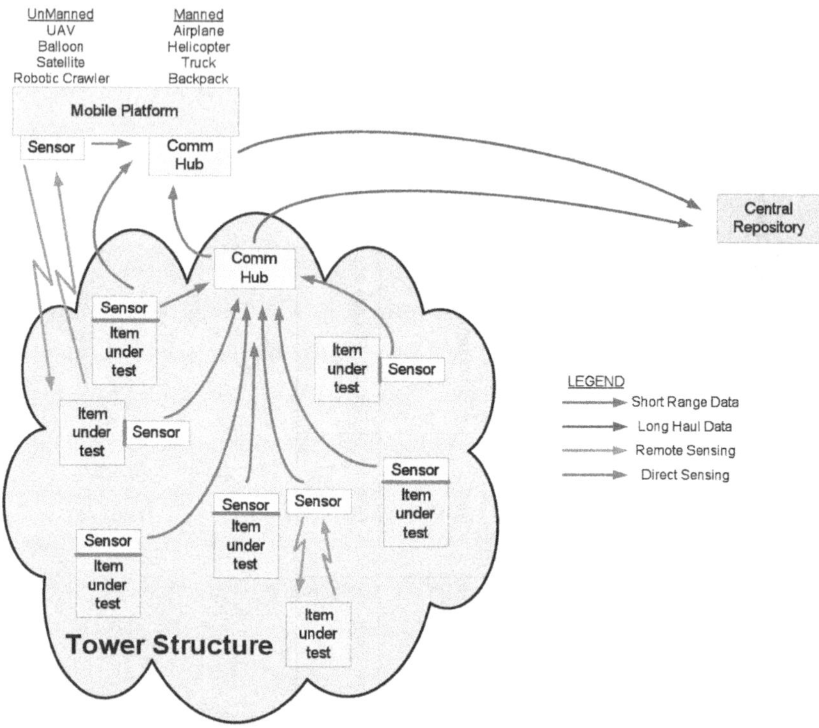

Figure 2-2. Sensor System Architecture (EPRI 1016921)
Examples of Transmission Line Sensors

Figure 2-3. Examples of Transmission Line Sensors (EPRI 1022519)

Figure 2-4. RF Conductor temperature and current sensor, offering power harvesting, live working install, and low cost. Cost is an order of magnitude lower than that of other available technologies. (EPRI 1022519)

Figure 2-5. Clamp-on RF leakage current sensor for transmission line applications installed on a 115-kV composite insulator. (EPRI 1022519)

Figure 2-6. An RF leakage current sensor installed on a post insulator in a substation. It also shows dry band arcing (the discharge activity due to contamination and wetting that causes the leakage currents) captured during a contamination event. (EPRI 1022519)

Figure 2-7. Antenna array installed on a portable trailer deployed in a 161-kV substation.

Short-Circuit Current Limiters (SCCL)

The short-circuit current limiter (SCCL) is a technology that can be applied to utility power delivery systems to address the growing problems associated with fault currents. The present utility power delivery infrastructure is approaching its maximum capacity and yet demand continues to grow, leading in turn to increases in generation. The strain to deliver the increased energy demand results in a higher level of fault currents. The power-electronics-based SCCL is designed to work with the present utility system to address this problem. It detects a fault current and acts quickly to insert an impedance into the circuit to limit the

Figure 2-8. A solar-powered tank-top temperature sensor installed on a transformer.

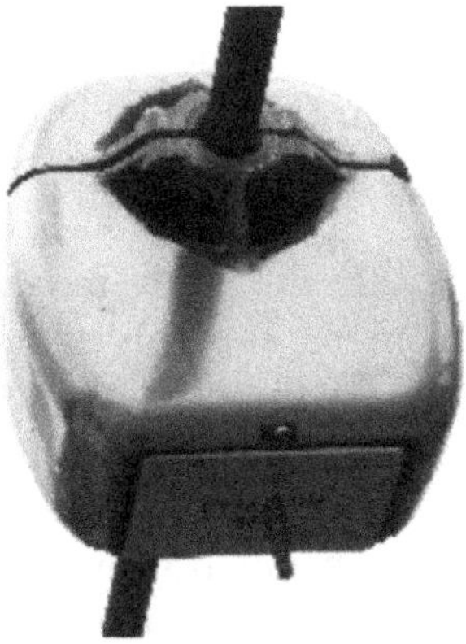

Figure 2-9. An MIS sensor for measuring acetylene gas levels in oil. The use of MIS gas-in-oil sensors can increase the number of transformers monitored due to their lower cost. (EPRI 1022519)

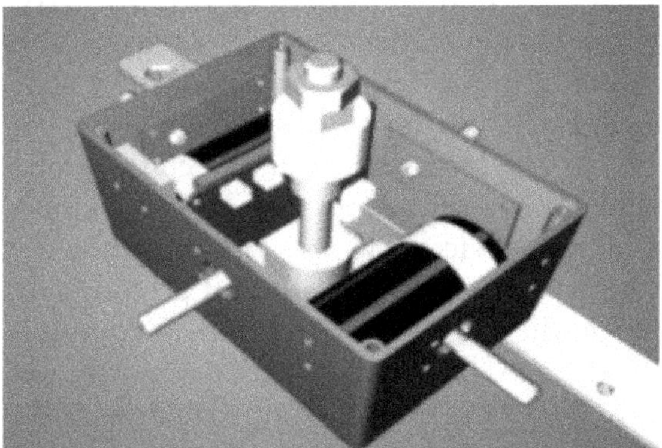

Figure 2-10. Conductor temperature sensors installed on a steel-reinforced aluminum cable (ACSR) conductor. The sensor communicates using a cell phone modem. (EPRI 1022519)

fault current to a level acceptable for normal operation of the existing protection systems.

Flexible AC Transmission System (FACTS)

There are a number of flexible AC transmission (FACTS) technologies which are critical to the Smart Grid. These all incorporate power electronics and can be applied to the transmission system. These include both the control and operation of the power system and applications that will extend eventually to transformers themselves.

FACTS devices can be used for power flow control, loop flow control, load sharing among parallel corridors, voltage regulation, enhancement of transient stability, and mitigation of system oscillations. FACTS devices include the thyristor controlled series capacitor (TCSC), thyristor controlled phase angle regulator (TCPAR), static condenser (STATCON), and the unified power flow controller (UPFC). AEP installed the first UPFC at their Inez substation in eastern Kentucky in 1998 (EPRI 1010633).

FACTS is a concept invented by EPRI in the 1980s, for which EPRI developed several patents licensed to Siemens. It involves the injection of a variable-voltage source, which adjusts the power flow across a transmission line, resulting in variable voltage, impedance, and phase angle. Six successful major FACTS installations were demonstrated, each with several key features. For example, one installation at the New York Power Authority's Marcy Substation resulted in the ability to increase power-transfer capacity into New York City by 200 MW, resulting in substantial savings. However, these FACTS devices did not spawn an anticipated revolution in the control of power flow, as was expected. They were plagued by three technical problems, the combination of which made FACTS 20% more costly than "conventional" solutions. These technical problems included the cost and performance of the control systems within the devices; the performance of the cooling systems used to cool the electronics; and the cost and performance of the power-electronic devices themselves. As of now, the first two of these issues have been resolved, and with the successful demonstration of advanced power electronics, there is relative certainty that the industry is poised for a rebirth of FACTS. This is especially important with the introduction of increasing amounts of variable generation—like wind and solar—located far from load pockets.

The advantages of FACTS technology are as follows:

- Increases the amount of power that can be imported over existing transmission lines.
- Provides dynamic reactive power support and voltage control.
- Reduces the need for construction of new transmission lines, capacitors, reactors, etc which mitigate environmental and regulatory concerns, and improves aesthetics by reducing the need for construction of new facilities.
- Improves system stability.
- Controls real and reactive power flow.
- Mitigates potential Sub-Synchronous Resonance problems.

FACTS for HVDC—The backbone of HVDC transmission is a FACTS device which is the converter station which converts AC to DC for transmission and then convert DC back to AC at the other end of the line. Power electronics do the heavy lifting in these applications. In the 400- to 1000-MW range, second-generation IGBT power-electronic devices are used, while in the 1000-MW and above range, older technologies are used (namely, thyristors). As load growth increases and the use of renewable power generation located far from the load pocket becomes the norm, there will be increased demand for DC technology. DC technology may be the only effective means of increasing power flow on an existing corridor that was originally built for AC transmission.

FACTS for Controlling Reactive Power—TCSCs are a derivation of the FACTS technology that uses power electronics to inject capacitance into the power system to improve power flow by controlling reactive power. There are roughly 100 TCSCs installed in the U.S., primarily in the west. Eastern utilities have been reluctant to take advantage of this technology due to concerns about sub-synchronous resonance. Further development of thyristor control can dampen those oscillations.

FACTS for Electronic Transformers—Power-electronic transformers will eventually become part of the transmission system. Existing mechanical switching is accomplished within about 6 AC cycles. This is rapid enough for most applications. However, there is a need for a power-electronic device that could reduce short-circuit currents and then act as power-electronic circuit breaker at the same time. The potential maximum for short-circuit currents is growing so as to exceed the maximum capacity of today's breaker fleet. A distribution device to limit short-circuit currents using power electronics has been demonstrated. Further development could allow this device to scale up to high voltage and, coincidentally, act as a fast-switching power-electronic breaker. This device would have the added benefit of eliminating the use of SF6. Considerable effort has been expended in the development of distribution transformers based on power electronics. Using power electronics in transformers can eliminate the majority of the inductance and, along with it, all of the oil used as coolant, resulting in a substantial reduction in the losses. In addition, it offers a great deal more flexibility in voltage control. The potential to scale up the distribution version of this transformer is promising. Further advancement in power electronics will be needed to realize for this transmission application.

Table 2-3.
DC Lines and Terminals Known to be Under Consideration (EPRI 1022519)

Renewable Studies	No. Lines	Type
EWITS	10	800 kV DC lines
WWIS	1	600 kV DC lines
HPX	2	500 kV DC lines
Santa Fe	1	500 kV DC lines
Clean Line	1	500 kV DC lines
Total	15	Lines or 30 HVDC Terminals

Voltage Source Converters

Voltage Source Converters (VSCs) are self-commuted high voltage direct current (HVDC) converters and contrary to "traditional" HVDC converters do not have to rely on synchronous machines in the AC system for its operation. The increased controllability improve harmonic performance as well as provide VAR support. VSCs, permit power flows to be reversed without reversing the polarity of the cable, thereby enabling the use of extruded cables (cables insulated with extruded polyethylene-based compound such as XPLE). It makes undergrounding (cables instead of overhead lines) more attractive.

Advanced Power Electronics for Transmission Applications

HVDC converters and other controllers are based on power electronic semiconductor devices. These devices exploit the properties of semiconductor material, principally silicon, germanium and gallium arsenide. Semiconductor material can be manipulated by "doping," the addition impurities and can be controlled by the introduction of an electric field, light or pressure. Power semiconductor devices are those intended for high-current and/or high-voltage applications. Silicon (Si) is the most widely used material in the manufacture of semiconductor devices. IT has low raw material cost, requires relatively simple processing, and offers good temperature range. Other, more advanced devices are converters. A new set of potentially "game-changing" switched has evolved. These switches fall into the category of "wide band gap" semiconductors. Among these, Silicon Carbide (SiC) and Gallium Nitride (GaN) have increased in reliability and dramatically decreased in cost. They hold a promise for significantly increased high-value applications across the electric utility industry in power generation, delivery, and end use.

Storage

Bulk storage is one of the major limitations in today's "just in time" electricity delivery system and one of the great opportunities for Smart Grid development in the future. Only about 2.5% of total electricity in the U.S. is now provided through energy storage, nearly all of it from pumped hydroelectric (PH) facilities used for load shifting, frequency control, and spinning reserve. System balancing is provided by PH, combustion turbines, and the cycling of coal power systems. (In contrast, some 10% of the electricity produced in Europe is cycled through a storage facility of some kind, and Japan stores 15% of the electricity it produces.) Deployment and policy have been instrumental in long-term resource planning and management. (EAC 2014)

Storage is essential for electricity consumers where power quality and reliability is critical, such as at airports, broadcasting operations, hospitals, financial services, data centers, telecommunications, and many finely tuned industrial processes. Such operations frequently install energy storage as part of an uninterruptible power supply. In the future, storage—as both an end user and electric utility energy management resource—will become possible due to a confluence of high TOU rates, dynamic pricing, and lower cost energy storage systems.

Compressed air energy storage (CAES), pumped hydro, and advanced lead-acid batteries are the primary options for utilities pursuing bulk storage for T&D grid support and system and renewables integration.

Communications and IT Infrastructure for Transmission and Substations

Smart substations require new infrastructure capable of supporting the higher level of information monitoring, analysis, and control required for Smart Grid operations, as well as the communication infrastructure to support full integration of upstream and downstream operations.

The substation of the future will require a wide-area network interface to receive and respond to data from an extensive array of transmission line sensors, dynamic-thermal circuit ratings, and strategically placed phasor measurement units. The smart substation must be able to integrate variable power flows from renewable energy systems in real time, and maintain a historical record or have access to a historical record of equipment performance. Combined with real-time monitoring of equipment, the smart substation will facilitate reliability-centered and predictive maintenance.

The core IT infrastructure will be able to coordinate the flow of intelligence from critical equipment, such as self-diagnosing transformers, with downstream operations, and be able to differentiate normal faults from security breaches. It will be able to distill and convey critical performance data and maintenance issues to back office systems.

The smart substation will build upon the existing platform. There is already a significant installed base of sensors at substations, but there is still limited bandwidth connecting the substation to the enterprise. Historically, the communications channel to the substation was justified as part of the installation of the energy management system (EMS) and supervisory control and data acquisition (SCADA) systems. A key consideration for the future is that these legacy systems have limited bandwidth.

Intelligent Electronic Devices (IEDs)

Intelligent Electronic Devices (IEDs) encompass a wide array of microprocessor-based controllers of power system equipment, such as circuit breakers, transformers, and capacitor banks. IEDs receive data from sensors and power equipment, and can issue control commands, such as tripping circuit breakers if they sense voltage, current, or frequency anomalies, or raise/lower voltage levels in order to maintain the desired level. Common types of IEDs include protective relaying devices, load tap changer controllers, circuit breaker controllers, capacitor bank switches, recloser controllers, voltage regulators, etc.

With available microprocessor technology, a single IED unit can now perform multiple protective and control functions, whereas before microprocessors a unit could only perform one protective function. A typical IED today can perform 5-12 protection functions and 5-8 control functions, including controls for separate devices, an auto-reclose function, self monitoring function, and communication functions etc.

Phasor Measurement Technology

Phasor measurement units (PMU) or synchophasors provide real-time information about the power system's dynamic performance. Specifically, they take measurements of electrical waves (voltage and current) at strategic points in the transmission system 30 times/second. These measurements are time stamped with signals from global positioning system satellites, which enable PMU data from different utilities to be time-synchronized and combined to create a comprehensive view of the broader electrical system. Widespread installation of PMUs will enhance the nation's ability to monitor and manage the reliability and security of the grid over large areas.

Synchrophasor technology has demonstrated the potential to enhance grid planning and operations processes. Recent industry R&D efforts have focused on developing a variety of applications including situational awareness, small signal stability behavior, event analysis, model validation, state-estimation enhancement, and on-line voltage stability assessment. While the industry continues to explore the use of PMU data in real time and off-line environments, the lack of killer applications has impeded the widespread use of synchrophasor technology. A concerted industry R&D effort is warranted among the research community, end users (grid operators and planners) and EMS vendors to produce production-grade PMU data applications for the users.

PMUs provide system operators with feedback about the state of the power system with much higher accuracy than the conventional SCADA systems which typically take observations every four seconds. Because PMUs provide more precise data at a much faster rate, they provide a much more accurate assessment of operating conditions and limits in real time.

The ultimate link between PMUs and other Smart Grid technologies is only now beginning to be revealed. PMUs are themselves an enabling technology that may make investments in advanced communication infrastructures and IEDs more desirable. The full potential benefits of PMUs will not materialize by simply installing PMUs. Wide area measurement systems (WAMS) or wide area control systems (WQACS), which include PMUs, communications infrastructure, other control devices and software application algorithms, will be required to fully realize the potential for PMUs.

Advanced HVDC Technologies

Several new HVDC technologies could be developed which would further enhance the functionality of that technology. First will be a family of devices which facilitate flow control and for direct voltage control in a meshed HVDC grid as well as technology which offer better overall control and protection of HVDC grids. Companion developments would address technical requirements and specifications of state-of-the-art HVDC switching equipment. Other developments include:

- Fixed series capacitors
- Large capacity (>60 kA) circuit breakers
- >1500 MVA transformers
- Controllable reactors
- Low Noise conductors
- Double circuit technology
- Plug and play multi-terminal HVDC devices
- Hybrid HVDC/AC systems

Next Generation Grid Operating System

Both the U.S. Department of Energy (DOE) and the Electric Power Research Institute (EPRI) are developing roadmaps for the development of a next-generation grid operating system or energy management system (EMS). The nation's grid faces significant transformation in supply, demand, consumer expectations, and markets. Given the long lead time required to develop and implement fundamental changes in the way the grid is managed, it is imperative to anticipate this transformation. It is envisioned that this EMS system of the future will contain enhanced state estimation and visualization technologies. Development should also include consideration of leveraging evolving high performance computing (HPC), analytics and power system analysis algorithms to take advantage of HPC capabilities.

Superconducting Cables

New high—temperature superconducting—based transmission cable technology could offer an important alternative to conventional cables in the longer term. These cables use high-temperature superconducting materials instead of copper or aluminum and have substantially higher power handling capabilities than conventional cables. When used for HVDC, these cables exhibit zero electrical losses aside from parasitic refrigeration losses. When used on land, they can be buried in a small width right-of-way. They are currently being demonstrated in small distances. (EPRI Superconducting 2009)

Materials for Electric Utility Industry Research

There is an ongoing need for research in innovative technologies both for transmission and distribution applications. This should include things like nanotechnology applications for use in breakers, insulators, conductors and even fluids (for transformers and other devices). It may be argued that these are not Smart Grid technologies, but these material advances promote a smarter responding, more resilient grid. Materials research would include better power electronics devices and new insulating materials that would be a replacement for SF6 or other items facing upcoming pressure, higher efficient materials and more resilient applications.

Cyber Security

Electric utilities have been incorporating cyber security features into their operations since the early 2000s. In recent years as the Smart Grid became increasingly popular, cyber security concerns have increased significantly. While there have to date been few reliable reports of cyber attacks on power systems, there is a great deal of urban lore which suggests alleged attempts to disrupt the reliability of U.S. electricity supply.

Cyber security is an essential element of the Smart Grid. It is the protection needed to ensure the confidentiality and integrity of the digital overlay which is part of the Smart Grid.

The North American Electric Reliability Corporation (NERC) has created Critical Infrastructure (CIP) Standards. These include standards for Critical Cyber Asset Identification (CIP002) and Security Management Controls (CIP003) as well as others. Meeting these standards are part of Smart Grid costs. See Chapter 9.

At present, utilities are considering cyber security as part of information technology (IT) projects for:

- Advanced metering infrastructure
- Plug-in electric vehicle (PEV) management systems
- Distribution automation
- Substation automation
- Transmission upgrades

Interviews with industry suppliers by the EPRI team indicate that as a percentage of IT project costs, cyber security costs range from 10 to 15% of SCADA and distribution automation and approximately 20% for AMI.

Cyber security costs vary by the size of the utility. Urban utilities are likely to be more aggressive in Smart Grid deployment than suburban or rural utilities. For purposes of estimating cyber security costs, three utility sizes were used—small, medium, and large—corresponding to rural, suburban, and urban.

It is assumed large utilities will use applicable industry standards and best practices, including emerging security standards like NIST's Smart Grid Interoperability Standards Framework and AMI-SEC System Security Requirements, for end-to-end security of the Smart Grid. Most will implement intrusion detection and prevention services (IDS/IPS) as well as security information event management (SIEM). They will likely use a system-of-systems approach to cyber security by deploying International Organization for Standardization and International Electrotechnical Commission (ISO/IEC), National Security Agency InfoSec Assessment Methodology (NSA IAM), Information Systems Audit and Control Association (ISACA), and International Information Systems Security Certification Consortium (ISC2).

Many utility AMI systems will likely use a certificate-based solution for identifying and authenticating trusted devices, authorizing commands and encrypting communications between user entities (people, programs, devices). This solution applies industry standard cryptography to privatize all data transmissions and ensures that communications between authorized entities are confidential, trusted and legitimate.

Enterprise Back-Office Systems

All large utilities already have enterprise back-office systems which include geographic information systems (GIS), outage management systems (OMS), and distribution management systems (DMS). To enable the Smart Grid, additional features will be required, including an historic data function in conjunction with analytic tools to take in data streams, compare and contrast with historical patterns and look for anomalies in the data.

Impacts on System Operators

Independent system operators (ISOs), transmission system operators (TSOs), and other independent operators (referred to as ISOs here) are making investments in an increasingly robust communications infrastructure as well as an enhanced analytical and forecasting capability. These investments are being made in response to requirements for ISOs to incorporate increasing functionality in order to maintain reliability, meet load growth, and to comply to new regulations which are increasing grid compliance with FERC rules, increasing the use of distributed resources, demand response and energy efficiency. This at the same time as market operations are increasingly more complex, the threat of cyber security increases, and pressures to maintain costs and improve the use of assets increases.

All ISOs initiated these investments, as part of sustaining core capabilities even before the nation began to evolve the concept of a "Smart Grid." For example, the development of techniques for real-time simulation and enhanced visualization have been under development since the 1990s. Today, they are considered part of the Smart Grid—but would likely have been viewed simply as necessary improvements a decade ago.

Three key ISO functions which are considered part of the Smart Grid in this study are:

1. Enhancing the visibility of the grid (transparency), increasing reliability, and energy efficiency.
2. Enabling the effective integration of increasing amounts of distributed resources including renewables, energy storage, and demand response.
3. Enabling effective response to increasingly sophisticated cyber and physical security threats.

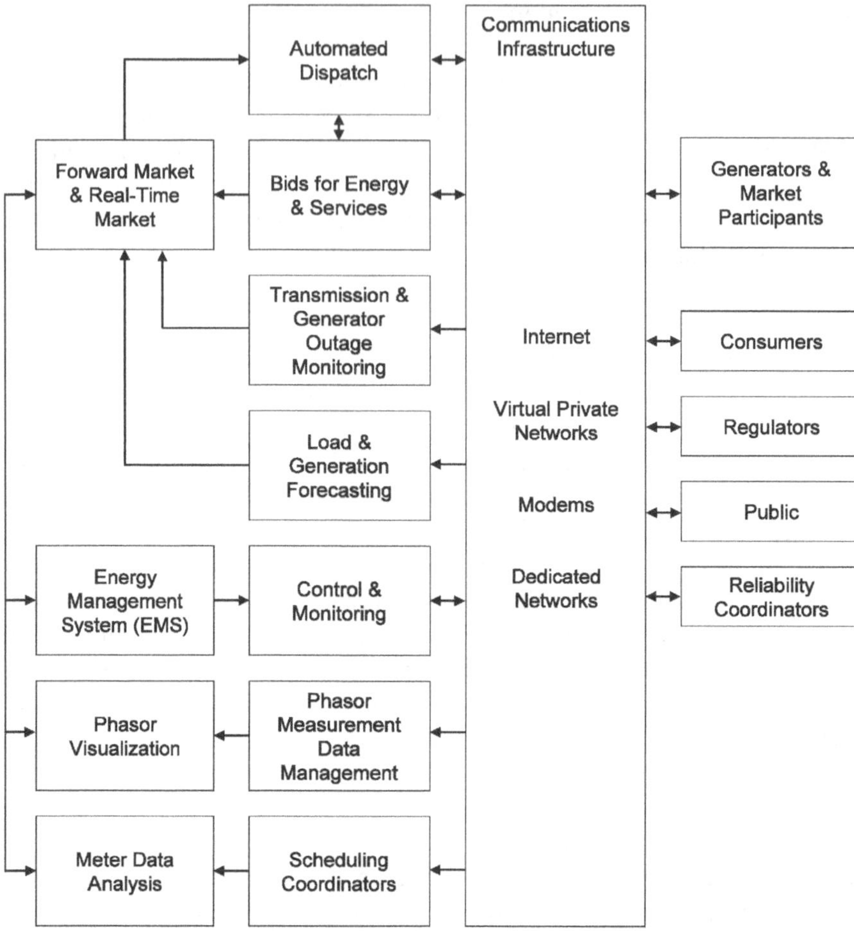

Figure 2-11. Key Components of an ISO Infrastructure (EPRI 1022519)

Figure 2-12 illustrates the key applications that form an ISO. The functional enhancements which the Smart Grid is expected to yield will require changes in these applications in the following areas:

- Communications—The increasing number of external interfaces will necessitate that telemetry data is compatible with IEC 61850. This will require retrofits and upgrades. In addition, the Smart Grid will contain many more transmission line sensors and sensors from wind and solar installations.

- Prices-to-Devices—Truly enabling consumers to respond to variations in prices and system constraints will require the ability to broadcast that information to consumers both through the Internet and through secure, automated area networks.

- Upgrade to NIST-Proposed Standards—The National Institute for Standards and Technology (NIST) has established a Smart Grid Interoperability Panel (SGIP), and the panel is developing Priority Action Plans (PAP) to encourage adoption of standard business practices and interface specifications. It is likely that these interfaces will require widespread adoption of the Common Information Model (CIM).

- Forecasting—Developing far more accurate data to forecast renewable energy production (particularly wind and solar) including very short-term forecasts and ramp rates. In addition, it will be necessary to forecast demand response participation.

- Cyber Security—The North American Reliability Council is developing Critical Infrastructure Protection Standards (CIP) to meet Federal Regulatory Commission (FERC) and Congressional requests.

- Integrate Synchrophasors—Integrating synchrophasor data sources with visualization and to enable advanced state estimation and dynamic stability analysis incorporating the data from numerous new distributed generation sources.

- Evolve New Markets—The advent of widespread consumer connectivity coupled with increasing participation of third parties allows for new markets to evolve. The ISO's system will need to be sufficiently flexible so as to allow the evolution of new markets.

- Other Applications—Individual ISOs may add other functionality like the ability to quickly add new market participants, to improve outage management data and other enhancements.

Each of these and other features will require an ongoing investment in the ISO's applications.

TRANSMISSION OPERATIONS—ADVANCED GRID MANAGEMENT TOOLS

Transmission system operators (TSOs), which is inclusive of traditional utility transmission owners/operators, independent system operators (ISOs), and regional transmission organizations (RTOs) and others (all referred to as TSOs) are making investments in an increasingly robust communications infrastructure as well as an enhanced analytical and forecasting capability. These investments are being made in response to requirements for TSOs to incorporate increasing functionality in order to maintain reliability, meet load growth, and to comply to new regulations which are increasing grid compliance with Federal Energy Regulatory Commission (FERC) rules, increasing the use of distributed resources, demand response and energy efficiency. At the same time, market operations are becoming increasingly more complex, the threat of cyber security is increasing, and pressures are continuing to mount to maintain costs and improve the use of assets.

A number of these investments were made as part of sustaining core capabilities even before the nation began to evolve the concept of a "Smart Grid." For example, the developments of techniques for real-time simulation of transmission operations and enhanced visualization have been under development since the 1990s. Today, they are considered part of the Smart Grid, but would have simply been viewed as necessary improvements a decade ago.

In order to further the ability of TSOs to take advantage of these investments, there is significant research and development underway in the area of AGM, or Advanced Grid Management. This involves the areas of data analytics, advanced mathematical algorithms, data management, and others that will allow system operators to use "faster than real-time data" to manage the transmission system.

DISTRIBUTION SYSTEMS

While a small percentage of electricity customers are served directly from the transmission system, the vast majority of the 165 million customers in the U.S. are served by the distribution system, which is comprised of a complex network of substations, lines, poles, metering, billing and related systems to support the retail side of electricity delivery. (EPRI 1022519)

Electricity distribution utilities are under pressure to improve reliability and system performance while dealing with the ongoing challenge of an aging infrastructure, conventional designs, and increasing customer demands for higher reliability and power quality. Budget and investment constraints require electric utilities to manage their distribution systems ever more efficiently, In addition, present expectations require that distribution systems support increased automation, new load types, increased distributed generation and storage, increased demand-side controls, environmentally friendly (green) technologies and more obligations to come. (EPRI 1022335)

Utility distribution systems are challenged by an aging infrastructure, conventional designs, and increased demands for digital-quality power. Compared to the transmission system, the greater complexity, exposure, and geographic reach of the distribution system results in inherently lower reliability, reduced power quality, and greater vulnerability to disruption of any kind. Using a reliability measure of average total duration of the interruptions experienced by a customer in a year, over 90% of the minutes lost by consumers are attributable to distribution events. In 2004, EPRI estimated that a fully automated distribution system could improve reliability levels by 40%. Advances in Smart Grid technologies may well prove that estimate to be overly conservative. From a reliability standpoint, feeder automation by itself only deals with outage symptoms, not root causes. For example, it is not a substitute for good maintenance practices, inspection, and vegetation management.

Investment in the distribution system has averaged $12 to $14 billion per year for last few decades, primarily to meet load growth, which includes both new connects and upgrades for existing customers. An urban utility may have less than 50 feet of distribution circuit per customer, while a rural utility can have more than 300 feet of primary distribution circuit per customer. Assuming a rough average 100 feet of line for each of the 165 million U.S. customers indicates the U.S. has an installed base of more than 3 million miles of distribution line. Upgrading a system this extensive to the level of performance required in a Smart Grid will require a substantial investment.

Distribution Planning for the Integrated-Grid Approach

It is becoming more common that utility distribution planners are faced with accommodating a large penetration of local generation and storage on their power distribution circuits. In many states, the renewable portfolio standards and incentives from various sources have resulted in larger solar PV installations than previously experienced.

The distribution planning process has traditionally been focused on determining the least cost alternative for meeting the peak load demand projected for some date in the future. The analysis is often simplified because loading patterns have been the same for many years, and there is much experience with dealing with these loading patterns such as for the rating of transformers. By looking at how the system behaves at one loading point, the planning engineer has a good idea of how it will respond at other times. The basic process can be summarized as follows:

- Define a distribution planning area (DPA) and model it for power flow analysis.
- Develop a load forecast for a selected planning horizon.
- Determine when planning limits on voltages and current capacities will be violated based on the load forecast.
- Identify one or more alternatives for correcting the violations.
- Determine the least cost alternative over the planning horizon using approved economic evaluation methods.

Research and development is needed in order to enable effective distribution planning. Advances are needed in:

- Distribution real-time modeling and simulation. Modeling the performance of smart inverters. Identifying how today's distribution management systems need to evolve.
- Identifying how best to utilize automated metering infrastructures (AMI) in a flexible load shape environment.

Distribution System Responsiveness, Resiliency, Flexibility and Customer Connectivity

The distribution system must become flexible; able to respond to changing system conditions; able to communicate real-time information; easily and efficiently accommodate; varied generation and load types; able to easily accommodate the installation of new equipment and equipment types; and to support system and end-use energy-efficiency and demand-response initiatives. The following are key components for achieving this future state:

- Utilities will be prepared to accommodate distributed generation, storage, and plug-in hybrid electric vehicles.
- Intelligent distribution devices will support plug-and-play capability.
- The distribution system will be designed with devices to manually or automatically respond to meet changing system conditions (demand needs, fault conditions, and so on).
- Utilities will use the Smart Grid infrastructure to accomplish core distribution tasks

The industry is developing and evaluating advanced distribution system applications for reliability improvement, system optimization, asset management, and integration of distributed resources. These applications involve implementation of monitoring equipment (sensors), communication infrastructure, and advanced protection and control functions. The results will support utilities in the migration to distribution management systems with model-based management of the system. The Distribution Management System (DMS) of the future will need to integrate many functions to optimize system performance, reduce losses, optimize voltage and VAR control, improve reliability through system reconfiguration and fast restoration, and integrate distributed resources. For some utilities, this relies on plans which build on the analytical capabilities of the Open Distribution Simulator Software (OpenDSS) for analytical assessment of advanced distribution-management functions. Early demonstrations of advanced functions for development of application guidelines and identification of gaps in the technologies are still needed.

Value and Risk

Results can be used to help specify voltage-optimization systems as part of the overall smart distribution development, develop the business cases for voltage-optimization functions, and evaluate performance of systems being implemented.

The results will provide the resources for needs assessment, development of business cases, and specification of advanced system reconfiguration functions that could be implemented as part of a distribution-management system, including:

- Performance assessment for advanced reconfiguration functions.
- Benefits that can be achieved with advanced reconfiguration functions.
- Requirements for advanced reconfiguration functions.
- A foundation for implementing voltage optimization systems as a function of distribution system characteristics and existing infrastructure.

Many challenges are impeding the implementation of Smart Grid technologies, including (1) existing designs and conditions of distribution systems that are not yet ready to support smart applications, (2) the need for plans for an orderly application of smart technologies, (3) standards and protocols for the technologies, (4) the need for plans and methods for investment cost allocation, and (5) justification for capital investments by utilities. In addition, new load and generation types are being added to the grid, creating an urgent need for the dynamic system response enabled by smart applications.

Integration of Local Generation

The existing electric distribution system is ready to accommodate a reasonable demand for deployment of local generation. Practices to establish hosting capacities, procedures for making interconnections, and methods for mitigating problems are well established. Operating the distribution system with distributed renewable generation does not compromise system reliability, power quality, or safety. However, as the use of local generation proliferates, a number of key developments will be required. The following are key components for achieving this future state:

- Distribution planners will have the data, experience, and needed modeling tools to plan for different renewable-deployment scenarios.
- Methods to determine feeder hosting capacities for variable distributed generation will be available.
- Distribution operators will know what distributed resources are

connected to their systems will be able to forecast and manage variability and will be able to employ mitigation strategies when needed.

- Distributed PV systems will include power and communication interfaces that are able to provide grid support.
- Communication protocols and interfaces will be available for distribution management systems and will be able to take advantage of advanced functions available from electronic inverters.
- Effective business models for utility ownership and operation of distributed PV will be available.
- Utilities that own and operate distributed PV systems will have established practices to minimize operation costs and avoid downtime.

Closing these gaps will require that the industry invest resources in developing and implementing new processes, tools, techniques, and technologies. Value to industry is derived by addressing the key drivers:

- Changing end-user preferences for electricity from clean and diversified generation sources, state renewable portfolio standards, and potential for a federal climate policy.
- A significant growth trend for distributed PV is bringing operational concerns and the need for new methods in distribution planning and design to maintain grid reliability and safety.
- Utilities recognizing the need to extend communications and to define functional interactions from substation to customer-sited local generation.
- Growing interest of utilities in ownership models, risk, and economic analysis and better decision-making techniques and development of strategies for engagement.
- New opportunities for rate-base recovery and leveraging the value of the utility distribution system to support renewable distributed generation.

The increased application of distribution generation and the application of demand-response initiatives as additional tools to meet capacity requirements will further challenge the distribution system. As the composition of the system changes, utilities will maintain the responsibility to meet demand with increasing efficiency, reliability, and quality.

Distribution System Communications Infrastructure

Communications constitute the critical backbone for integrating customer demand with utility operations. Detailed, real-time information is key to effectively managing a system as large and dynamic as the power grid. Each smart meter in the advanced metering infrastructure (AMI), described later in this section, must be able to communicate with a wide range of user control systems, as well as reliably and securely communicating performance data, price signals, and customer information to and from an electric utility's back-haul system.

No single technology is optimal for all applications. Among the communications media now being used for AMI applications are cellular networks, licensed and unlicensed radio, and power line communications. In addition to the media, the type of network is also an important part of communications design. Networks used for Smart Grid applications include fixed wireless, mesh networks, and a combination of the two. Several other network configurations, including Wi-Fi and Internet networks are also under investigation.

Interoperability remains one of the most critical success factors for Smart Grid communications. There is growing interest in the potential use of the Internet Protocol Suite (TCP/IP) as a networking protocol that could run over many different communication technologies. In June, 2009, EPRI submitted a Report to NIST on the Smart Grid Interoperability Standards Roadmap that became the starting point for NIST's own roadmap, released in September, 2009 (www.nist.gov/smartgrid).

Communication architectures remain diverse for integrating residential devices with the grid. Some utilities envision using the meter as a gateway to the home for price and feedback information, whereas others envision using the Internet or other communication channels. Radio frequency (RF) networks communicating in both licensed and unlicensed radio bands and have been chosen by the majority of Smart Grid deployments in the U.S. Mesh networks incorporate multi-hop technology where each node in the network can communicate with any other node. Star networks utilize a central tower that can communicate with a large number of end devices over a wide area. Each type has certain advantages and disadvantages and are selected based on the unique needs and circumstances of the utility. Power line networks where communications are carried via the grid network are also used by a large number of utilities both in the U.S. and abroad.

Distribution Automation

Distribution automation (DA) involves the integration of System Control and Data Acquisition (SCADA) systems, advanced distribution sensors, advanced electronic controls and advanced two-way communication systems to optimize system performance. The SCADA system collects and reports voltage levels, current demand, MVA, VAR flow, equipment state, operational state, and event logging, among others, allowing operators to remotely control capacitor banks, breakers and voltage regulation. Substation automation, when combined with automated switches, reclosers, and capacitors, will enable full Smart Grid functionality.

This includes not only building intelligence into the distribution substations and into the metering infrastructure but also into the distribution feeder circuits that link these two essential parts of the grid. This means automating switches on the distribution system to allow automatic reconfiguration, automating protection systems and adapting them to facilitate reconfiguration and integration of DER, integrating power-electronic based controllers and other technologies to improve reliability and system performance, and optimizing system performance through voltage and VAR control to reduce losses, improve power quality and facilitate the integration of renewable resources.

Intelligent head-end feeder reclosers and relays. Replacing electromechanical protection systems with microprocessor-based, intelligent relays and reclosers are an integral part of Smart Grid operation. Advantages include multiple functionality, including both instantaneous and time-overcurrent protection, greater sensitivity, better coordination with other devices, and the ability for self diagnosis. Approximately 70% of all feeders will include intelligent reclosers and relays by 2030, at an estimated unit cost of $50,000.

Intelligent reclosers. The use of intelligent switching and protection devices on feeders (referred to as "tie-reclosers") to allow isolation of segments of feeders and to thereby enhance reliability.

Remotely controlled switches. Remotely controlled switches contain distributed intelligence and use peer-to-peer communications to take actions without the need for central control intervention in order to isolate faults and restore power quickly in the event of an outage. As a result, distribution system operators will no longer be the only ones that can perform that function

Power electronics, including distribution short circuit current limiters. Advances in power electronics allow not only greater fault protection but flexible conversion between different frequencies, phasing, and voltages while still producing a proper ac voltage to the end user.

Voltage and VAR control on feeders. Voltage/VAR controls are a basic requirement for all electric distribution feeders to maintain acceptable voltage at all points along the feeder and to maintain a high power factor. Recent efforts by distribution utilities to improve efficiency, reduce demand, and achieve better asset utilization, have indicated the importance of voltage/VAR control and optimization. Utilities continue to face system losses from reactive load, such as washing machines, air conditioners. By optimizing voltage/VAR control through voltage reduction, power factor optimization, and conservation voltage reduction, great efficiencies can be realized.

Smart Grid software will be able to combine information flowing from the automated substations with SCADA data points throughout the distribution system to analyze and recommend re-configuration of the distribution system for optimum performance. Circuit optimization will minimize line loses and integrate customer data from AMI to regulate voltage while still maintaining acceptable levels for customers. This functionality will help support conservation voltage reduction (CVR) strategies to achieve energy savings.

Intelligent Universal Transformers

Conventional transformers suffer from poor energy conversion efficiency at partial loads, use liquid dielectrics that can result in costly spill cleanups, and provide only one function–stepping voltage. These transformers do not provide real-time voltage regulation nor monitoring capabilities, and do not incorporate a communication link. At the same time, these transformers require costly spare inventories for multiple unit ratings, do not allow supply of three-phase power from a single-phase circuit, and are not parts-wise repairable. Future distribution transformers will also need to be an interface point for distributed resources, from storage to plug-in hybrid electric vehicles.

Regional Energy Grid Interface (REGI)

The intelligent universal transformer (IUT) is a first-generation, power-electronic replacement of conventional distribution transformers. EPRI has developed an IUT which can serve as a "Renewable Energy Grid Interface" (REGI). The new concept includes a bi-directional power interface that provides direct integration of photovoltaic systems, storage systems, and electric vehicle charging. It will also incorporate command and control functions for system integration, local management, and islanding.

REGI will become a key enabler in the overall Smart Grid development strategy. It plays a transformational role by combining the traditional functions of a power transformer with new interface capabilities. It can seamlessly integrate widespread renewable energy technologies, including energy storage, electric vehicles, and demand response, while also providing an architecture that allows the operation of reliable local energy networks. The controller will interface with distribution management systems, energy management systems, and demand response systems to optimize overall grid performance and improve reliability.

Extending the Grid's Boundaries

Traditionally, the grid is viewed as the power delivery system which connects bulk power generation with consumers. Its boundary starts at the central generator's busbar and proceeds through the transmission and distribution system and typically ends at the meter. That model has been extended with the adoption of local generation, storage, plug-in electric vehicles and through the use of smart meters. In effect the boundary of the grid now reaches "beyond the meter." The grid now extends all the way from the central generator's busbar through the transmission and distribution system to the meter and is now poised to go beyond into buildings. This now enables a transactional relationship to be established between the bulk power system, local generation, storage and end-use appliances and devices.

For example, if the grid is extended so as to include inverters and the use of technologies like the intelligent universal transformer (IUT) and then the grid boundaries could be viewed as migrating further into the consumer's domain. Figure 2-12 illustrates this evolution and highlights the benefits it may bring to consumers.

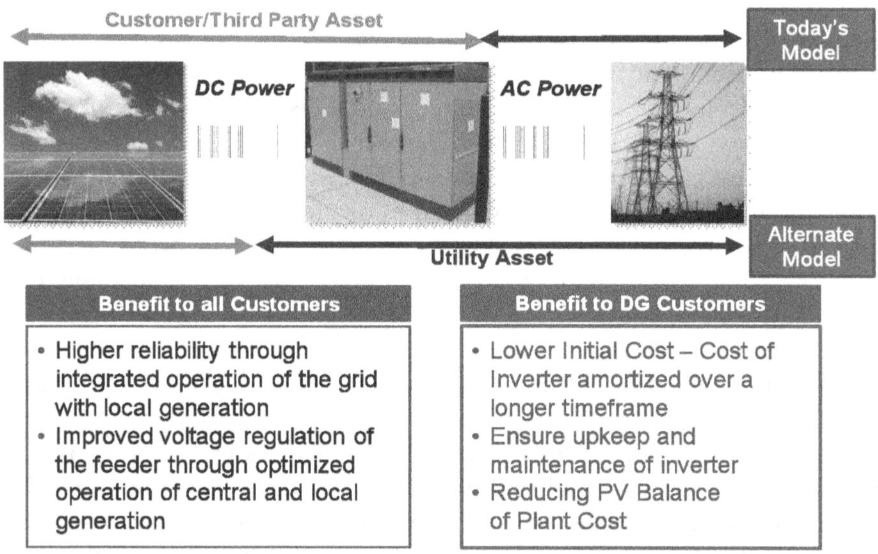

Figure 2-12. Extending the Distribution Grid to Include Inverters

Conservation Voltage Reduction and Volt-VAR Optimization

Volt/volt ampere reactive (VAR) control is an operating requirement for all electrical distribution utilities. However, in preparing to serve tomorrow's customer needs, it must now incorporate conservation voltage reduction (CVR) as utilities begin looking for ways to increase the efficiency of the power distribution system, to reduce peak demand, and to conserve energy. In tomorrow's world, not only CVR, but Volt VAR optimization (VVO), will be needed to reduce greenhouse gas emissions, improve system losses, reduce generation requirements, and reduce critical peak impacts to the system. The advanced communications that have been developed by the industry and implemented by the utilities have changed the capabilities of the CVR and VVO systems that are available today. (EPRI 1024482)

To assure the functionality required by tomorrow's grid, the industry must move from individual voltage or VAR controllers operating independently on the feeders to systems that communicate feeder readings back to a central VVO controller. These central controllers can then calculate optimal operating set points or commands for the system equipment and send them out over the system to create the optimal scenario for CVR and/or VVO on the feeders.

Each utility needs to assess the benefits of CVR and VVO on their system using modeling tools to analyze circuits for CVR/VVO. The effects of implementing various degrees of CVR/VVO on a circuit can be evaluated by using the Open DSS (Open Distribution System Simulator) modeling tool

The U.S. Department of Energy (DOE) has supported the development of additional technologies that rely on solid-state power electronics to provide more precise Volt-VAR optimization and greater efficiency. However, these solid-state technologies are in the later stages of development and demonstration.

Additionally, the development of analytical tools and information to facilitate the analysis and adoption of Volt-VAR optimization technologies are needed.

ADVANCED METERING INFRASTRUCTURE (AMI)

An advanced metering infrastructure (AMI) involves two-way communications with smart meters, customer and operational data bases, and various energy management systems. AMI, along with new rate designs, will provide consumers with the ability to use electricity more efficiently and to individualize service, and provide utilities with the ability to operate the electricity system more robustly.

Smart meters are the main component of AMI and generally the first technology deployed by an electric utility in a Smart Grid program. Although smart meters have been used by commercial and industrial customers for decades, only recently have they become economical for widespread use in residential settings. The broader AMI system in which smart meters operate involves the two-way communication network to exchange energy usage, price and curtailment signals, and operational control signals. Integral to AMI is a common enterprise bus network architecture linking all key enterprise systems including meter data management, customer care, auto-demand response system, and energy management. The goal is to provide a highly secure, resilient and flexible technology upgrade to the core business of electric utilities and to integrate electricity usage into Smart Grid dynamics. Three basic functions are involved:

Smart meters: Smart meters are capable of two-way communication with the utility, remotely programmable firmware, and, optionally, a remotely manageable service disconnect switch. In addition to consumption measurements, smart meter functionality includes: voltage measurement and alarms that can be integrated with distribution automation projects to maximize CVR benefits, and interval data to support dynamic pricing and demand response programs.

Communications: A communications system that is highly secure (encrypted), redundant and self-healing, and related hardware and software systems to communicate between smart meters, substation and distribution automation equipment, customer energy management systems, and head-end software applications/meter data management systems.

Meter data management system: A meter data management system capable of storing and organizing data, allowing for advanced analysis and processing, and interfacing AMI head-ends with a range of other enterprise software applications.

Controllers for Local Energy Networks

Local energy networks (LEN) are means by which consumers can get involved in managing electricity by reducing the time and effort required to change how they use electricity. If usage decisions can be categorized so they are implemented based on current information, and that information can be readily collected and processed, then consumers will purchase and operate such a system, install and operate a home area network (HAN), an electronic information network, connected to an energy management system (EMS), a decision processor. The HAN accommodates the flow of information to and from network nodes. Each node is associated with a device or element of the household's electric system. Nodes can be hard-wired devices that account for substantial portions of electricity used like the HVAC, a pool pump, lighting circuits, or smaller plug loads like TVs, entertainment centers, and a multitude of chargers. Communication among devices and the EMS is accomplished through wireless, wired, or power line carrier media that define and make operational the HAN.

Energy Management Systems (EMS)

An EMS is an intelligent device that acts as the coordinator for the devices that comprise the home area network. It maintains user-defined rules for when appliances and other household loads should turn on or off. These rules can be based on the price of electricity at a particular instance of time (e.g., when it exceeds some threshold), on current conditions (e.g., the time of day a household service is typically expected to run), or in response to a command to do so from an external agent (e.g., a curtailment order from a curtailment service provider).

Energy Management Systems (EMS) are the brain, making decisions based on exigent conditions viewed in light of a predefined instruction set, and the HAN is the neural system that conveys information about the state of the nodes and delivers commands and verifies their receipt and enactment. It is an electronic device whose purpose is to manage household electricity consumption better than the household can do so in its absence. Achieving that result requires understanding how the household members use and value electricity, establishing ways for them to negotiate differences in value systems, and establishing a holistic household utility function that establishes the relative value under different system states and executes pre-established operational decisions.

Architectures are evolving for marrying the Smart Grid with low-carbon central generation, local energy networks (LEN) and electric transportation. LEN includes a combination of end-use energy service devices, distributed generation, local energy storage, and integrated demand-response functions at the building, neighborhood, campus or community level. These architectures to facilitate a highly interactive network based upon a distributed, hierarchical control structure that defines the interactions of LEN, distribution systems and the bulk power system (Gellings, 2010).

These architectures facilitate the inclusion of multiple centralized generation sources linked through high-voltage networks. The design implies full flexibility to transport power over long distances to optimize generation resources and to deliver the power to load centers in the most efficient manner possible. In particular, these architectures enable the inclusion of inherently less controllable variable resources such as wind, solar and certain kinetic energy sources by offering a variety of balancing resources. To enable integration of these elements, these architectures must address the key transformative technical challenges shown in Table 2-4.

The concept of these distribution architectures is to optimize performance locally without complete dependence on the bulk power system infrastructure by taking advantage of the overall infrastructure

Table 2-4. Key Technical Challenges for Tomorrow's Distribution Architecture EPRI 1022335)

Operational Area	Modeling, Simulation, and Control	Monitoring, Data Management, and Visualization	Advanced Control Architecture	Control Structure
System Operations	Physical and distributed models, real time simulation, local and global system constraints	Widespread sensor integration with simulation tools and expert systems. visualization tools for decision making	Distributed control architecture – local optimization integrated with system management integration with simulation tools and expert systems. visualization tools for decision making	Autonomous Control Devices
Market Operations	Aggregate resource models, local vs. global optimization	Market and participant awareness	Market structures to support distributed control architecture	Autonomous Control Devices

to optimize energy efficiency and energy use. A key transformative element will be the development of distributed, hierarchical control devices that will be able to constantly balance generation and load at the device, home, neighborhood, city, area and regional levels. To achieve this vision, specific controllers will need to be designed and prototyped, tested and demonstrated in field applications to verify their interactions.

Planning Tools

While the industry is rapidly expanding its knowledge and experience in planning and operating large HVAC and HVDC systems with increasing levels of DETs, there are several areas where further research and development is needed. Power systems planners use computer models of components to design power systems. As new technologies evolve, there is the opportunity to develop robust data sets on the performance of those technologies under an array of conditions. This data is then used to develop models for use in mimicking the new power system. In the case of many of these new technologies, there isn't sufficient data to develop such models. This hampers the planner's ability to design tomorrow's grids for reliability and availability and to assess the range of configurations available so as to establish optimal designs.

Micro-Grid Architecture

Recent storms and the resultant extensive blackouts have raised the question as to what impact would expanding the use of micro- grids have and what would their preferred architectures be so as to enable their functionality and integration into the grid during normal and emergency operation.

Advanced Power Electronics

Increasing penetration of distributed energy technologies (DET), especially photovoltaic (PV) systems on the distribution power grid is introducing newer grid integration challenges for utility engineers. Over voltage, reverse power flow, excessive switching of capacitor banks and/or line tap changers are often experienced in circuits with higher penetration of variable generation sources like solar PV. Some of these technical challenges can be resolved or at least can be minimized by employing the full potential of power electronics inside the inverters interfacing these sources with the electric grid. Inverters with grid supportive functionality, including reactive power support, low/high voltage ride-through, watt-frequency, watt-voltage, and real power curtailment, can contribute to grid stability and hence assist to increase higher rate of renewable adoption.

DOE through its ARPA-e program has supported the initial development solid-state power electronics that can be deployed on distribution systems to manage power flows, optimize voltage and reactive power to reduce generation requirements and improve power quality, and in data centers and industrial facilities to ensure power quality and reliability for sensitive applications.

Next-generation Integration

A critical gap lies in the capability of the Distribution Management Systems (DMS) used to monitor and control distribution systems. While the state of DMS technology available to electric utilities is evolving from existing components, it is not clear that such an evolutionary development will keep pace with emerging requirements to manage distributed energy technologies.

DMS contain a number of elements. Critical among these is an Outage Management System (OMS). OMS systems are used to manage and restore service to customers following major storms and other events. Reliance on traditional Outage Management Systems (OMS) has become more inefficient as utilities deploy more Supervisory Control and Data Acquisition (SCADA) systems across their distribution grids. The need for an integrated, real-time distribution management system with computer-aided tools for switching, load flow analysis, fault location and isolation is prompting utilities to move toward DMS. DMS will support and enable the Smart Grid by becoming the single repository of certain real-time and near-real-time data, as well as, the power system connectivity model (asset characteristics, connectivity, and renderings). Current DMS systems are not tracking net load at the customer level in near-real time; and are not providing an integrated near-real-time model of transmission and distribution power flows; or

Dispatching or supporting distribution level markets to manage the operation of distributed energy technologies. The DMS in conjunction with the OMS becomes the information foundation to support the Smart Grid with:

- *Faster Simulation and Modeling*: Developing simulation tools that could provide faster simulation and modeling for improving situational awareness and decision support.

- *Self-Configuring Grid: Fault location information,* in combination with knowledge of electric connectivity and switch status/remote controllability, will allow the DMS to develop a new configuration to minimize the extent of outages.

Condition-based Maintenance

Usage and response data inform a condition-based maintenance application to analyze and develop different maintenance strategies. These strategies can be tested in a study environment within the DMS on the as-operated or as-built model.

There exist many subtle differences between DMS products: DMS systems have developed from different starting points—either SCADA or OMS—and evolved to their current forms. These different origins have colored product capabilities. Products derived from SCADA started with schematic one-lines and remain device-centric. These provide switch execution linked to supervisory control systems in native mode and advanced supervisory control device configuration capabilities. These DMS provide real-time architecture for tracking device configurations and managing centrally controlled distribution devices. Alternatively, products developed from OMS remain more customer and outage oriented. This path has provided more experience with distribution processes and provides crew management capabilities in native mode, but their SCADA/tagging capabilities are often somewhat cumbersome. These differences are beginning to blur as DMSs mature and offerings become more comprehensive.

Current DMS offerings serve to integrate certain information and operational systems and to handle the significant quantity of data characterizing the real-time management of the system. A fully functional DMS, given current technology, would consist of a single user interface for SCADA, OMS, distribution system state and switching information, distribution automation applications, and advanced analytics.

As previously described, the Smart Grid includes numerous new actors previously absent from the electric grid. A capable solution is needed to effectively integrate into transmission and distribution planning and operations the rapid proliferation of distributed energy technologies including intelligent energy using devices and other responsive demand, distributed and customer-sited generation and storage, and both utility and third party microgrids. Distributed energy technologies include the use of integrated or stand-alone generation and storage equipment. These can be owned by customer, utility or a third party and can provide power directly to the customer. When connected and integrated into the distribution system, distributed generation can potentially defer additional utility transmission and distribution system upgrades while purportedly improving owner quality and reliability. Compact, technically advanced units in a variety of technologies include diesel engines, lean burning gas-fired internal combustion engine generator sets, cogeneration packages, small "micro turbines," or renewable resources, such as photovoltaic or wind resources. For various reasons including renewable portfolio standards and other mandates, many utilities are exploring renewable generation initiatives. Additionally, many local generation initiatives are undertaken by the consumer without the involvement of the utility. Distributed technologies also encompasses storage technologies. The storage resources can be batteries, capacitors, or a variety of emerging technologies such as plug in hybrid electric vehicles.

Systems that include dynamic transmission and distribution topologies and distributed energy technologies will require more transparent and integrated planning and more detailed real-time operational management and coordination. The operation of the system in which distributed technologies can impact both distribution and transmission system operations may require:

- A federated control architecture connecting transmission and distribution operations;
- Integrated modeling and state estimation to give both transmission and distribution operators real-time awareness of power flows across transmission and distribution systems;
- A flexible, advanced information architecture to manage a large expansion in operational data, integrate an evolving set of information systems and applications, while maintaining cyber security;
- The ability to commit and dispatch or forecast and coordinate the operation of large numbers of distributed technologies, dynamically manage the topology of mesh or microgrid based distribution networks, optimize voltage, and simultaneously maintain phase balance across the distribution system; and
- Ultimately distribution level market structures that can coordinate settlement of transactions.

These requirements, in part, parallel the types of systems that were developed for the operation of RTOs and ISOs. However, efficient operation of a more highly distributed system will have to accommodate a larger number of control points and manage greater complexity. The extent to which control of such systems should be distributed and how such a distributed control system might operate are open research questions. The development of such systems will require significant, multi-year RD&D investments. And, there is a limited market to support entirely private sector development of such systems. A variety of economic factors and policies will continue to drive the deployment of distributed energy technologies. There could be significant reliability and economic costs if control systems do not advance at pace sufficient to efficiently manage a more distributed energy system. Moreover, if the U.S. were to cede a leadership role in the development of such technologies, this could increase the cyber-security vulnerability of domestic power systems. Therefore, RD&D is needed for information and control systems to support the efficient integration of a more dynamic transmission and distribution grid and for larger scale deployment of distributed energy technologies.

CUSTOMERS' TECHNOLOGIES

The Smart Grid allows the electricity customer to become fully integrated with the traditional electricity supply system. Such integration began decades ago with commercial and industrial accounts, but with more advanced and lower cost IT and communications technology, it is beginning to gain traction with residential customers. In 2008, FERC estimated 4.7% of the U.S. customers had advanced meters, and that 8% of U.S. customers were engaged in some form of demand response (DR) program. These programs increased the nation's demand-response potential to 5.8% of peak demand by 2008, or more than 40,000MW. DR potential is expected to grow rapidly over the next 20 years as AMI fully saturates the market.

There are more than 142 million customers in the U.S., of which 13% represent commercial and industrial accounts. The customer base is expected to grow 16% over the next 20 years to more than 165 million as shown in Table 2-5. However, since most consumer appliances will by then be DR-ready, the actual number of individual communication-connected end nodes will more than double.

Table 2-5. U.S. Electricity Customers (EPRI 1022519)

Number of Electric Customers	2007	2030	Load Growth
Residential	123,949,916	143,928,676	19,978,760
Commercial	17,377,219	20,178,151	2,800,932
Industrial	793,767	921,709	127,942
Transportation	750	750	0
Total	**142,121,652**	**165,029,286**	**22,907,634**

Phasor Measurement Units for Distribution Systems

With the increasing number of Distributed Energy Resources, many of which are variable and low inertia, PMUs applied at the Distribution System is an area that needs research around the applications and modeling required for real time applications. Micro-PMUs would be used to provide real-time data and visibility for electric distribution participants and systems to use for diagnostic and control applications to support the integration of distributed energy resources. Because of the relatively small angles and high noise-to-signal- ratios on the electric distribution system, the demands on micro-PMU technology design and performance at the distribution level are about an order of magnitude higher than those for transmission. Research and development is underway on micro-PMUs and demonstrations of prototypes are in startup phases at universities and selected utilities in the US. For example, under ARPA-E Award #DE-AR0000340, Micro-Synchrophasors in Distribution Systems, the California Institute for Energy and Environment (CIEE)/University of California Berkeley, in conjunction with Power Standards Lab (PSL) and Lawrence Berkeley National Lab (LBNL), are conducting a three-year research project to develop a high-precision micro-synchrophasor, or μPMU, and to study its applications for diagnostic and control purposes in distribution systems.

Integrating Customers Enables New Functionality

Over the next 20 years, integrating the customer into the Smart Grid will enable new functionality to take hold:

- **Increase demand response capabilities** and enable time varying rates to reduce costs, improve load factors, and optimize the economic performance of the grid.

- **Facilitate integration of DER,** including a variety of customer-owned systems, such as rooftop photovoltaic (PV) systems.
- **Integrate the transportation system into the grid** via electric vehicles (EVs) and plug-in hybrid electric vehicles (PHEVs), which can provide the distributed means for large scale electricity storage for the grid, balance daily load cycles, and substantially reduce the nation's oil dependency.
- **Improve energy management** in the home and commercial buildings, reducing peak demand and improving energy efficiency.
- **Optimize grid performance** by using the demand side as a resource for stabilizing the grid, for shaving peak demand, and for reducing the capital investment requirements for peaking generation.

PV Inverters

Inverters are microprocessor-based units used to transform dc to ac power that can be used to connect a photovoltaic (PV) system with the public grid. The inverter is the single most sophisticated electronic device used in a PV system, and after the PV module itself, represents the second highest cost. It is also considered the weakest link. Whereas, solar panels are very robust and carry 25-year warranties, inverter warranties have traditionally been in the 5 to 10 year range. Inverter reliability, however, has been trending up.

There are many types of inverters. Some are stand-alone units isolated from the grid and used to support a stand-alone rooftop system; others are grid-tied, in which case the microprocessor circuits are more elaborate and require additional functionality, including lightning protection. Central inverters are used in large applications. Many times they can be connected according to "master-slave" criteria, where the succeeding inverter switches on only when enough solar radiation is available. Module inverters are used in small photovoltaic systems, such as household rooftops.

A new generation of micro-inverters holds promise to increase PV performance. With current PV design, all solar panels are connected in series, so that if any panel in the series is shaded, it brings down the performance of the entire system. Moreover, for a series module to work, all panels have to have the same orientation and tilt, which limits roof top configuration. The micro-inverter scheme, on the other hand, allows each panel to be connected to its own micro-inverter, increasing overall system performance and providing flexibility for the staggered roof designs of many modern homes. Austin Energy, among others, is testing new micro-inverter designs.

Residential Energy Management System (EMS)

A residential EMS is a system dedicated (at least in part) to managing systems such as building components or products and devices. Residential EMS systems are not typically called "portals" in today's parlance. Portal is a term commonly applied to a web portal. This subdivides into several components including resident EMS and intelligent home devices (IHD). In addition the system may handle customer preferences and occupancy via a schedule, on-demand, or occupancy sensing automation. The line between a residential management system that handles lighting, family calendars, shopping or replenishment, and an EMS has become fuzzy. While proponents of a dedicated device propose that a homeowner will eventually purchase such a device, we are seeing parallel development of other approaches where the core of the system is a software application bundled on a server located at a third party data center.

Online energy management portals offer customers insight into their energy usage and automatic management of energy efficiency. Through a central view on a web page, for example, customers can access current energy usage statistics, historical usage patterns, and the amount of carbon dioxide emissions avoided by utilizing a renewable energy source. The portal can also display price signals and tie a customer's energy consumption and production patterns into their utility's rate schedule. Current standards developments may also enable effective aggregation and third party information sharing that will impact the adoption of a residential EMS. As of the writing of this document, preliminary information regarding consumer purchases of advanced residential EMS systems show promise but adoption to date has been low.

These aspects make it difficult to pin the price tag onto the residential EMS. Many components have a dual purpose and exist under separate financial justifications. Consumer reluctance to purchase an EMS may be driven by on-line options that could to replace key parts of the functionality of an EMS. These issues could either imply that the cost per customer is low, or the penetration rate is low. However, the end result should be similar regardless of which way we apply this observation.

In-home Displays and Access to Energy Information

Providing real-time feedback on energy consumption holds significant promise to reduce electricity demand. Several studies over the past 30 years have evaluated the effectiveness of energy savings from home energy displays of varying sophistication. Most of these studies verified savings between 5% and 15% with a longer-term sustained impact toward the lower end of this scale. Other studies have found that information alone does not appear to be sufficient to achieve appreciable reductions. People need a strong motivation to change, such as compensation, confidence they can change, and feedback that changes they do make are having an impact. As such, most successful approaches provide more frequent feedback, as well as feedback on specific behaviors. (EPRI 1022519)

As the Smart Grid unfolds, various methods to provide energy, cost, and environmental information are beginning to emerge. A specific class of stand-alone devices has been utilized extensively and is referred to as the in-home display (IHD). Typically, IHDs present basic information, such as real-time and projected hourly electricity cost and electricity consumption (kWh). Some can display additional information, such as electricity cost and consumption over the last 24 hours, the current month (and/or prior month) consumption and cost, projected usage, monthly peak demand, greenhouse gas emissions, and outdoor temperature. A similar approach is a component of a prepayment system, also known as a pay-as-you-go system, since these also have a display. The very nature of the pay-as-you-go billing encourages consumers to keep an eye on the display to monitor their usage and know when they will need to replenish their energy account.

In contrast, there are simpler approaches that may not be quantitative. That is, they do not include feedback on electricity consumption or electric demand. Such devices do not require direct attention, but effectively communicate information peripherally. For example, a small glowing ball has been used to indicate a higher electric price or energy demand by changing colors. Other implementations have developed a simple plug-in device with red, green, and yellow lights as simple indicators of energy demand or price.

Small-scale demonstrations have utilized more sophisticated home energy displays. Typically, they provide much greater detail about electricity consumption broken down by different end uses, or circuits, and use richer display graphics. In most cases, these advanced displays are part of a more comprehensive system that may have many features beyond energy management. This makes allocation of the Smart Grid component cost more difficult to pin down since energy management may or may not be the driving force behind the decision to interact with the device.

Alternative methods to provide energy information to the consumer continue to emerge. As the standards development processes move forward, additional innovations in this area will continue to become available. Some of these products offer an alternative to the dedicated IHD device or at least a subset of functionality. Although the consumer must have a physical means to view the information, the means may already exist in some form such as the PC/laptop, cell phone and PDA. Additional developments might use any products with a consumer facing display as a location to display energy information. This includes appliances, security systems and any new consumer product categories that may be on the drawing board.

Market and Rate Design

Accelerating end-use energy efficiency, increasing consumer adoption of energy generation technologies, the emergence of intelligent energy using devices, and rapid adoption of new technologies such as all forms of distributed energy technologies, including PEVs and energy storage imply a changing business environment. This may necessitate changes in utility rate design, particularly for those customers who choose to take advantage of these new technologies. Rate design changes may also need to accommodate flat or decreasing kWh sales.

Understanding Consumer Behavior

To meet the needs of tomorrow's retail customers served by distributors, it is critical that the industry develop an enhanced understanding of consumer behavior regarding electric technologies. Specifically, the industry needs a comprehensive and accurate evaluation of consumer perspectives toward Smart Grid technologies in order to further the understanding of motivators for consumer adoption. Increasingly, behavioral economics and other social science disciplines are providing insights that can help improve the designs of interfaces between the consumer and Smart Grid technologies. Some smart technologies, including intelligent thermostats, have already been adopted by millions of electric customers often without receiving any utility incentive. In other cases, utilities may seek to test Smart Grid technologies that interface with users and programs approaches aimed at consumers using valid experimental design methodologies prior to providing adoption incentives. Better understanding consumer preference will help coordinate activities ranging from designing informational materials to assisting stakeholders in relating to consumer attitudes in order to improve engagement (EPRI 1024565).

This includes a desire to understand the value drivers in consumer terms and to answer questions such as "How would a consumer experience the technology and benefits of the Smart Grid?" Some intelligent energy using devices will automatically implement customer preferences after purchase of the smart device or its initial online or point-of-purchase program enrollment and require no further engagement with the customer. Other Smart Grid technologies may involve greater consumer engagement. In these cases, utilities must be able to go beyond acceptance and create a "pull" whereby consumers are asking to become involved in Smart Grid technology.

Among the research needs are the following:

Motivation: What are the key drivers for adoption and ongoing appropriate use of technology which facilitates consumer connectivity to the Smart Grid? This research needs to recognize that consumers are motivated in a broad range of ways and to take advantage of the growing body of behavioral and social science research.

Understanding: Understanding technology and how to relay this in consumer terms through an understanding of consumer perception, awareness and interest level. Approaches should be tested under experimentally designed pilot conditions.

Education: Identify keys to effective education and a common industry message to enable the same story to be told, which can promote a consistent level of understanding.

Incentives and Program Designs: Education alone is often insufficient to motivate consumer action. Consumers are motivated by financial and other incentive or program design aspects such as convenience, trust, competition, social norms, etc. Additional research is needed to refine these areas.

Competitive Market Development: Given appropriate wholesale pricing and settlements, utilities and competitive retail electric suppliers may package electric supply, intelligent devices, and/or other services in a service offering that helps the customer manage their energy bill. Additional research may help utilities and regulators better understand the market structures and rules that provide economic incentives for suppliers to compete in providing energy services and improve power markets.

Grid-ready Appliances and Devices

Grid-ready appliances do not require truck rolls to retrofit with remote communications and control capabilities. Grid-ready appliances and devices, which are often referred to as "DR-ready," are manufactured with demand-response (DR) capabilities already built in. The universal entry of grid-ready devices into the marketplace, which is fully anticipated to start to take shape in the next several years, will lead to ubiquitous demand response capability.

The average American home has 4.67 appliances per home, with the refrigerator being the most universal (99.8% of U.S. households have a refrigerator based on U.S. Census data). The number of households in the U.S. is projected to reach 143,928,676 by 2030. The study team assumed that the first grid-ready appliances will start to appear in 2011. The penetration of DR-ready appliances is expected to approach 40% over the next 20 years. To account for homes with electric water heating and air conditioning "appliances," the team increased the average to 5.67 appliances per home. The reader could argue that many water heaters are electric and not every home has AC. There is a balancing argument that the number of appliances per home may also grow, leaving the team comfortable with the 5.67 appliances per home as being a reasonable estimated average.

By using data from AHAM (Association of Home Appliance Manufacturers) for the life expectancy of major appliances, the team used the average life expectancy of 13.91 years for each appliance. By dividing this into the 20-year span of the study, the average appliance would be replaced 1.44 times during this time.

Sales figures indicate that a single year of appliances sales has trended toward 10% of the current installed base. This number includes both new construction and replacement sales. Since we are using the projected number of households for the year 2030 which includes new construction, the replacement from appliance life expectancy was used against the year 2030 households projection to avoid double counting. (EPRI 1022519)

Plug-in Electric Vehicle Charging Infrastructure and On-vehicle Smart Grid Communications Technologies

Plug-in electric vehicles (PEVs) are defined as any hybrid vehicle with the ability to recharge its batteries from the grid, providing some or all of its driving through electric-only means. Almost all of the major automotive manufacturers have announced demonstration or production programs in the 2010 to -2014 timeframe, and their announced vehicles feature all-electric, plug-in hybrid-electric and extended-range electric vehicle configurations. Notable and earliest introductions among these are globally targeted production vehicles from General Motors (Chevrolet Volt, extended-range EV) and Nissan (Leaf, battery-only EV). It should also be noted that Tesla remains the manufacturer today producing their two-seater roadster and continues to work on their lower-priced, family (Model S) sedan.

EPRI is deeply involved in developing and demonstrating on- and off-vehicle PEV to Smart Grid connectivity technologies. Automotive manufacturers are bringing their early generations of PEVs designed for grid rechargability with uni-directional electric power flow capability (from grid to vehicle). These first generation of vehicles will be relying on the consumer inputs via in-vehicle interactive touch panel display or through cellular/Telematics connectivity, to program vehicle charging, input pricing or other information related to on-board energy management. Given this state of technology on board the PEVs, the initial impetus for any type of demand response and load control as well as critical peak pricing program implementation has been on the off-board "charging" station equipment for which the technical term is electric vehicle supply equipment (or EVSE), given that the off-board equipment is just a glorified 240 V outlet while the actual AC/DC power conversion and energy management actually occurs on-board the PEVs. The most predominant costs associated with the Smart Grid infrastructure capability will, therefore, be in developing Smart Grid to PEV and Smart Grid to "charging" station (aka, electric vehicle supply equipment or EVSE) connectivity and communications.

All applications which may provide reverse flow power capability such as vehicle-to-grid (V2G) are unproven. Their impact on battery durability, utility/automotive/consumer acceptance, and economics are yet to be demonstrated. In addition, it is uncertain what services they will enable and whether policies and incentives will be needed to bring them to reality. V2G, therefore, remains on R&D agenda of several automotive manufacturers, EPRI, some ISOs/RTOs, and several R&D institutions. It is, however, too early to predict the direction and magnitude of this technology's installed base any time in the near future.

PEV Load Management

In terms of load management or time shifting of the load due to PEV charging, both utilities and automotive manufacturers have agreed to jointly pursue standardization activity that will enable the PEVs to act as just another appliance on the AMI or HAN. Two activities within SAE, under J2836 (use cases) and J2847 (data specification) with extensive automotive and utility participation, are in the process of defining the requirements for PEV to Smart Grid communications, which will enable the PEVs to be utility-controllable distributed resources for load shifting, demand response, and pricing-signaling purposes. ZigBee Alliance and HomePlug Alliance have created the Smart Energy Initiative, which is crafting the Smart Energy 2.0 (SE2.0) specification for AMI and HAN applicability. SAEJ2836/J2847 and SE2.0 teams are working together to coordinate the data exchange requirements between Smart Grid and PEVs, and a draft Marketing Requirement Document (MRD) and Technical Requirements Document (TRD) are currently being refined for late 2010 ratification. SE2.0 and J2847 are expected to define a consistent set of data specifications for the charging load of the PEVs to be controllable by the same utility load management systems that determine demand response and load control signals for other loads, such as air conditioners, smart thermostats, or other smart appliances. The SAEJ2836 use cases include enabling PEV owners to enroll into utility demand response, load control and special incentive pricing programs and then program their vehicles to accept or reject utility requests for participating in demand response, load control and critical peak pricing-related events. (EPRI 1022519)

For the first generation of PEVs, the technology options for integrating PEVs with the Smart Grid will reside off-board, in the form of the closed, proprietary networks of charging station operators such as Coulomb Technologies, ECOtality, and Silver Spring Networks. Whereas significant public funding to the tune of $300M through stimulus awards from federal government and state and local authorities has been directed towards focused regional charging infrastructure buildout, the focus has been on enabling PEV technology adoption in early adopter markets, rather than on scalability and cost competitiveness of these technologies longer term. EPRI's collaborative research with the automotive industry indicates that for PEVs to be widely deployed, the infrastructure overhead for them would need to get down to minimal to none, with each PEV carrying its own required technology on-board that can connect to the nearest Smart Grid node, using either AMI/HAN to connect to the Smart Grid through the "front-end" or the on-board Telematics-based technology to connect through the "back end" to the utility back office systems and meter data management systems through standardized server to server communications.

To this end, EPRI envisions the PEV manufacturers to quickly integrate the standards-driven communication technologies on-board the PEV. The only significant costs for "Smart Grid-enabled" PEVs will, therefore, be the cost-plus for incorporating the communications hardware necessary to send/receive data from the utility based on applicable standards. The automotive and utility industries have agreed for PLC- (power line carrier-) based wired interface to be the physical interface between the PEV and the AMI/HAN, with the PLC(HomePlug AV or IEEEP1901 are the currently adopted technologies) transceiver chipset and associated Smart Grid communications "application layer" software with requirements defined by SAEJ2836/J2847 and SE2.0, residing on-board, as well as a PLCto X bridge residing off-board, with X being the transport layer of the AM I/HAN network, which also implements SE2.0-based messaging as the application layer.

Communication Upgrades for Building Automation

Today, over one-third of the conditioned and institutional buildings in the U.S. have some form of energy management and control systems installed (EPRI 101883). Automated demand response (ADR) can be accomplished by communicating to the building energy management system using an Internet-communicated signal or some other form of direct link. Open automated demand response (OpenADR) involves a machine-to-machine communication standard that provides electronic, Internet-based price and reliability signals linked directly to the end-use control systems or related building and automated control systems (EPRI 1016082). The building automation system is pre-programmed to reduce load according to the messages it receives, and it may also provide real-time energy consumption information back to the utility or service provider.

Experiences with Open ADR presumes the building has an existing

EMS system. There are two cost components that enable the building to respond to DR signals. The first is enabling the existing EMS to receive the DR signals. In some cases, this might mean upgrading the software, and in other cases, this might mean installing a "simple client" whose only purpose is to receive the DR signals and pass them on to the EMS system. One of the features of OpenADR is to allow very simple and inexpensive clients to be built that can interface to existing EMS systems via dry-relay contacts. Dry relay contacts seem to be the near-universal interface mechanism for EMS systems.

The second and perhaps largest cost component is the programming of load control strategies in the EMS. The cost is primarily one of manpower that involves audits of loads in the facility and specialized knowledge of how to convert the EMS to implement load control strategies. In this regard, the simple response levels sent as part of an Open ADR signal can be used. In many cases, it is more convenient for the facility manager to think in terms of "normal, moderate, and high" response levels instead of prices or specific dispatch commands. Also, it is not insignificant that if the engineers set up their load control strategies based upon simple levels, then they can more easily move between different programs without the need to reprogram their EMS system.

Customer-side—Electric Energy Storage

Advanced lead-acid batteries represent the most prevalent form of electric energy storage for residential, commercial and industrial customers wanting to maintain an uninterruptible power supply (UPS) system. As shown in Table 7-9, commercial and industrial systems can supply power for up to 8 hours at 75% efficiency, and maintain performance through more than 5000 cycles. Residential versions typically involve two hour duration at 75% efficiency and 5000 cycle performance.

Both standby and online UPS technologies are available. The online UPS is ideal for environments where electrical isolation is necessary or for equipment that is very sensitive to power fluctuations. Although once previously reserved for very large installations of 10 kW or more, advances in technology have permitted it to now be available as a common consumer device, supplying 500 watts or less. The online UPS is generally more expensive but may be necessary when the power environment is "noisy" such as in industrial settings, or for larger equipment loads like data centers, or when operation from an extended-run backup generator is necessary.

In an online UPS, the batteries are always connected to the inverter, so that no power transfer switches are necessary. When power loss occurs, the rectifier simply drops out of the circuit and the batteries keep the power steady and unchanged. When power is restored, the rectifier resumes carrying most of the load and begins charging the batteries, though the charging current may be limited to prevent the high-power rectifier from overheating the batteries and boiling off the electrolyte. The main advantage to the on-line UPS is its ability to provide an electrical firewall between the incoming utility power and sensitive electronic equipment. While the standby and Line-Interactive UPS merely filter the input utility power, the Double-Conversion UPS provides a layer of insulation from power quality problems. It allows control of output voltage and frequency regardless of input voltage and frequency.

References

- "Technical Architecture for Transmission Operations and Protection: Envisioning the Transmission Smart Grid," EPRI, Palo Alto, CA: 2008. 1016055
- "8th FACTS Users Group Meeting: Presented Material at the 8th FACTS Users Group Meeting," EPRI, Palo Alto, CA: 2005. 1010633.
- "Report to NIST on the Smart Grid Interoperability Standards Roadmap," (Contract No. SB1341-09-CN-0031), D. Von Dollen, EPRI, June 17, 2009
- Future Inspection of Overhead Lines, EPRI, Palo Alto, CA 2008. 1016921
- Distribution Research Area Strategic Plan: September 2010. EPRI, Palo Alto, CA: 2011, 102335
- A National Grid Energy Storage Strategy: Offered by the Energy Storage Subcommittee of the Electricity Advisory Committee (EAC)," January, 2014
- Estimating the Costs and Benefits of the Smart Grid: A Preliminary Estimate of the Investment Requirements and the Resultant Benefits of a Fully Functioning Smart Grid. EPRI, Palo Alto, CA: 2011. 1022519
- EPRI Program on Technology Innovation: A Superconducting DC Cable, Electric Power Research Institute (EPRI), Palo Alto, CA, 2009
- Grid Strategy 2011: Conservation Voltage Reduction and Volt VAR Optimization in the Smart Grid, EPRI Report 1024482, November 2011

Chapter 3

Smart Grid Roadmaps

TECHNOLOGY ROADMAPS

Roadmaps, or more correctly for this book, Technology Roadmaps have many different applications. As explained by a publication of The International SEPT Program at Universitat Leipzig (SEPT 2009), technology roadmaps can be applied for eight areas:

1. Planning of Products
2. Planning of Services and Capabilities
3. Strategic Planning
4. Capabilities and Knowledge Planning
5. Process Planning
6. Long Run Planning
7. Integration Planning

Roadmapping as applied to the electric power sector was popularized by EPRI in its 2003 Report titled "Electricity Technology Roadmap: Meeting, The Critical Challenges of the 21st Century.

The original Electricity Technology Roadmap ("the Roadmap") provides a global vision for realizing electricity's essential value to 21st century society, a plan to set strategic technological priorities, and an outline of the associated research, development, and delivery requirements needed to achieve this vision. The Roadmap's goal was to encourage the debate, consensus, leadership, innovation, and investment that will enable electricity to realize its potential for increasing the quality of life on a global scale. (EPRI 1010929)

Given the central importance of electricity to enhancing the quality of life, EPRI launched the Electricity Technology Roadmap initiative with the fundamental societal concerns of the 21st century's five essential destinations:

1. Strengthening the power delivery infrastructure
2. Enabling the digital society
3. Boosting economic productivity and prosperity
4. Resolving the energy / environment conflict
5. Managing the global sustainability challenge

The Roadmap Destination flows from a set of logical propositions that begins with the current state of the domestic power infrastructure and continues through to the benefits of global electrification.

The roadmap identified critical development and investment decisions which need to be made to transform the electricity system for the pressing needs of the new century. That effort identified three overriding imperatives:

- Meet the precision-power requirements of the emerging digital economy
- Create the clean-energy portfolio required for long-term environmental sustainability
- Accelerate access to the benefits of electricity among all the people of the world.

SMART GRID ROADMAPS

Roadmapping can be applied to the Smart Grid in order to facilitate planning which can take the generic concept of the Smart Grid through to implementation plans. Along the way, roadmapping can facilitate outlining conceptual architecture, conducting benefit assessments, defining a technology stage-gate process, testing, demonstration and deployment. The Smart Grid roadmap should link regulatory constraints, corporate strategy and consumer needs with vendor technology and standards. Properly executed it will produce a time ordered sequence of actions. Smart Grid Roadmapping is activated after the utility has defined its corporate strategy.

Methodology

A number of utilities have employed EPRI's Smart Grid Roadmap Methodology (SGRM). The SGRM is a hybrid of strategic, technology,

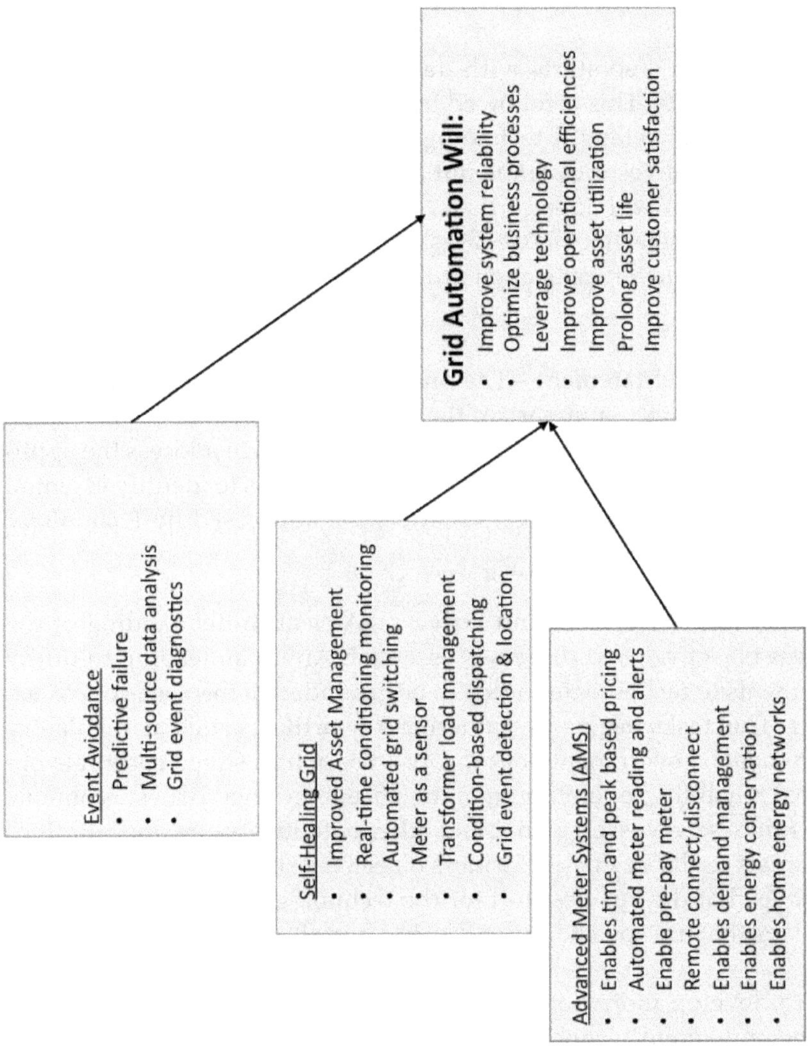

Figure 3-1. CenterPoint Energy Roadmap Summary

business and implementation planning. (EPRI 1025470). It utilizes five key steps: Vision; Requirements; Assessment; Planning; and Roadmap Implementation. Within each step there are three or four recommended tasks. The SGRM is shown in Figure 3-2.

Vision

The Vision step starts with developing the utility's Smart Grid Vision Statements. This is followed by obtaining the business objectives and drivers that relate to technology planning and deployment. The business imperatives dictate the initial selection of the applications and technologies that will address them. This list of applications is used as an input for the selection of use cases. An application chosen at this stage may be a higher level application utilizing one or more technologies and standards.

The Vision Statement—The purpose of a Smart Grid vision statement is to succinctly summarizes the utility/ISO's goal to both leverage existing and adopt new technologies and standards to address the applicable business objectives and drivers. The purpose is to identify essential business objectives and drivers that can be addressed by technology investments.

Business Objectives and Drivers—A clear understanding of the business objectives and drivers is essential. Any plan leading to utility investments in technologies needs to be grounded in the needs of its customers. This tasks begins by gathering key artifacts such as regulatory requirements; government policy; corporate policy; stated corporate objectives; reliability; operating; maintenance; efficiency issues; economic opportunities; risk assessments; and other potential future opportunities and threats to the business. Business objectives should include both key objectives that the business has for the technology investments and the objectives that are specific to the Roadmap itself.

To develop more detail around the business objectives use cases and recommended technologies it is helpful to identify these imperatives in the form of drivers with internal and external sources. These drivers can then be easily mapped to technologies, use cases and technology investment recommendations for future traceability.

Figure 3-2.
EPRI Smart Grid Roadmap Methodology (SGRM) (EPRI 1025470)

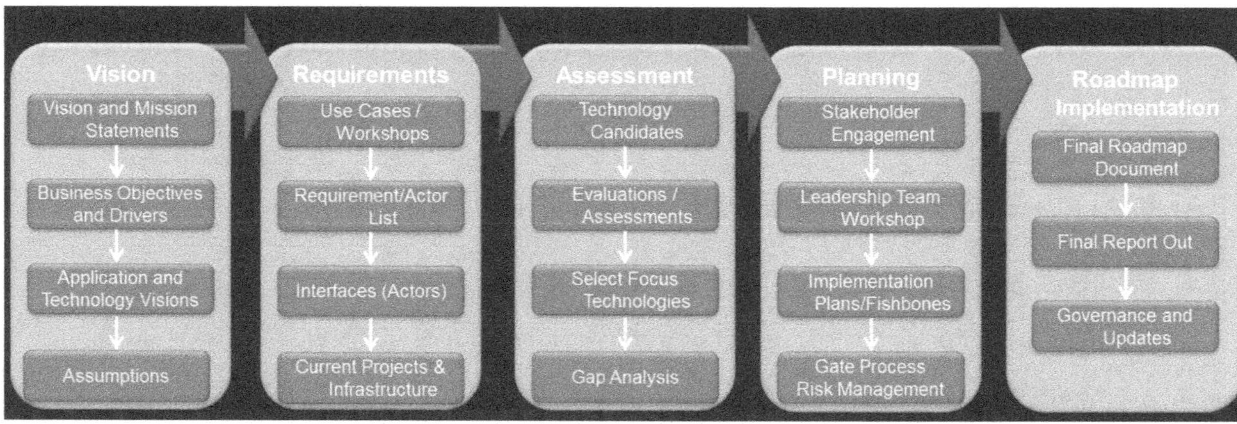

Application and Technology Vision Statements—The initial set of high level applications are chosen to address the priority business objectives. These applications are defined at a high level without stating specific references to technologies or standards. The application and technology vision statements are defined in the context of the business objective(s) to be addressed and are able to be mapped to at least one business objective or driver.

Assumptions—Documenting assumptions is a useful way of communicating to the rest of the organization, the basis on which the Roadmap is developed, the effort and support needed for a good Roadmap result and the key elements necessary for successful longer term technology adoption and deployment.

Requirements—The Requirements Step best starts with use cases including use case workshops. Once the use case titles and narrative are developed and the workshops have been held and documented, the next task is to derive the requirements and so called "primary actors." As part of the workshops, the interactions between the primary actors are identified. This allows the documenting of the Actor Interfaces. The IntelliGrid Requirements and Architecture development process is shown in Figure 3-3.

Use Cases/Workshops—Use cases are a very useful methodology originated in other industries and adopted use in the power industry as part of the EPRI IntelliGrid Methodology. (IEC PAS 62559). This methodology is applicable in a wide range of situations for technology adoption and deployment. In simple terms, if the technology being considered is proven or presents a low business risk or is planned for a minimal deployment over a reasonable time frame then a reduced scope use case approach is suggested. However if the technology being considered is new or presents a high business risk or requires a thorough cost benefit analysis or is complex or will interface with many other systems, and will be deployed in a large and costly program on an accelerated schedule, then a rigorous use case approach is suggested. The detailed cross functional requirements development, risk mitigation, business value identification and life cycle benefits are invaluable.

A use case is a "story" that includes various "actors," and the "path" they take to achieve a particular functional goal. By considering

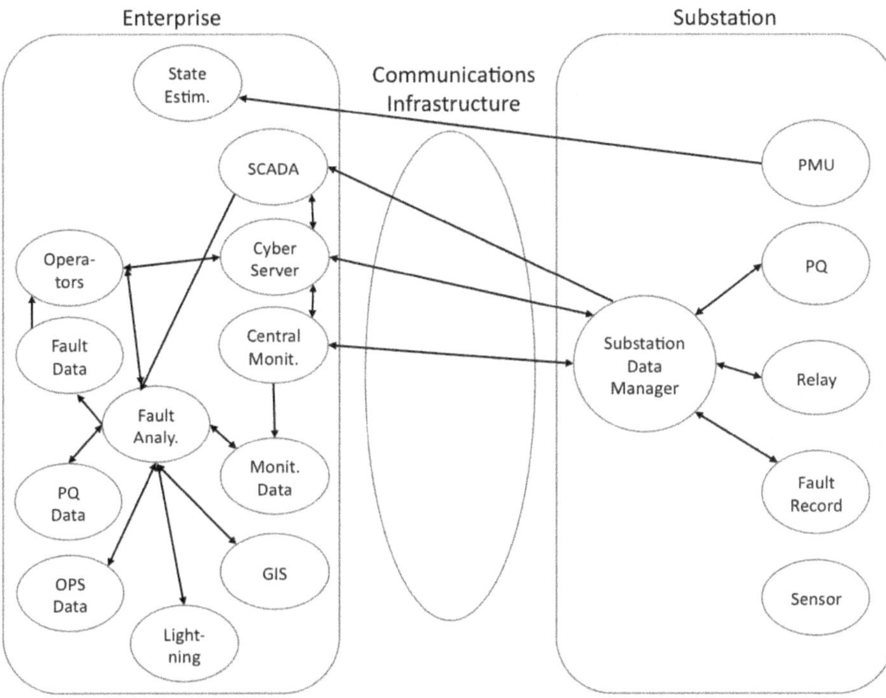

Figure 3-3. IntelliGrid Requirements and Architecture Development Process (EPRI 1025470)

the actions of the actors working to achieve functional goals, a completed use case results in the documentation of multiple scenarios, each containing a sequence of steps that trace an end-to-end path. These sequential steps describe the functions that the proposed systems and processes must provide, directly leading to the requirements. The goals of the use case methodology as part of the SGRM are:

- Collect all requirements that will have an impact on the architecture

Specific use cases are chosen on the basis of the following criteria:

- Address business objectives and drivers
- Address application and technology vision statement(s)
- Present the most architecturally significant applications
- Have a high probability of being justified

Requirements and Actors List—Once the use case titles and narratives are developed a series of workshops and webcasts should be scheduled with representation from multiple stakeholder departments.

The information gathered during the workshop can include:

- Existing technology and communications infrastructure
- Primary actors and interactions
- Technical requirements—Functional Requirements (FR) and Non-functional Requirements (NFR)
- Concerns and issues regarding the current system
- Business requirements and potential business values
- Views on necessary and preferred new applications, requirements, characteristics, integrations and architectures
- Current projects and technology pilots underway
- Regulatory, policy and tariff information that is applicable

Interfaces (Actors)—Primary actors (applications, equipment, staff) are identified and their relationship with other actors documented in the form of interface diagrams for each use case. Links between each actor are numbered. Message sequence diagrams (Unified Modeling Language—UML format) can then be created to indicate the data/information movement between actors.

Information for each of the numbered links on the interface diagram can be documented to provide the utility guidance on the technologies, standards and communications performance that should be considered. Figure 3-4 shows a typical interface diagram.

Current Technology Projects and Existing Communications &Technology Infrastructure—The documenting of the existing infrastructure along with known deficiencies is an important part of the SGRM. Understanding the current situation is essential in developing the gap analysis task in the Assessment step as well as the starting point (tail of the fish) for the implementation plans task in the Planning Step.

Assessment

The first task for this step consists of identifying the candidate technologies applications and standards. Candidates for this list are identified in Chapter 2.

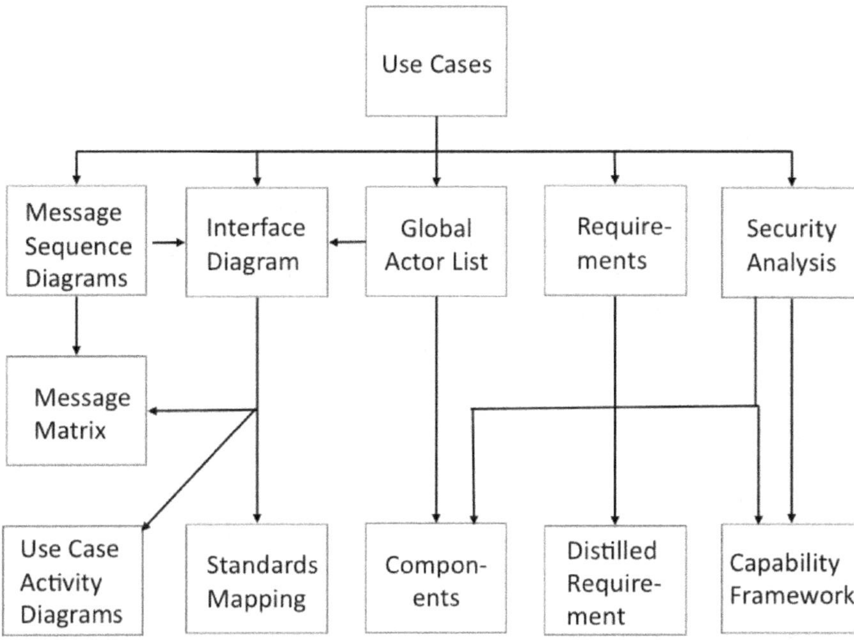

Figure 3-4. Typical Actor Interface Diagram (EPRI 1025470)

Technology Candidates—A list of technology candidates is developed by referring to a range of industry sources as noted above. A wide range of methods are used in the industry to evaluate candidate technologies, applications and standards. The method and criteria should be selected based on the needs of the utility but should be as broad as possible.

Evaluations and Assessments—Detailed Technology Assessment Criteria need to be applied to the candidate technologies. A highly ranked technology or standard will rank high in one or more of the following:

- A key enabler for a Smart Grid vision
- Encouraged by a regulatory authority
- Addresses a safety concern or risk
- Essential in maintaining or substantially improving reliability
- Provides a substantial improvement in operational or energy efficiency

- Yields a significant cost reduction
- Addresses an important strategic or business need and includes risk mitigation
- Supports a large enhancement in customer satisfaction

Select Focus Technologies—Once the above assessments are completed, the top candidate technologies, applications or standards can be selected for further focus. Upon selection, a new vision or objective statement is developed for each of the focus technologies. This statement will be used in the development of the gap analysis as well as the implementation plans.

Gap Analysis—Once the focus technologies are selected and the objective statements are developed, the next task is to assess the current status, issues, concerns, degree of deployment, plans underway for that technology. The gap analysis is then assembled by comparing the current situation to the objective statement for each focus technology.

Planning

Once the focus technologies are selected, the tasks in the planning step can be addressed. The first of these is to reengage key stakeholders, especially the subject matter experts to ensure that all needed input is received. The next task for the planning step is the development of diagrams which define steps in the development of individual or clusters of technology and its supporting systems, based on gap analysis for each technology. The final task is the revisiting of the technology adoption gate process which is an excellent method for managing risk.

Stakeholder Engagement—The first task of the planning step is to ensure that all stakeholders have the opportunity to have input in aspects of the decision.

Leadership Team Workshop—This is optional task of presenting the Roadmap findings to date to the utility's Smart Grid Leadership team for validation.

Implementation Plans and Ishikawa (Fishbone) Diagrams—These diagrams are called Ishikawa diagrams. They characterize needs and gaps. Ishikawa diagrams was created by Kaoru Ishikawa in 1968, also known as fishbone diagrams are intended to breakdown, in successive layers of detail, root causes that potentially contribute to a particular effect (Ishikawa 1968) categories typically include:

- People
- Methods
- Machines
- Materials
- Measurements
- Environment

The development of the Ishikawa diagrams occurs with close interaction with the utility's subject matter experts and starts with the outcome of the gap analysis for each technology. Refer to Figure 3-5 for an example fishbone diagram.

Stage-Gate Process and Risk Management—The last task for the Planning stem is the management of risk by revisiting the technology adoption gate process in order to manage risk.

A Stage-Gate Model* is a method by which a technology development process is divided into separate stages†. The stages are then said to be separated by gates. In the Stage-Gate Process, the development is evaluated at each gate by management and a decision made as to whether the development process should be modified, put on hold, or proceed as originally planned. The assessment is conducted by evaluating know development progress and comparing preliminary performance with originally expected targets. Typically, the business case is reviewed at the time of assessment along with a risk analysis, the investment needed, and the availability of staff and contractors.

The Stage-Gate Process allows management to assess the progress of technology development and make commitments in stages. This ensures that development with lower probability of success can be stopped, held or modified. It also allow reallocation of resources with the "R&D" portfolio. Dividing the technology development process into appropriate phases is the key to effectively applying Stage-Gate. Typically, technology development is divided into stages such as idea generation, establishment of feasibility, development of capability, testing and validation and product launch.

---

*Stage-Gate is registered trademark of the Product Development Institute, Inc.

†The Stage-Gate Model was developed by Robert G. Cooper of McMaster University and first described in his 1986 book "Winning at New Products" (Cooper, 1986).

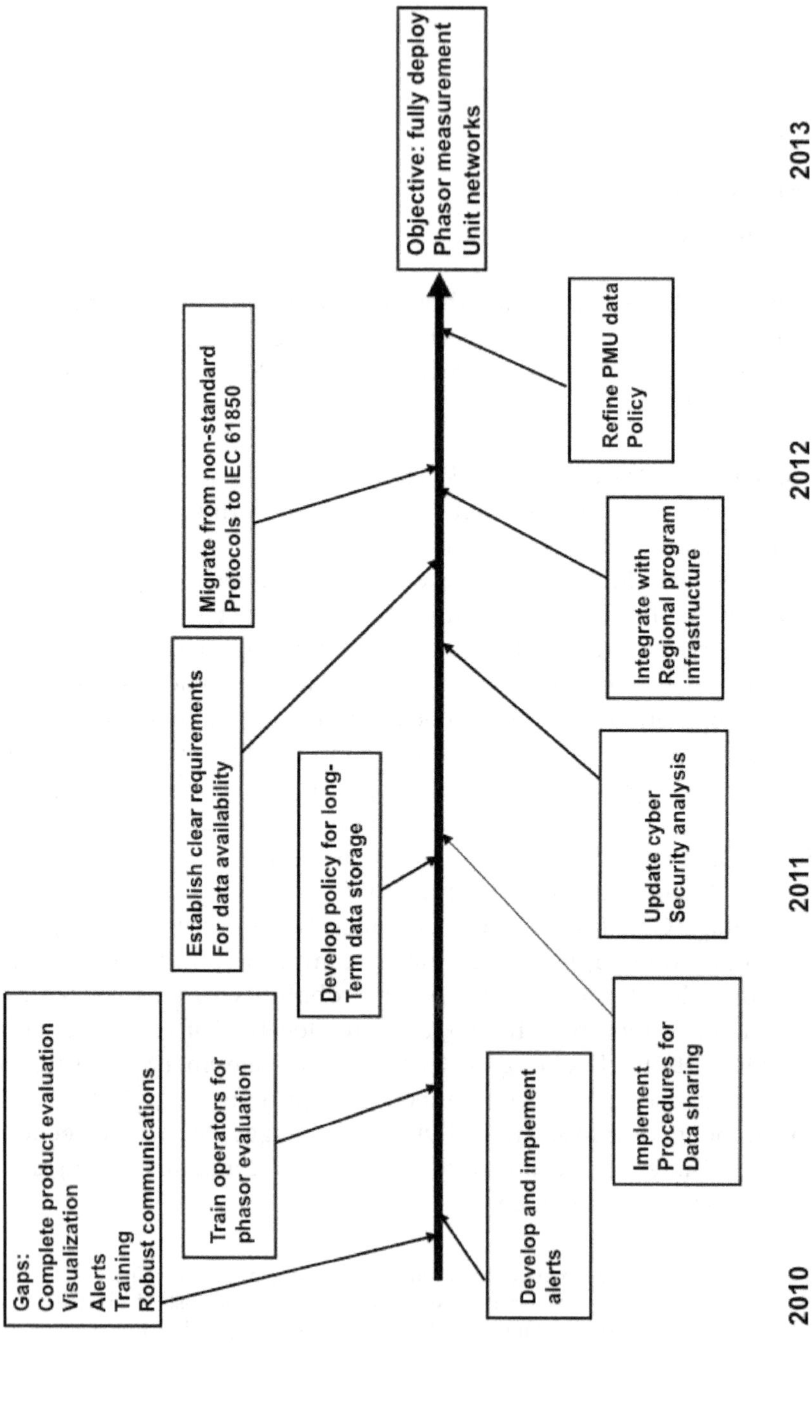

Figure 3-5. Example of Ishikawa (Fishbone) Diagram Showing Gaps and Objective (EPRI 1025470)

Technology Readiness—A well-accepted method used to identify the readiness of a technology for ultimate deployment in smart grids is the use of the Technology Readiness Level (TRL) process developed by the National Aeronautic and Space Administration (NASA) as a means to manage NASA's space related R&D. TRLs are a convenient means to describe the stage of smart grid technology by allowing a consistent comparison of maturity between different types of technology. The technology readiness of each of the smart grid technologies can be assessed in preparation for use of stage-gate. Figure 3-6 presents an approximate guide to how a specific TRL number corresponds to a specific stage of smart grid technological development.

What Do TRLs Include? TRLs include (a) basic (blue-sky) research in new technologies and concepts (targeted identified goals, but not necessarily specific systems); (b) focused technology development addressing specific technologies for one or more potential applications; (c) technology development and demonstration for each application; (d) system development; and (e) commercialization. Table 3-1 lists a short description of each level.

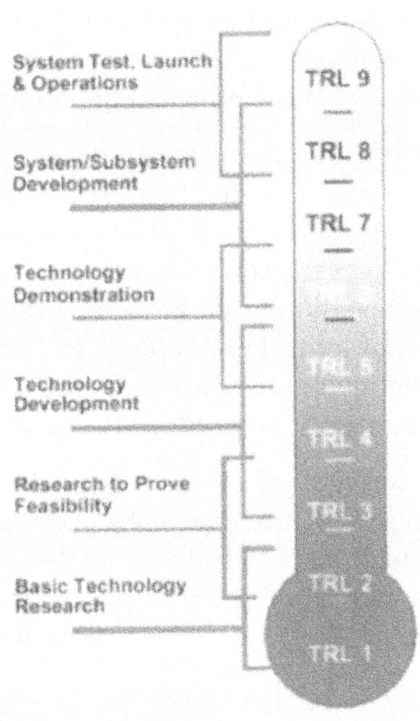

Figure 3-6. Technology readiness levels (NASA 1995 and DOE 2008)

Figure 3-7 illustrates where stages and gates could possibly be aligned with TRLs. This applies Stage-Gate to TRLs.

When management decides that a select project is to be subjected to the stage-gate approach, the project manager will follow the following steps:

- **Stage 1: Develop a Proposal**. Stage 1 involves development of a proposal and presentation outlining the intent of the potential development.
- **Gate 1 Review**. The Gate 1 Review involves management's concurrence as to the proposals' recommendation.
- **Stage 2: Develop a Business Plan**. During Stage 2, scoping of the technology and a business plan is developed.
    - **Scoping**: Scoping entails evaluating the potential performance of the technology, its possible applications, and intended markets. Possible strengths and weaknesses are also identified. It is also important to know the costs and performance of commercial technologies which are currently being applied.
    - Building the Business Case: In building the business case, a solid analysis is conducted to define the product, create the business case, build the project plan, and review the feasibility.
    - Product definition is where the product is defined and the development activity is justified. In this activity, the potential performance of the ultimate product is targeted, the potential value to utilities is assessed, and target clients are identified. Often this phase would entail discussions with advisors and construction of a market analysis. In this phase, the product concept is defined sufficiently so as to know the components and methods needed to produce the product.
    - *Building the business case* entails detailing the findings regarding the product performance, value, cost, and potential market. This phase may also include an analysis of legal, regulatory, and environmental performance requirements. The business case must state all uncertainties and assumptions.
    - *The project plan* is a list of activities, tasks, and events with milestones and timelines for the technology development. This includes an assessment of resources (money and personnel)

Table 3-1. Technology Readiness Levels (NASA 1995)

Technology Readiness Level	Description
1	Exploratory research transitioning basic science into laboratory applications
2	Technology concepts and/or application formulated
3	Proof of concept validation
4	Subsystem or component validation in laboratory environment to simulate service conditions
5	Early system validation demonstrated in laboratory or limited field application
6	Early field demonstration and system refinements completed
7	Complete system demonstration in an operational environment
8	Early commercial deployment (serial nos. 1, 2, etc.)
9	Wide-scale commercial deployment.

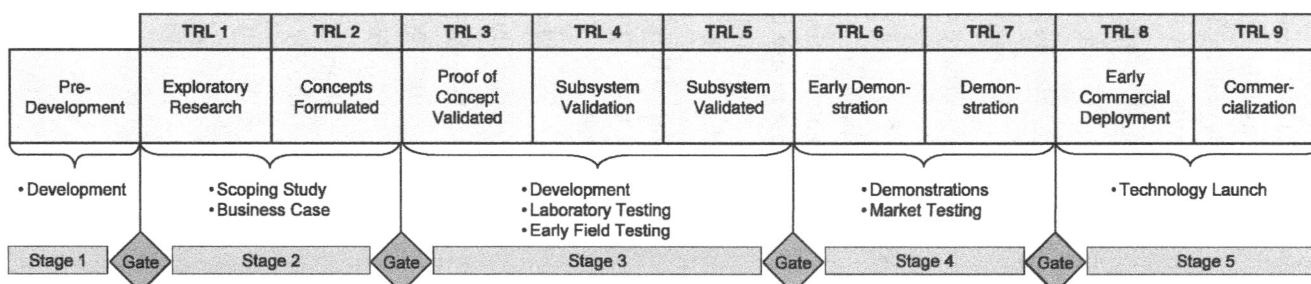

Figure 3-7. Mapping TRLs the Stage Gate Process

needed to complete each task.

- — **Gate 2 Review**. The Gate 2 Review is literally a review of the feasibility study and the business plan. The purpose of this review is to determine whether to move to the next stage. Questions to be answered in this review:
    - Do reports confirm basic science principles?
    - Where have scientific observations been published?
    - Have stakeholders and funders been identified:
    - Research plan formulated?
    - Where would R&D be undertaken?

- **Stage 3: Development**. In the development phase, the plans formulated in Stage 2 begin to be executed. The development team maps out realistic timelines with budgets and schedules, and executes the first phase of development. At this phase, the project begins to build momentum, and enthusiasm among potential clients is stimulated.

    — **Gate 3 Review**. The Gate 3 Review is to assess the development of the technology to date. Questions to be answered in this review:
    - What key process requirements have been identified?
    - Was the performance of components validated?
    - Where preliminary system performance characteristics estimated?
    - Did the laboratory experiments verify overall feasibility?
    - Were the principle process variables fully identified and their variability assessed?
    - Were individual components thoroughly tested?
    - Are all components assembled at laboratory scale in a working model?
    - Were any environmental hazards identified?
    - Have all system performance metrics been verified?
    - Have system integration requirements been identified and solutions determined?

    - Are all demonstration plans finalized?
    - Have customer demonstration commitments been received?

- **Stage 4: Demonstration**. Stage 4 includes TRL Level 6 and 7 activities. This phase provides validation for the entire development. The technology itself is tested and performance is validated, as well as the original performance and cost targets are reviewed. This phase typically consists of final laboratory testing, demonstration, and market testing.
  — **Gate 4 Review**. Gate 4 Review is to assess the technology's readiness for launch. Questions to be answered in this review:
    - Has performance, availability and reliability been fully assessed?
    - Have any design improvements been identified?
    - Have all operating parameters been identified?
    - Has a full engineering feasibility study been conducted?
    - Has the environmental impact assessment been completed?
    - Have all commercialization agreements been completed?
    - Have final design parameters been established?
    - Is the full-scale system validation completed?
    - Are the full documentation of installation and operations been complete?

- **Stage 4: Technology Launch**. At this stage, the product is ready for development

Roadmap Implementation

This final step needs to capture all the tangible results and leverage the intangible impact such as cross departmental cooperation that may have resulted.

- Final Roadmap Document—This task is the final reviews, editing and publication of the report.
- Final Report Out—To ensure the largest possible benefit is derived from the Roadmap effort, the final report presentations need to be

delivered to three key sets of stakeholders.

- Governance and Update of the Roadmap -Experience has shown that the leadership and governance policies and capabilities of an organization are by far the biggest determinants of the degree of impact of a Roadmap development. These steps are recommended:
  - Monitoring Progress Toward the Vision
  - Modifying the Roadmap Based on Technology Assessments and Trials

THE INTERNATIONAL ENERGY AGENCY'S (IEA) SMART GRID ROADMAP

IEA published a smart grid roadmap in 2011. IEA believed that "The development of smart grids is essential if the global community is to achieve shared goals for energy security, economic development and climate change mitigation." (IEA 2011) The agency acknowledged that Smart grids enable increased demand response and energy efficiency, integration of variable renewable energy resources and electric vehicle recharging, while reducing peak demand and increasing electric reliability.

IEA acknowledged that the physical and institutional complexity of electricity systems makes it unlikely that the market alone will implement smart grids on the scale that is needed. Governments, the private sector, and consumer and environmental groups must work together to define electricity system needs and determine smart grid solutions. According to the IEA, rapid expansion of smart grids is hindered by a tendency on the part of governments to shy away from taking ownership of and responsibility for actively evolving or developing new electricity system regulations, policy and technology. They believe that these trends have led to a diffusion of roles and responsibilities among government and industry actors, and have reduced overall expenditure on technology development and demonstration, and policy development. As a result, slow progress has been made on a number of European regional smart grid pilot projects that are needed.

In Europe, large-scale, system-wide demonstrations are urgently needed to determine solutions that can be deployed at full scale, integrating the full set of smart grid technologies with existing electricity infrastructure. These large-scale pilot projects would allow testing of various business models to allow their adaptation to local needs. European electricity stakeholders believe that Countries and regions will use smart grids for different purposes; emerging economies may leapfrog directly to smart electricity infrastructure, while countries who are members of the Organization for Economic Cooperation and Development (OECD) are already investing in incremental improvements to existing grids and small-scale pilot projects. In Europe, peak demand will increase between 2010 and 2050 in all regions. Smart grids deployment could reduce projected peak demand increases by 13% to 24% over this frame all across Europe.

IEA believes that current regulatory and market systems can hinder demonstration and deployment of smart grids. Regulatory and market models—such as those which induce system investment, prices and customer participation—must evolve as technologies offer new options over the course of long-term, incremental smart grid deployment. To encourage a revision in policies, regulators and consumer advocates need to engage in system demonstration and deployment to ensure that customers benefit from smart grids. Building awareness and seeking consensus on the value of smart grids should be a priority for regulators and utilities alike. In addition, greater international collaboration is needed to share experiences with pilot programs from around the world, to leverage national investments in technology development, and to develop common smart grid technology standards that optimize and accelerate technology development and deployment while reducing costs for all stakeholders.

IEA's roadmap acknowledges that smart grids can provide significant benefits to developing countries. Capacity building, targeted analysis and roadmaps—created collaboratively with developed and developing countries—are required to determine specific needs and solutions in technology and regulation.

Purpose, Process and Structure of IEA's Roadmap

To provide guidance to government and industry stakeholders on the technology pathways needed to achieve energy security, economic growth and environmental goals, the IEA is developing a series of global low-carbon energy roadmaps covering a range of technologies. The roadmaps are guided by the IEA Energy Technology Perspectives "BLUE Map" Scenario, which aims to achieve a 50% reduction in energy-related CO2 emissions by 2050. Each roadmap represents international consensus on milestones for technology development, legal and regulatory needs, investment requirements, public engagement and outreach, and international collaboration. The Smart Grid Roadmap aims to: "Increase understanding among a range of stakeholders of the nature, function, costs and benefits of smart grids; Identify the most important actions required to develop smart grid technologies and policies that help to attain global energy and climate goals."

IEA intends that their roadmap will develop pathways to follow and milestones to target based on regional conditions. The roadmap was compiled with the help of contributions from a wide range of interested parties, including utilities, regulators, technology and solution providers, consumer Financial Regulatory and policy.

IEA's roadmap does not attempt to cover every aspect of smart grids and should be regarded as a work in progress. As global analysis improves, new data will provide the basis for updated scenarios and assumptions. In addition, as the technology, market and regulatory environments evolve, additional tasks will come to light. The broad nature of smart grids requires significant collaboration with other technology areas, including transport electrification, energy storage, generation and end-use.

IEA's Conclusions: Near-Term Roadmap Actions for Stakeholders

IEA's road mapping process concluded that smart grids are a foundational investment that offer the potential to substitute efficient use of information for more conventional "steel-in-the ground" investments in the electricity system, at considerable cost savings to consumers, as demonstrated by early results of pilot projects. Smart grids will also change how power system planning is done, and how wholesale and retail electricity markets evolve. The information collected by smart grid sensors, meters and systems will not only empower customers to manage their electricity consumption but will enable electricity system operators to better understand and meet users' needs.

IEA believes that the roles of the government and the private sector are often misunderstood, at times by themselves and often by each other. The broadness and complexity of the electricity system, and its importance to society in general, increase the necessity to understand who should perform the actions outlined in this roadmap. Neither the government alone, nor the private sector alone, can accomplish the goal of modernizing the electricity system. IEA highlighted that collaboration is vital.

Their roadmap summarized the actions key electricity system stakeholders will need to take to implement the roadmap. Each action requires a broad range of actors. (IEA 2011)

Electricity Generators

- Utilize flexibility and enhancements delivered by smart grids to increase use of variable generation to meet demand growth and decrease emissions.

Transmission and Distribution System Operators

- Develop business models along with government and regulators that ensure all stakeholders share risks, costs and benefits. In addition, lead education in collaboration with other stakeholders on the value of smart grids, especially with respect to system reliability and security benefits.
- Promote adoption of real-time energy usage information and pricing to allow for optimum planning, design and operation of distribution and transmission systems in a coordinated fashion.
- Demonstrate smart grids technology with business models that share risks, benefits and costs with customers in order to gain regulatory approval and customer support.

Government and Regulators

- Collaborate with public and private sector stakeholders to determine regulatory and market solutions that can mobilize private sector investment in all electricity system sectors.
- Recognize that smart grid deployments should reflect regional needs and conditions—a "one-size-fits-all" does not apply to the deployment of smart grids.
- Plan for evolution in regulation along with technology development—new technologies will both offer and need new regulatory options.
- Invest in research, development and demonstration (RD&D) that address system-wide and broad-range sectorial issues, and that provide insights into behavioral aspects of electricity use.

Technology and Solution Providers

- Deliver full technology solutions to system operators through partnership with others in the value chain to address concerns with technology system integration, long-term post installation support, and security and reliability.
- Create a strategy and develop standards in participation with industry and government stakeholders on an international level to ensure interoperability of system components and reduce risk of technology obsolescence.

Consumers and Consumer Advocates

- Develop an understanding of electricity system reliability, quality, security and climate change benefits of smart grids.
- Help develop regulatory and market solutions that share investment risks, costs and benefits with all consumers.
- Actively engage in developing system demonstrations and deployments in order to ensure consumer contribution to and benefit from future electricity systems and markets, while ensuring consumer protection.

Environmental Groups

- Support the development of smart grids necessary for a range of clean energy technology deployments such as wind, solar and electric vehicles.

International Governmental Organizations

- Support the RD&D of smart grid solutions for developing countries through targeted analysis, road mapping exercises and capacity building.
- Support international collaboration on and dissemination of smart grid RD&D, including business and regulatory experiences.

OTHER ROADMAP SUMMARIES

The following are examples of a number of technology roadmaps developed by utilities, states or individual countries.

California ISO

Those are significant opportunities and challenges facing the California Independent System Operator (ISO) driven by California energy and environmental goals and emerging technologies that are expected to be proven essential in meeting these goals. As with most Independent System Operators, the California ISO had plans to invest significantly in communications infrastructures and a wide range of technologies to address corporate objectives for grid reliability, regulatory compliance, operating efficiency, market operations, cyber security, renewables integration, system integration and asset utilization.

The ISO and its roadmaps teams ranked five themes to arrive at a prioritized list of technology theme statements and recommendations as detailed below:

Phasor Measurement Units Gathering Recommendation

Establish comprehensive requirements addressing all aspects of phasor measurement unit data collection and communication in cooperation with the utility equipment owners and the Western Electricity Coordinating Council (WECC).

Advanced Forecasting Recommendation

As California approaches the state legislated renewable generation targets of 20% of energy usage by 2013 and 33% by 2020, the impact of both variable resources and loads such as (PEVs) on ISO operations should continue to grow. ISO Operations should enhance its operational tools.

Advanced Grid Applications Recommendation

Advanced grid applications working in conjunction with other key applications such as advanced forecasting, external entities supplying data, field data sources, and the communications infrastructure should be essential in providing ISO's staff with the ability to maintain system reliability in the midst of significant change, while facilitating effective market services.

Cyber Security Recommendation

The ISO should expand its cyber security policy to include application layer security to address this growing area of vulnerability. Also, new policy elements addressing data integrity on the basis of data source and class should be developed.

DER Enabling Recommendation

Enabling distributed energy resource (DER) participation in the ISO market requires that ISO adopt a robust and flexible system architecture that builds upon recent Market Redesign and Technology Upgrade (MRTU) design.

The California ISO Roadmap

The California ISO along with EPRI, PG&E, SCE, and SDG&E with assistance from The California Energy Commission Public Interest Energy Research (PIER) Program undertook developing a ten year plan for California's Smart Grid Roadmap in 2011 looking forward to 2020. The California 2020 Roadmap represented the California utility vision of the technical infrastructure and corresponding operating environment that were needed to meet the policy goals. The Roadmap Plan identified as series of visions as the most critical elements in establishing the Smart Grid in California by 2020.

2020 Vision Communication/Architecture Domain

The 2020 Vision for Smart Grid communications infrastructure and architecture was envisioned to integrate diverse, single-purpose communications networks into an efficient and flexible unified information infrastructure.

2020 Vision for Customer Systems Domain

The Smart Grid should enable customers to actively support the reliable, sustainable, economic delivery of power by providing feedback to help customers manage the timing and quantity of their energy usage; enabling third-parties, including utilities, to manage energy usage on behalf of customers; and allowing customers to adopt environmentally-friendly technologies without sacrificing grid reliability.

2020 Vision for Grid Operations Domain

By 2020, the California transmission grid should be well on its way to being fully monitored in real time with an integrated set of advanced sensors and monitoring devices and robust communication systems.

2020 Vision: Renewables & DER Domain

The 2020 vision for this domain is to utilize, integrate and develop bulk renewable and distributed energy resources (DERs) to meet the varying customer and market demands.

2020 Vision: Grid Planning and Asset Efficiency Domain

The Smart Grid of 2020 should operate assets and resources more efficiently by dynamically controlling voltage and optimizing power flows.

2020 Vision: Workforce Efficiency Domain

Smart Grid Workforce Effectiveness involves preparing the workforce to support Smart Grid technologies and tools while enhancing safety and productivity.

Additional Considerations

Security considerations must be addressed in an integrated statewide Architecture, in the communications infrastructure, and in all the information exchange requirements at points of interoperability.

At the transmission level, the WECC synchrophasor initiative and coordination should provide the foundation for significant expansion of synchrophasor applications and development of advanced system management and control applications that take advantage of synchrophasor information.

Asset management and maintenance strategies should also take advantage of widespread sensors and automation. Modeling and forecasting improvements were needed to better integrate variable resources with markets and grid operations.

Duke Energy's Transmission Roadmap

Duke Energy has developed vision and mission statements for the definitions and development of a power delivery communications and automation infrastructure, which included the communications upgrade to substations, the implementation of select technologies and applications and the integration of enterprise applications, as well as databases and systems as an enabler for Duke to optimize the business.

The five most critical use cases for Duke Transmission going forward were identified as: Secure Remote Access to Intelligent Electronic Devices (IED)s, Transmission Fault Analysis, Integrating Equipment Monitoring & Diagnostic into EA Management, Dynamic Rating Integration with EMS, and Distribution Load Reduction. From these, the following nine key technology visions were evaluated and recommended for implementation:

1. Implement secure remote access to substation data.

2. Deploy new substation data gathering technologies.
3. Improve on Duke's existing cyber security strategy.
4. Created and deploy a substation protocol integration strategy.
5. Implement a strategy for secured and controlled enterprise-wide sharing of applications and databases.
6. Migrate from planned T&D Facility Ratings Project to an on-line continuous asset monitoring approach.
7. Implement automatic fault record retrieval and analysis.
8. Deploy substation phasor measurement units (PMUs) at select sites.
9. Expand the deployment of distribution feeder automation (DFA).

Southern Company (SoCO) Smart Grid Roadmap

Starting from SoCo's existing Smart Grid Strategy work that had been completed previously, this Smart Grid Roadmap process began with the identification of a set of distribution-focused Smart Grid use cases and scenarios. Smart Grid use cases/scenarios were selected based on Southern Company's interest in pursuing functionalities implied by these use cases/scenarios.

Use Cases 1-2: Distribution Operator Locates Outage Using AMI Data and Restores Service

Outage location and service restoration is facilitated by AMI data. Outages reported by AMI meters should substantially augment customer-reported outage data, resulting in faster collection of more extensive outage data and faster, more accurate outage predictions. The AMI system should also facilitate the identification and resolution of nested outages. AMI meters should report the presence of line-side voltage, allowing "false" outages to be identified by customer service requests (CSRs), eliminating unnecessary truck rolls.

Use Cases 3-4: Distribution Uses AMI System to Optimize Network

Distribution Planning should utilize historical feeder load information to optimize load balance on the distribution system. By capturing actual peak load data on feeders rather than using estimates, Distribution Planning is better able to target and plan capital improvements to the distribution system: optimize capital investments, void feeder loading problems (:over/underload detection), and improve asset utilization.

AMI data should facilitate determination of phasing of single-phase meter thereby enabling balancing of voltage among phases.

SoCo's Key Findings

The following is a summary of key findings from the roadmapping process:

- Upgrade of communications infrastructure
- Transition to open standards to support future technology investments (no new legacy systems). Continue growth in AMI installations
- Integration and organization-wide coordination of Enterprise Applications such as Outage Management System (OMS) with AMI (RNI) system that should deliver substantial benefits to SoCo as well as SoCo customers
- Application integration using open standards on a prioritized strategy basis was recommended for all Southern operating companies
- Implement substation Local Area Network (LAN)s using open standards and associated devices to enable improved substation data retrieval
- Continue investments in high value applications: utilizing greater penetration of power system monitoring to derive key system knowledge yielding improved efficiencies and reliability

Salt River Project (SRP) Roadmap

SRP developed vision and mission statement for the definition and development of a fully architecturally-integrated power delivery communications infrastructure, which included the integration of all related applications, databases and systems.

SRP and Roadmap Use Cases

Roadmap use cases were developed in the following seven most pressing areas:

- Enterprise Strategy for cyber-security: system wide and fully integrated from HAN, AMI, WAN, DFA, SCADA, back-office, EMS, etc.

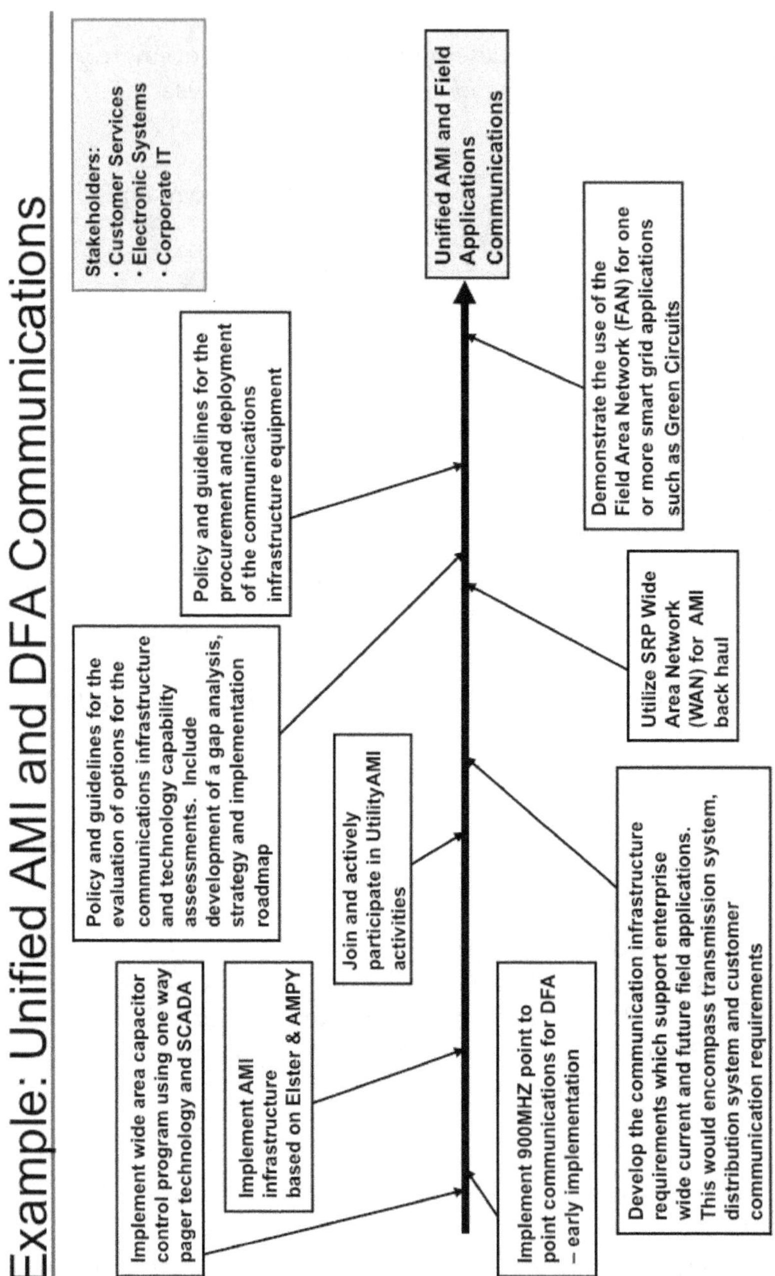

Figure 3-8. SRP Ishikawa Diagram for Unified AMI and DFA Communications

- Automated Tools for WAN Monitoring: in preparation for significant growth in all categories of devices, automated tools for the monitoring and control of all aspects of network traffic and optimization.
- Integrated Substation LAN-DNP & IEC 61850: detailed plans for migration to IES 61850 in large-scale deployments at both T&D levels.
- Unified AMI & DFA Communications: develop policy and specific guidelines to assure that future deployed communication systems are scalable and versatile enough to handle not only current AMI issues (e.g. DR, TOU, outages, HAN, PEVs, DG, etc.) as well as unforeseen future applications.
- Expanded Distribution Feeder Automation Policy: primarily focused on outage management issues and strategies, locally intelligent but globally optimized systems for power restoration and management.
- Electric System Data Acquisition & Data Management: improve system efficiency, safety, and reliability through wide deployment of enhanced sensor and data aggregation and management systems at the substation, feeder, and transmission line levels.
- Enterprise Application Integration: create the framework, policies, and unifying strategies for the foundation of an enterprise-wide data exchange

Tennessee Valley Authority (TVA) Roadmap

TVA's Roadmap is focused on Grid Modernization investments through 2020 and beyond. The TVA Grid Modernization Roadmap provides a high level framework for implementing the tools and capabilities that are required to modernize the grid.

The Roadmap Planning recommendations cover four targeted Roadmap Initiatives describing activities TVA should undertake to lay the foundation for grid modernization over the next ten years.

Information Communications and Technology Infrastructure

The applications and technologies that TVA will implement to modernize the grid generate significant information flows. To realize full value from the applications, the data should be available throughout the TVA organization including customers.

Modeling and Analytical Tools for Planning and Operations

To continue improving reliability, cost-effectiveness and security/compliance, TVA will invest time and resources into advanced planning, modeling and operational tools. TVA will evaluate and deploy advanced tools to improve planning, modeling and system operations, especially in areas such as customer load management where new uncertainties complicate the operating environment.

Advance Control Strategies

Advanced tools technologies and systems are improving operator control over uncertainties (e.g. voltage, reactive power, frequency, reserves, distributed energy resources and renewables). Consequently, TVA will investigate and implement strategies that improve operator analysis and control of the power system with a particular emphasis on preparation for greater diversity n generation portfolio.

Strategies and Systems to Optimize Asset Management

TVA continues to make significant investment in transmission assets. Adopting state of the art strategies and systems to manage those assets cost effectively offers the best return on TVA's grid modernization investment over the next decade.

The State of Kentucky's Roadmap

Kentucky is one of the most richly endowed U.S. states in terms of energy but also one of the state's most dependent on just one single resource. Coal provides nine-tenths of the state's electrical power, and coal mining is the state's largest industrial employer. In a sense, the state is like a Third World country almost wholly decided to the cultivation of just one crop, so that the country will prosper as long as the world market for that crop remains robust but will be in serious trouble if the crop ever goes out of fashion. With an eye on that vulnerability, in 2008, Kentucky's governor issued a challenge to make the state a "national leader in energy technology and production," and to "help the country move toward greater energy self-reliance. I intend to put us on such a path." (IEEE 2011)

In 2010, the Kentucky Public Service Commission, the University of Louisville Speed School of Engineering and the University of Kentucky's Department of Electrical and Computer Engineering formed a collaboration to produce a working roadmap for the development and deployment of smart grid technology in Kentucky. Out of this collaboration came the Kentucky Smart Grid Roadmap Initiative (KSGRI), with stimulus funding from the Department of Energy's National Energy Technology Laboratory (NETL).

Smart Grid Classes

The KSGRI created the Kentucky Smart Grid Assessment Mode (KSGAM) to measure the extent of Smart Grid deployments, operations, and planning in Kentucky, to determine priorities for areas of improvement within the state electrical grid, and to include stakeholder input as to what our societal priorities for a modern grid should be. The KSGAM divided Smart Grid operation into 10 groupings of related characteristics defining Smart Grid operations. These groupings, called Smart Grid Classes, are: (KSGRI 2012)

1. Strategy and Management (SM): Smart Grid vision and strategic planning, internal governance and management processes, and collaboration with internal and external stakeholders.
2. Organization and Structure (OS): workplace structure, training, communications, and knowledge management within the utility.
3. Technology (TECH): deployment evaluation and strategic planning of advanced technologies.
4. System Architecture and Operation (SAO): operation of the power grid as an automated system with a high degree of situational awareness.
5. Demand and Supply Management (DSM): dynamic management of supply and demand based on real-time information, particularly for load management and distributed energy resources.
6. Work and Asset Management (WAM): optimal management of grid assets and workforce resources, with decisions based on real-time data.
7. Physical and Cyber Security (SEC): protection of equipment and data from cyber and physical security attacks through security architectures, risk assessments, and cyber security standards.
8. Government and Regulation (GR): the extent to which regulators authorize Smart Grid investments, approve innovative regulatory strategies, and facilitate intra-utility optimization.

9. Customer (CUST): the role of customer participation and experience, regarding pricing, education, and advanced services.

Environment and Society (ENV)

Contributions of the utility to achieving societal goals regarding reliability, safety, security, energy sources, energy source impacts, and quality of life.

Smart Grid Characteristics

The KSGAM also identified seven technological independent abilities of the Smart Grid. These abilities, called Smart Grid Characteristics, are:

1. Active Participation by Consumers (APC): increased interaction of consumers with the grid, characterized by the use of price signals and demand response programs to give customers choice and control regarding power purchasing.

2. Accepts All Power Generation and Storage (AAPGS): integration of diverse resources with "plug-and- play" connections to multiply the options for electrical generation and storage, including centralized power plants, distributed energy resources and energy storage devices.

3. Enables New Products and Services (ENPS): Direct linking of the buyers and sellers of electricity, the evolution of new commercial goods and services and a restructuring of power markets.

4. Improved Power Quality (IPQ): The delivery of "clean" digital-grade power characterized by reduction in voltage sags, surges, frequency harmonics, and phase imbalances.

5. Efficient Operation and Use of Assets (EOUA): Use of real time information from advanced sensors to allow operators to better understand the state of the system and respond accordingly.

6. Self-Healing: The grids ability to identify, isolate, and restore problematic sections of the grid with minimal human intervention.

7. Defend against Attack and Natural Disaster (DAAND): The grids ability to protect against physical and cyber-attacks.

Stakeholders were also asked to allocate $100 million amongst improvements in the electric power system. The results indicated that Kentucky stakeholders would prefer to see improvements regarding the environmental impact of the electric grid. The results are elucidated in Table 3-2.

Priorities

Table 3-2. Economic Valuation of Smart Grid Benefits (KSGRI 2012)

Attribute	Value Assigned ($ Million)
Reliable	18.55
Secure and Safe	17.13
Economic	19.95
Efficient	19.63
Environmentally Friendly	24.75

The final result of KSGRI's Smart Grid Roadmap development process was the actions outlines in Tables 3-3 and 3-4.

The Pacific Northwest Roadmap

The Bonneville Power Administration (BPA), sponsored and managed the creation of a smart grid roadmap referred to as the Pacific Northwest Smart Grid Regional Business Case (RBC). Is was developed with input from a number of regional entities including the Northwest Power and Conservation Council (NPCC), and the Pacific Northwest Smart Grid Demonstration Project, which is being managed by the Battelle Memorial Institute (Battelle). The RBC incorporated input from regional utilities and other regional stakeholders as well as findings from smart grid (SG) studies nationwide. (BPA 2013)

BPA defined the smart grid, as using two-way communications and automated intelligence to enhance the traditional electricity delivery system. They believe that the smart grid promises many benefits for the Pacific Northwest, including better reliability, more efficient and flexible operation of the grid, lower rates, and reduced carbon emissions.

Table 3-3. Kentucky Smart Grid Roadmap Initiative (KSGRI) Roadmap Results; actions in the near term—1 to 3 years (KSGRI 2012)

Organization	Actions
KY Smart Grid Task Force	Identify Founding Taskforce Members
	Develop Organization Structure
	Recruit Voting/ Advisory Members
	Evaluate and Endorse Smart Grid Roadmap as Appropriate
	Identify and Endorse Smart Grid Metrics
	Form Evaluation Plan
	Perform 1st KY Smart Grid Evaluation
	Release Results on SG Evaluation w/ Implementation
Industry	AMI Deployments Phase 1-3 (for early adoption)
	IT Deployments Phase 1-3 (for early adoption)
	ADO Phase 1-3 (for simultaneous AMI/ADO deployment – ALT1)
	Work w/ Government to Identify Funding Mechanisms to Support KY Smart Grid Clearinghouse
	Launch Initial Smart Grid Educational Campaigns
	Industry
Government	Work w/ Industry to Develop Requirements for SG Education
	Programs
	Support Development of Smart Grid Clearinghouse
	Develop Recommendations for Demand Response Tax Credits
Industry and Government	Identify Funding Mechanisms for KY Smart Grid Research Center and KY SG Test Lab.
	Identify Infrastructure and Human Resources for Research Center and Test Lab.
	Open Research Center and Test Lab
	Identify and Hire Smart Grid Consultant for KY AMI Pilot
	Perform Initial AMI Pilot Design and Participant Recruitment
	Perform 1 YR AMI/Dynamic Pricing Pilot

Table 3-4. Kentucky Smart Grid Roadmap Initiative (KSGRI) Roadmap Results; actions in the midterm—4 to 11 years (KSGRI 2012)

Organization	Actions
KY Smart Grid Task Force	Collect Industry/Stakeholder Input on 1st KY Smart Grid Evaluation Re-Evaluate Smart Grid Metrics and Implementation Recommendations as Necessary Perform KY Smart Grid Evaluations on 3 Year Intervals • Update Metrics and Implementation • Recommendations as Necessary w/ Input from Stakeholders. Release Evaluation/Progress Reports
Industry	AMI Deployments Phase 4-6 IT Deployments Phase 4 DER Phase 1 -5 ADO Phase 4-7 – ALT 1 ADO Phase 1-6 – (for Delayed ADO Deployment – ALT 2)
	Industry· Bring KY Smart Grid Clearinghouse Online All Utilities Have Smart Grid Education Programs In Place According to Government Requirements Continue to Fund KY Smart Grid Clearinghouse
Government	Evaluate AMI Pilot. Expand if Necessary to Form Recommendations. Implement KY DER Pilot Implement DR Tax Credits Evaluate Utilities SG Consumer Education Programs.
Smart Grid Research Center and Testing Lab	Work w/ SG Taskforce to ID Research Areas • Focus on Cost Reduction and Energy Efficiency • Focus on New Technology Development Support Needs of KY SG Taskforce Support Industry in SG Deployments as Needed

Table 3-5. Kentucky Smart Grid Roadmap Initiative (KSGRI) Roadmap Results; actions in the long term—11 to 25 years (KSGRI 2012)

Organization	Actions
KY Smart Grid Task Force	Perform KY Smart Grid Evaluations on 3 Year Intervals • Update Metrics and Implementation Recommendations as Necessary w/ Input from Stakeholders • Release Evaluation/Progress Reports
Industry	ADO Phase 7 (ALT 2) AAM Full Deployments ATO Full Deployment
Government	Evaluate Effectiveness of Tax Incentives for Demand Response to Consider Program Expansion or Termination Based on Efficacy.
Smart Grid Research Center and Testing Lab	Support Transfer of New SG Technology to Market Continue to Support KY SG Task Force and Industry in SG
	Deployment and Analysis for Future Needs

The Pacific Northwest Roadmap assesses the benefits, costs, and risks of a comprehensive regional smart grid deployment in the Pacific Northwest. This summary highlights several of the key conclusions from BPA's road mapping efforts. (BPA 2013)

T&D Optimization

The roadmap highlighted transmission and distribution optimization as encompasses smart grid capabilities that improve the controllability or utilization of electrical infrastructure assets, leading to more efficient delivery of electricity. Example capabilities include conservation voltage reduction (CVR) and power factor control.

Grid Reliability

Technologies were included which enhance end-use energy reliability encompassing smart grid capabilities that reduce the likelihood, duration or geographic extent of electricity service interruption, and maintain or improve the quality of delivered power. Example capabilities include fault location, isolation, and service restoration (FLISR), enhanced fault prevention, and wide area monitoring (WAM).

Dynamic and Responsive Demand (Smart DR)

BPA chose to refer to Demand Response as dynamic and responsive demand. In their definition it encompassed smart grid capabilities that allow short-term influence of end-use consumption by signals provided through the electricity supply chain. Their roadmap included seven end use categories of demand response (DR): lighting; space heating; space cooling; appliances/plug loads; water heating; refrigeration/industrial processes; and agricultural pumping. They mapped these capabilities across four sectors: residential; commercial; industrial; and agricultural. And used three program approaches: pre-enrolled participant events; price signals; and fast-acting/ancillary services.

Smart End-Use Energy Efficiency

Smart end-use energy efficiency was included and encompassed smart grid capabilities that reduce energy consumed by customers through enhanced information feedback, identification of poorly performing equipment as candidates for replacement or maintenance, and other enhancements to end use energy efficiency that require smart grid functionality. Capabilities considered encompass consumer behavior change, automated energy management, efficiency equipment upgrades, and improved maintenance.

Grid Storage Integration and Control

Grid storage integration and control was included which encompassed all smart grid capabilities were included that provide the ability to store electrical energy in battery systems. This includes battery systems sited at end-use facilities, on the distribution system, on the transmission system, and electric vehicle batteries when connected to charging stations.

Utility Operational Efficiency

Utility operational efficiency was also included in the BPA roadmap encompassing smart grid capabilities that improve the ability of a utility to deliver energy with the same reliability and efficiency, but with lower operations and maintenance costs. This included automated meter reading and billing, reduced truck rolls from targeted repair work, and improved planning and forecasting.

Smart Grid Capabilities

Using the technologies listed above, the RBC analysis characterizes a number of smart grid capabilities, three of which are summarized here.

Conservation Voltage Reduction

Conservation voltage reduction (CVR) is a reduction of voltage levels aimed at reducing energy consumption on the distribution feeder. CVR is a key area of interest in the Pacific Northwest, with a number of regional utilities exploring opportunities for CVR through pilots and full-scale program rollouts

PMU Applications

Phasor measurement units (PMUs), often called synchrophasors are highly precise sensors that communicate grid measurements from across a transmission system. The measurements are taken at high speed and synchronized to give a precise and comprehensive view of a broad transmission geography as compared with conventional technology. Synchrophasors enable a better indication of the condition of the in real time, and can be used to trigger corrective actions to maintain reliability. Several PMU applications were mapped in BPA's roadmap including the following:

- 1. **Dynamic Capacity Rating**—enabling transmission facilities to be used to optimal capacity.
- 2. **Renewable Monitoring and Integration**—enabling major renewable sites to be more fully utilized.
- 3. **Wide Area Monitoring**—making it possible for the condition of the bulk power system to be observed and understood in real-time.

Demand Response

The changing mix of supply resources in the Pacific Northwest is driving a renewed interest in DR in the region. This change is due, in part, to continued limitations in available traditional regional generation capacity resources and increasing penetrations of renewable generation. Several forms of DR are possible without the smart grid like interruptible contracts for industrial customers or direct load control programs for residential customers. The smart grid can bring important benefits to DR programs that traditional DR mechanisms cannot provide, such as improved response rates, deeper curtailment, increased participation, and use of DR for ancillary services and oversupply mitigation that require more advanced communication and control capabilities than traditional DR can provide.

New York Power Authority (NYPA) Roadmap

To facilitate New York State's renewed energy vision to re-imagine the grid and to facilitate the increasing pace of industry transformation, NYPA believes that an investment in smart grid technology across their Generation and Transmission infrastructure is an important and necessary step. NYPA undertook development of a roadmap, referred to as a "Strategic Initiative" for Smart Generation and Transmission. (BYPA 2014) This initiative is part of NYPA's Strategic Vision 2014-2019 and is aimed at providing both NYPA and New York State with what NYPA believes is the most advanced grid in the industry and to ensure that the most modern industry solutions are leveraged to deliver capability in key areas. Their objective is to pave the way for increased benefits to customers by providing New York State with a market leading platform for future technologies and services, better management of bulk and distributed renewable generation and storage sources, near-real-time access to information, improved outage management and overall reduced electrical losses and increased system efficiency.

This initiative has been designed to align with the Blueprint for New York State's Energy Highway ("EH"), issued in 2012 and the 2014 launch of the New York Public Service Commission, Reforming the Energy Vision (REV). The REV initiative aims to ensure that the New York electricity market transforms to accommodate the changing nature of the market, and the actions in this

New York Energy Industry in Transition

Smart Generation and Transmission roadmap will be entirely complementary of that vision, which is "The energy industry is in transition. Technological innovation and increasing competitiveness of renewable energy resources, combined with aging infrastructure, extreme weather events, and system security and resiliency needs, are all leading to significant changes in how electric energy is produced, managed and consumed. NYPA asserts that New York State must lead the way to ensure these trends benefit the State's citizens, whose lives are so directly affected by how electric energy is manufactured, distributed, and managed." (NYPSC 2014)

NYPA had already initiated several developments before this roadmap was developed, including Life Extension and Modernization (LEM) Upgrades, Generation Automation Control deployments, sensor deployment continuation, Supervisory Control and Data Acquisition (SCADA) upgrades and other system improvements. This roadmap proposes to build on that foundation through the advancement of an R&D schedule and deployment of a number of projects. Through the deployment of advanced components, advanced control methods, sensing and measurement, improved interfaces as well as leveraging analytics and big data), and integrated communications, NYPA will lead the market and support the transition in the NY state marketplace.

Capability Areas

As Figure 3-9 outlines, NYPA's smart grid journey will largely fall into three major steps, and as noted in step one, building a smart grid foundation has already begun at NYPA with the Transmission Life Extension and Modernization (TLEM) program, annual research and development schedules, and the proactive and continuous maintenance and improvement programs across the NYPA's grid. Building on this foundation the Smart Generation & Transmission roadmap will look to develop capability in a series of areas.

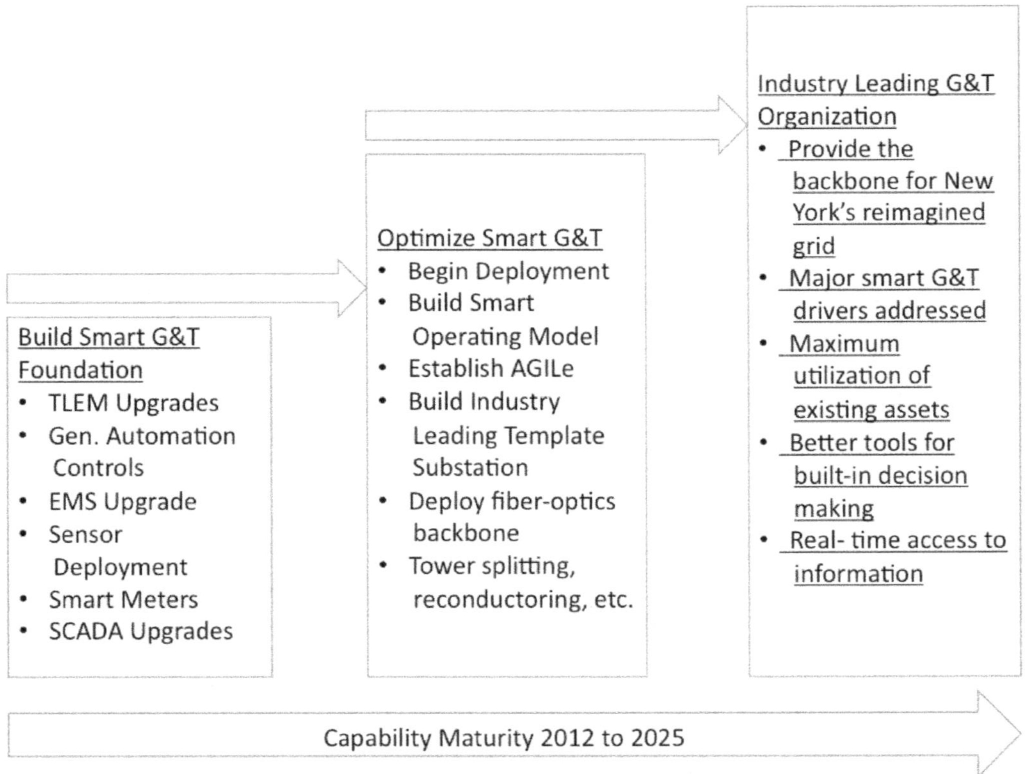

Figure 3-9. NYPA's 2014-2015 Smart Generation & Transmission journey of transformation (NYPA 2014)

Increased reliability and resiliency

- Advanced transmission monitoring, control and protection systems to decrease likelihood of cascading failures and wide-area blackouts.
- Robust security measures to reduce likelihood of catastrophic bulk system failures from human-caused and natural disasters.

Enhanced situational awareness

- Advanced analytical tools to convert data from grid sensors into insight, leading to wide-area situational awareness and control capabilities.
- Improving operator effectiveness and enhance system protection and restoration.

Optimize transmission assets

- Ensure flexibility and efficiency by optimizing the utilization of transmission assets.
- Reduced congestion and bottlenecks, Reduced costs for operation and maintenance tasks,
- Reduced risk due to old equipment or necessary downtime, increased system efficiency.

Optimize generation assets

- Automatic controls, predictive maintenance cycles on existing generation facilities to maximize hydro flow and other generation capabilities.

Integration of bulk renewables

- Develop bulk renewables to meet environmental policy demands, using intelligent monitoring, climate micro-forecast, protection and control technology, storage tech, and advanced information and operational technology integrated with the underlying assets.

Integration of distributed generation

- Ability to manage distributed generation and storage to help balance the intermittency of renewable resources and provide grid support.
- Advanced system protection to manage intermittency and bidirectional power flow.

- Advanced energy management system to truly integrate distributed with central resources

NYPA designed their roadmap to give customers, regulators and other interested stakeholders a detailed look at how smart generation and transmission will work at NYPA, how NYPA plans to deploy grid technology across its network, and why this approach makes sense for the Authority's customers and the state as a whole. Sine NYPA believes that the energy market is changing and the evolution of smart grid technologies is maturing at various speeds, this plan and its underlying assumptions will need to be periodically be reassessed regularly.

High Level Timeline

Across the Smart Generation & Transmission initiatives NYPA identified a number of categories of projects. In their documentation (NYPA 2014), NYPA details each of the capability areas and the individual projects, pilots and areas of study that align to these groups. In general there are four areas of deployment across the initiative over the next 10-15 years. These can be largely grouped into the following;

- Work that is in-flight already and needs to be completed as per schedule
- Projects that can be integrated right away
- Projects that will need to be piloted
- Further R&D Studies and evaluation

In terms of work that is already underway, NYPA's Transmission Life Extension and Modernization (TLEM) program is a key initial step for creating a smart generation and transmission grid at NYPA as this initiative is replacing and upgrading existing transmission system components that are at the end of service life or at a high risk of failure.

In addition to the TLEM program, other ongoing efforts include an annual research and development program and the continuous maintenance and improvement of its generating assets.

References

- IEC PAD 62559—"IntelliGrid Methodology for Developing Requirements for Energy Systems"—http://webstore.iec.ch/preview/info_iecpas62559%Bed1.09%DEN.pdf
- Technology Roadmapping Handbook, L. Bernal, et al., International Sept. (Small

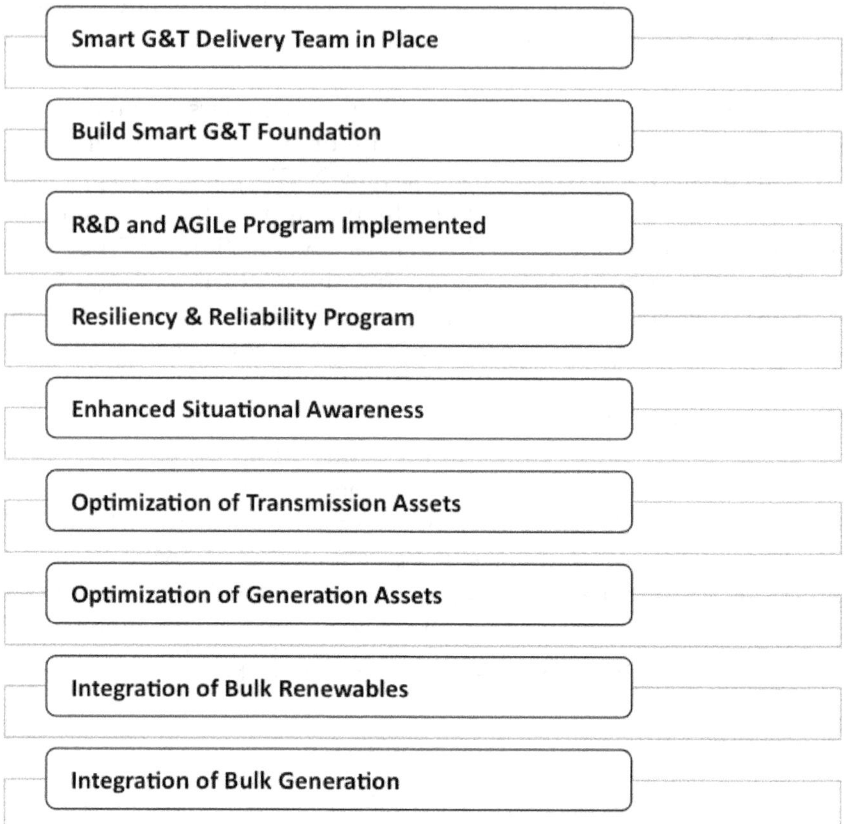

Figure 3-10. New York Power Authority (NYPA) High Level Smart Generation & Transmission Initiative Deployment Timeline (NYPA 2014)

Enterprise Promotion + Training) Program, Universitat Leipzig, June 2009

- Smart Grid Roadmap Guidebook, EPRI, Palo Alto, CA: 2012. 1025470
- Electricity Technology Roadmap: Meeting the Critical Challenges of the 21st Century, 2003 Summary and Synthesis, EPRI 2003
- Cooper, Rogert G. (1993), "Winning at New Products: Accelerating the Process From Idea to Launch" (2nd Edition), Cambridge, Mass: Addison-Wesley
- Cooper, Robert G. (1986), "Winning at New Products," Addison-Wesley.
- Technology Readiness Levels—A White Paper," John C. Mankins, National Aeronautic and Space Administration (NASA), April 6, 1995.
- "Technology Readiness Assessment (TRA)/Technology Maturation Plan (TMP) process Guide," U.S. Department of Energy Office of Environmental Management, March 2008.
- Technology Roadmap: Smart Grids, International Energy Agency (IEA), Paris, France. 2011.

- Kentucky's Smart Grid Roadmap: Recommendation on a Vision and Direction for the Future of the Electric Power Grid in the Commonwealth, Kentucky Smart Grid Roadmap Initiative (KSGRI), 2012.
- Electricity Technology Roadmap: Meeting the Critical Challenges f the 121st Century—2003 Summary and Synthesis, EPRI 1010929, 2003.
- Guide to Quality Control, K. Ishikawa, JUSE, Tokyo, 1968
- Session: Smart Grid, Presentation by J. Nowaczyk, Salt River Project at the 2009 World Energy Engineering Congress, November, 2009, Washington D.C.
- World Energy Engineering Congress (WEEC), Presentation by D. Cortez, CenterPoint Energy at the 2009 World Energy Engineering Congress, November, 2009, Washington D.C.
- Strategic Initiative Business Plan for Smart Generation & Transmission, New York Power Authority (NYPA), White Plains, NY, 2014.
- Reforming the Energy Vision (REV): NYS Department of Public Service Staff Report and Proposal, Case 14-M-0101, New York State Public Service Commission (NYPSC), April 24, 2014.
- The Smart Grid Regional Base Case for the Pacific Northwest: Interim Results and Analysis, Prepared by Navigant Consulting, Bonneville Power Administration (BPA), December 17, 2013.

Chapter 4

The Smart Grid as an Integrated Grid

The Smart Grid needs to be an Integrated-Grid so as to recognize the optimal combination of local generation, energy storage, energy efficiency and new uses of electricity integrated with central generation and storage to provide society with reliable, affordable and sustainable electricity. This approach enables resources and technologies to be deployed operationally to realize all potential benefits. It requires a modern grid characterized by connectivity, rules enabling interconnection and innovative rate structures that enhance the value of the power system to all consumers.*

The Integrated-Grid Approach to the Smart Grid allows society to enjoy the benefits of a blend of the most valuable generation, storage, power delivery and end-use technology tailored to meet local circumstances. It does not favor any one technology beyond its unique contribution. Instead the approach includes objective and thorough analysis to identify optimum architectures considering end-user requirements, delivery system, grid operation and available energy and storage resources.

At no time since the first power systems were invented and deployed has there been such a variety of new energy technology available or related breakthroughs on the horizon. These have potential to fundamentally and profoundly change the generation, delivery and utilization of electricity and related electric energy services. Applying the Integrated-Grid Approach to the Smart Grid supports the evolution of roles for suppliers, producers, consumers, utilities and regulators.

---

*Operational integration does not necessarily imply that utilities must control local generation. However, it does require monitoring and overall balance between resources.

THE INTEGRATED-GRID APPROACH

Several key considerations must be taken into account to define an Integrated-Grid Approach that is universally applicable and robust under foreseeable conditions, including recognition that:

- Consumers of all sizes are increasingly adopting local generation.
- Most consumers with local generation are choosing to remain connected to the grid.
- The grid provides substantial value to consumers with local generation.
- Extending the conceptual and operationally boundaries of today's grid by modernizing the grid and by properly integrating local generation with utility generation creates an integrated grid.
- The Integrated-Grid Approach optimizes central and local resources to ensure grid reliability at lowest cost.
- The Integrated-Grid Approach will require:
  - — Grid modernization and upgrades to meet current and foreseeable requirements for accommodating all forms of localized self-supply of electric services
  - — Interoperability standards to ensure that new technologies can be used anywhere with equal effectiveness
  - — Updating interconnection rules to ensure that they offer incentives compatible with established goals
  - — Innovative rate structures and market accommodations so that distributed resources are fully and optimally utilized.

Flexibility

The electricity sector will increasingly rely on flexibility in power systems and operations—generation and power delivery—over an interconnected delivery system. Interconnecting the grid with local generation, storage and other devices will result in a power delivery system that requires more than wires and connectors to deliver electricity effectively. It requires that it become an interactive, dynamic system of distributed resources defined by two-way power flow and expanded functionality, including real-time grid security assessment and recovery.

Resiliency

Power providers must deal with the fact that extreme weather events are increasing in frequency and severity even as "wired" society becomes less tolerant of widespread or extended power outages. Hurricane Katrina, Super storm Sandy, and the tsunami leading to the Fukushima accident inflicted considerable damage and left citizens without power, clean water, food, and fuel for heating or transportation. In several of these events, a digital-dependent public responded unfavorably to the loss of cell phone service. A public sense of vulnerability is affected also by manmade outages and cyber security incursions. Combined, these represent a new reality for the industry and its stakeholders.

Prevention employs engineering designs and advanced technologies to "harden" both generation and the transmission and distribution (T&D) system to limit damage. Recovery utilizes tools and techniques to restore service quickly to as many customers as practical. Survivability anticipates and deploys technologies to aid customers, communities and institutions in continuing some level of normal function without complete access to the grid.

Connectivity

Significant opportunity is arising for the electricity sector to use connectivity to improve resiliency, connect digitally with customers, and improve operational efficiency, as well as to leverage more effectively the advantages of the local generation by using the Integrated-Grid Approach to the Smart Grid.

How will the Integrated-grid Approach Change the Power System as We Know it Today?

Figure 4-1 illustrates today's power system. As shown, it is largely comprised of large central-station generation connected by a high voltage network grid to local electric distribution systems that, in turn, serve homes, business and industry. In today's power system, electricity flows predominantly in one direction using electro-mechanical controls.

Figure 4-2 illustrates the elements which will be part of a fully functional Smart Grid deployed as part of applying the Integrated-Grid Approach to the Smart Grid. Tomorrow's power system will still depend heavily on large central-station generation, but if the Integrated-Grid Approach is embraced it will include a substantial number of installations of electric energy storage and of renewable generation facilities,

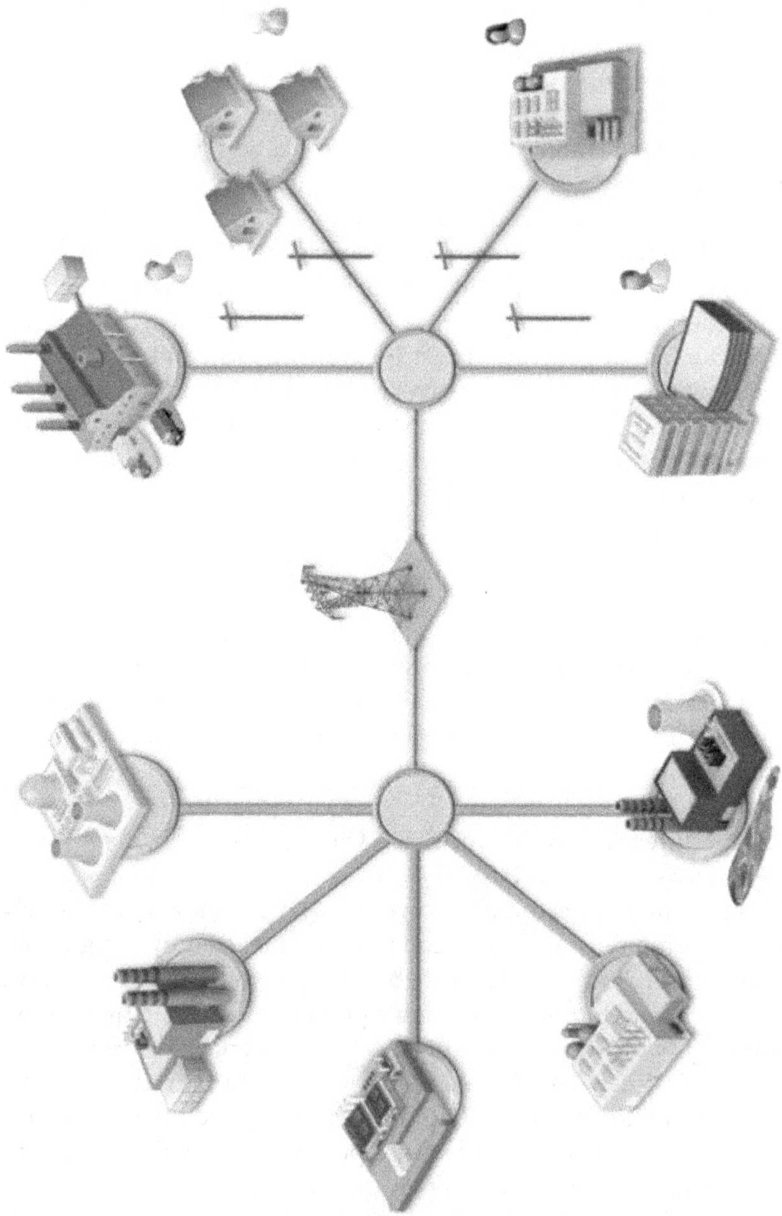

Figure 4-1. Today's Power System

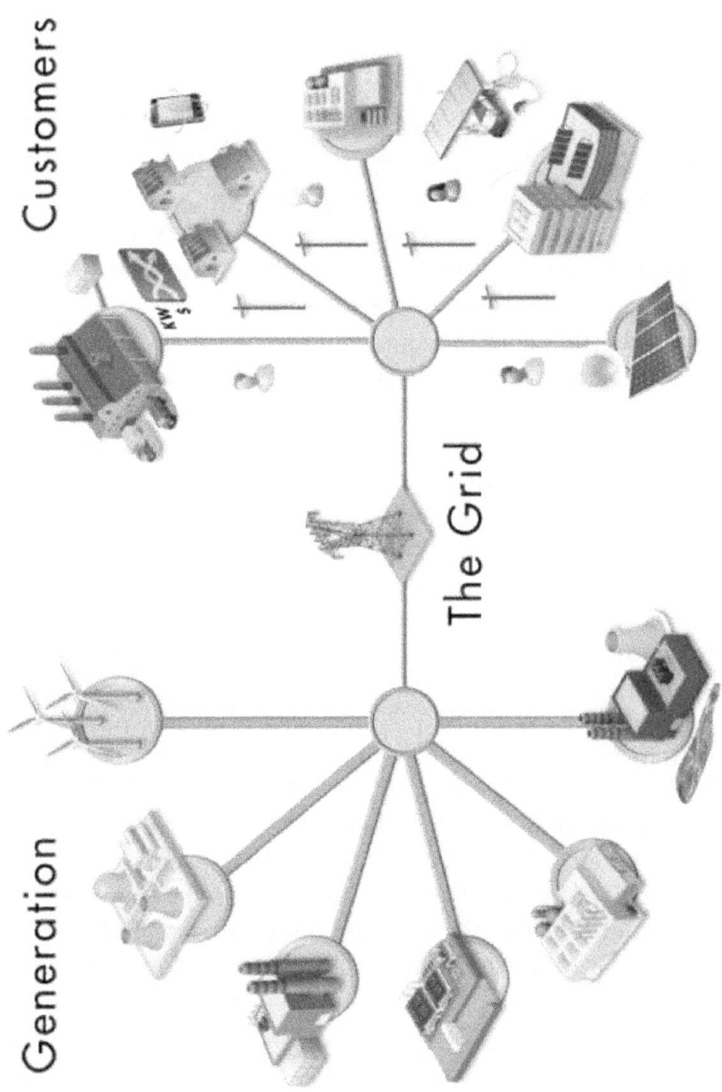

Figure 4-2. Realizing the Integrated-Grid Approach: Tomorrow's Power System

both at the bulk power system level and distributed throughout the system. The Integrated-Grid Approach recognizes that accomplishing diversification requires greatly enhanced sensory and control capability configured to accommodate these distributed resources as well as electric vehicles, direct consumer participation in energy management and efficient communicating appliances.

What Grid Cost is incurred to serve a Typical Consumer?

Broadly speaking, in discussions and deliberations about the changes in consumer demands, the industry—utilities, consumers and regulators alike—are underestimating the importance of using an Integrated-Grid Approach to optimize the value of a combination of local generation, energy storage, energy efficiency and new uses of electricity integrated with central generation and storage.* A robust and ever-present grid is essential for those that partially meet their electricity requirements from on-site generation and for those that seek to self-generate all of their needs, at least in a collective sense. Without the grid, a distributed generation consumer would experience reduced reliability and, without the grid or without substantial investment in supporting technologies, would not have the ability to power their building or premise when the sun doesn't shine or have the electrical support to operate certain devices such as conventional residential air conditioners. The benefits of interconnection go both ways. The grid benefits substantially from consumers who generate some or all of their electricity locally because, as long as they are connected to a modernized grid, that power can be used as a system resource to the benefit of all electricity consumers.†

It has been 130 years since Thomas Edison energized the first electricity supply system on September 4, 1882 at Pearl Street in New York. Networks of electrical systems subsequently evolved to what the National Academy of Engineering referred to as he "supreme engineering achievement of the 20th century." In many respects, today's power system looks similar to Edison's. Electricity flows predominantly in one

*Local generation is often referred to as distributed generation (DG). Photovoltaic (PV) power generation installed by customers is a form of DG. Most practitioners also include DG in a basket of technologies called distributed energy resources or DERs which include DG, responsive loads, and electric energy storage.

†The concepts discussed herein are equally applicable for: campus-style microgrids; for modernization of transmission systems with wide area management systems (WAMS); and combinations of other technologies such as phasor measurement units (PMUs).

direction, from generators to end-users, using mechanical controls to balance Dispatchable Generation with Forecastable Customer Demand (need to differentiate one way flow with respect to transmission where it is two way and distribution where it is predominantly one way—accuracy cannot be sacrificed because of simplicity). Today's power system is designed around the needs of 1,000 GW of dispatchable generation and 2.4 million miles of one-way power flow that provides electricity to customers, whose involvement is considered passive, in that they only consume electricity.

Today's power system has served society well, in large part because it's so far very reliable. It has allowed economies to grow and the standard of living to continue to increase while keeping electricity rates affordable. However, with the advent of local generation and soon to follow—storage, it is imperative that the traditional system be operated in an integrated fashion to accommodate these new resources without sacrificing the standards that have been set.

Figure 4-3 illustrates the functional costs of providing grid services to the average US residential consumer. The average residential electricity customer consumes 982 kWh per month, which unbundled amounts to an average of $70 for energy consumed, $30 for the distribution system and $10 for the transmission system that deliver the energy when and where it is wanted. These represent the average costs of providing energy and grid services. These reflect the price consumers actually pay—but do not represent the real break-down into generation and delivery. Those differences are a result of customer characteristics, the pattern and amount of customer demands, and the technology required

Figure 4-3. The Cost of Today's Grid Connectivity (Note that the consumers often do not see this cost breakdown. There costs are often allocated differently depending on rate design.)

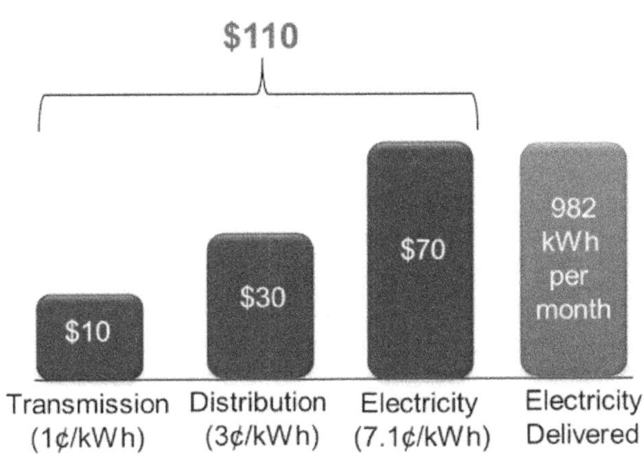

to meet those demands. Costs also vary among utilities. In some cases, consumers pay considerably less for the transmission and distribution systems and somewhat more for the energy component. This breakdown is the consequence of the Rate Design principles that are applied by each utility to allocate costs to customer classes.

The average cost does not do justice to the diversity of the costs to serve residential customers of varying sizes, Figure 4-4 illustrates this breakdown for transmission and distribution costs to serve residential electricity customers of different types and sizes.

A careful examination of the importance of these functions from a power system engineering perspective suggests that a portion of the "strength" of the grid is provided by the availability of central station power plants connected throughout the grid that provides for "inertia" which supports the overall delivery system. Figure 4-5 illustrates the breakdown of grid costs in to capacity and energy-related costs.* This figure acknowledges that the combination of transmission, distribution and the portion of generation, which provides grid support, is an average of $51/month while the energy costs alone are approximately $59/month. Again, these costs are average and will vary widely across the U.S. and among various sizes of consumers.

An investment in central generation capacity is still needed to support grid connectivity, albeit those investments may be somehow included in the cost of generation, transmission and/or distribution depending on the presence of retail or wholesale markets and the regulatory paradigm present. This cost is estimated and called out separately here for illustrative purposes.

It should be noted that distributed generation does not necessarily reduce the need for central generation or transmission capacity. As

*Generation, Transmission, Distribution Breakdown are provided from EIA estimates in ($/kWh) assuming an average customer usage of 982 kWh/month. Generation is broken down into two components Energy and Capacity/Ancillary) based on PJM market estimates of the price breakdown: "2010 PJM Market Highlights: A Summary of Trends and Insights." 2011. http://www.pjm.com/~/media/documents/reports/20110513-2010-pjmmarket-highlights.ashx. Distribution and Transmission are estimated based on cost breakdown from Southern California Edison (SCE).
SCE Implied Cost breakdowns scaled to $40/month:
Cost Breakdown Fixed % Variable %
Customer 100% 0%
Distribution 90% 10%
Sub transmission 60% 40%
Transmission 100% 0%

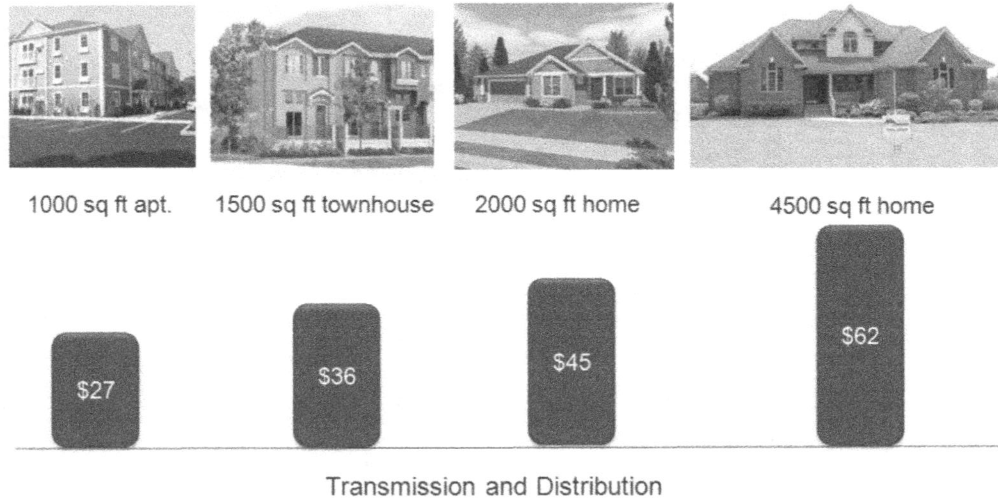

Figure 4-4. Ranges of Monthly Cost of the Grid for the average residential customer (Source: AEO 2013)

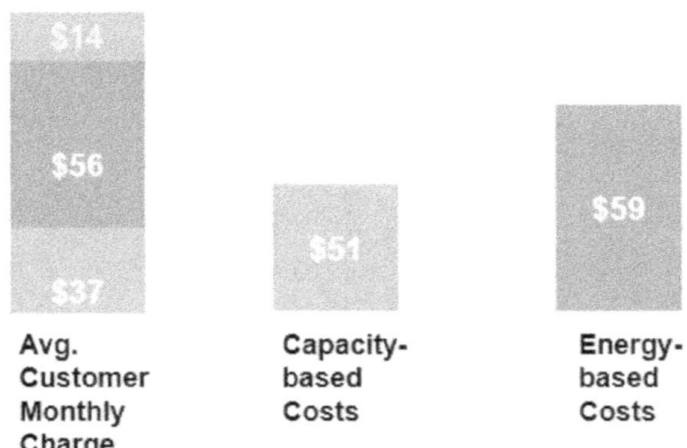

Figure 4-5. Average Residential Customer Costs broken into capacity based and energy based costs.

a result, an analysis would not necessarily allocate any capacity credit to local generation. Some of the recent grid operators support this idea based on their recent studies and actual experience. This since even if there is value of capacity credit higher penetration of PV type variable generation will increase the need for ramping resources, inertia and other ancillary service so overall the fixed cost for generation to support grid connectivity of local generation may actually increase. Figure 4-6 illustrates two cases that support this. Each illustration represents the projected system—wide demand for electricity in California and Germany, respectively considering both the total demand net of renewable power generation. The left side is from the California Independent System Operator (ISO) illustrating their now famous duck curve which depicts how California's net power generation will change dramatically beginning in 2015 and the right side depicting the load profile of the German power system. Each system illustrates the substantial impact which photovoltaic power generation can have as penetration increases substantially. Two issues are most notable: 1) there is significant potential for over-generation at peak sun periods which without a modernized grid and smart inverters will cause system instability and potential call apse; and 2) the rapid increase in power generation required when the sum sets and evening demand builds (the ramp) will require a substantial amount of central generation.

Increasing Importance of the Distribution System

Although the demand for electric energy supplied from the grid may be flat or only increases slightly, the importance (and cost to provide) of the distribution system will still increase. This is due to four factors:

- The growth in the use of distributed generation (photovoltaic, small wind, micro-generation, combined heat and power (CHP), etc.) which is to be connected to the grid and both draws power from the grid and supplies power to it.
- The growth in the deployment of plug-in-electric vehicles (PEVs) and other new, yet unforeseen appliances.
- The potential (future?) use of micro-electric energy storage.
- The increase in the end-use efficiency and the digital nature of electric energy-consuming devices and appliances.

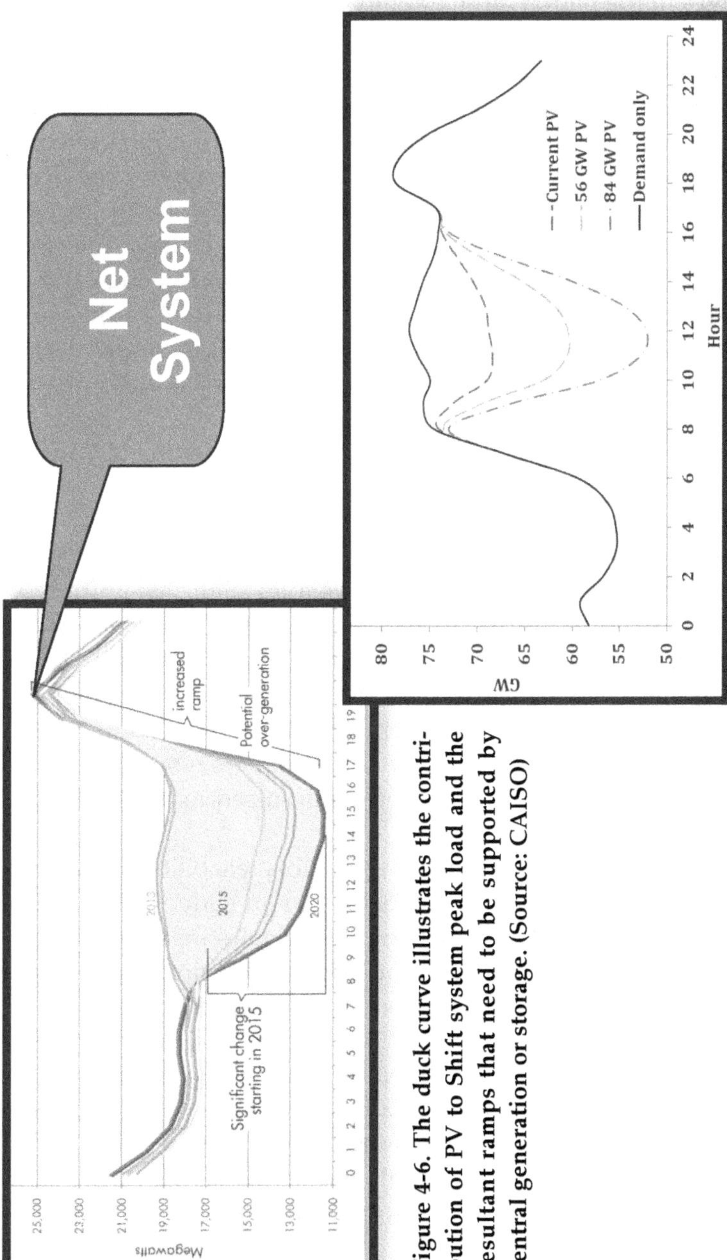

Figure 4-6. The duck curve illustrates the contribution of PV to Shift system peak load and the resultant ramps that need to be supported by central generation or storage. (Source: CAISO)

These four factors, taken individually or in combination, will result in changes in the pattern and amount of electricity demanded from grid resources, as well as in the electrical characteristics imposed on elements of the grid. In the case of PEVs and small electric energy storage installations, these may devices impose a substantial demand when vehicles are highly concentrated in radial portions of the distribution system—often over a limited period. Only a robust distribution system can accommodate these new demands on the grid. In addition, evolving new end-uses are increasingly "digital" in nature and no longer have the frequency response, short-circuit current characteristics, and voltage response of their analog ancestors. They are much more sensitive to changes in the physical characteristics of electricity delivery. This increases the role and importance of a strong electrical backbone to which buildings and microgrids can connect.

The Integrated-Grid Approach will be an essential ally in addressing these issues.

The Value of Local Generation to <u>All Consumers</u>

The **Integrated-Grid Approach** enables high penetration of local energy resources and broadly benefits all consumers of electricity who are connected to the grid. Figure 4-7 illustrates the primary benefits of local generation to all consumers provided these consumers remain connected to the grid. Table 4-1 elucidates these attributes. It enables the following benefits to accrue to all consumers who are connected to the grid provided the grid has been modernized to accommodate local resources.

The Integrated-Grid Approach enhances reliability and affordability by reducing the risk of grid instability due to sudden drops in a large amount of central and/or local resources, enables the integration of both local generation and local or distributed storage, minimizes outages, and provides ancillary services (frequency response, non-spinning reserve and participation in Demand Response (DR) programs. It can reduce cost by optimizing from a combination of local and central resources. It also enables a higher penetration of local resources. It can be configured to provide incentives for local resources to provide services to the grid. Further examples of this value are illustrated below: The Integrated-Grid Approach* enables true synergy between consumers and the grid.

*The **Integrated-Grid Approach** has been referred to as part of the concept ElectriNet^{SM} in previous literature.

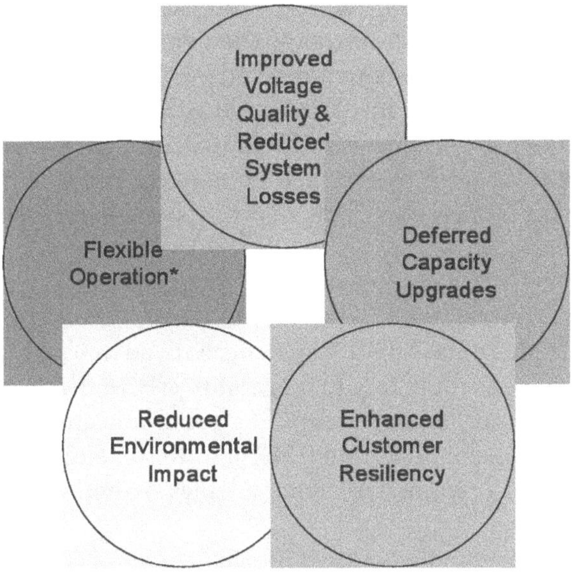

*Enhances reliability and impacts distribution operations, DER integration, minimizes re-conductoring, provides ancillary services (frequency response, non-spinning reserve and DR)

Figure 4-7. Primary Benefits of Local Generation to all Consumers (Provided they are Grid-Connected)

Grid Support

Photovoltaic (PV) systems offer natural support for distribution systems. For example, Figure 4-8 illustrates the coincident demand from PV as compared to a typical distribution load profile. The system can support utility needs and mitigate distribution system demand. However, the extent to which this is possible is very dependent on local circumstances. The overall need for generation capacity on the power system may not be impacted substantially until there is large penetration of local generation and depending on the modernization of the grid. Local generation, however, will still reduce the need for the operation of central generation, even if it does not reduce capacity requirements.

Figure 4-9 further illustrates the effectiveness of central voltage control-enabled by connectivity which interconnected local generation systems can yield. The figure on the left represents a system with local generation installed with no voltage control. The relative high voltage and fluctuations can interfere with the effective operation of customer and distribution equipment, leading to premature failure of components and appliances. The figure on the right represents a system where the voltage can be controlled to mitigate the over-voltage problem, and hence reduce the adverse costs that result.

The Integrated-Grid Approach would enable coordinated control of high-penetration PV systems as illustrated in Figure 4-10. Here advanced inverters are used to enhance voltage control and balance the ratio of real and reactive power needed to reduce losses and improve system stability.

Providing Ancillary Services

Figure 4-11 illustrates how PV systems with advanced inverters can offer reactive compensation. Key to this ability is both the deployment of advance inverters and the availability of a communications system and a distribution management system (DMS). In this example, the synergy between utility and consumer systems enables overall optimization of the system.*

*Depending on how the system is designed, consumers with local generation can provide the utility with various ancillary services, including operating reserves, regulation and load balancing, and voltage support. While the value of these kinds of services can be significant, it is highly variable. (King, 2006)

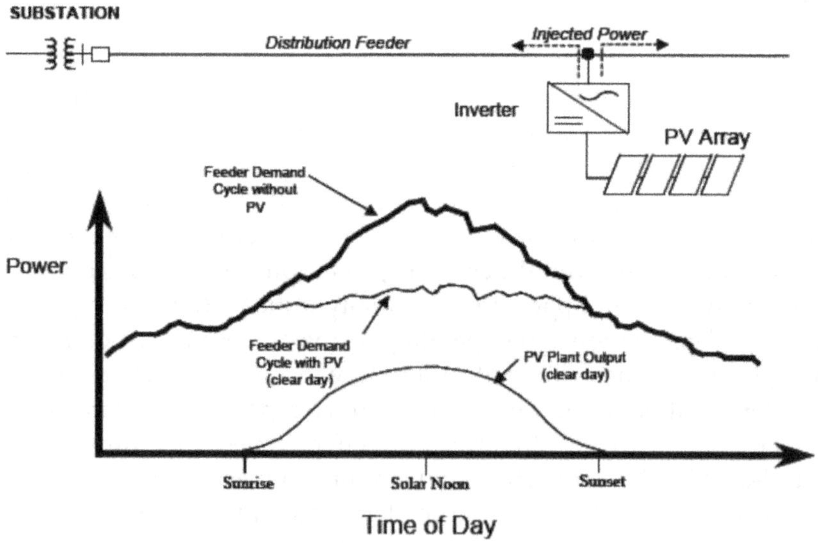

Figure 4-8. Grid Support Value of Distributed PV (Source: EPRI 1018096) Optimizing Distribution Operations

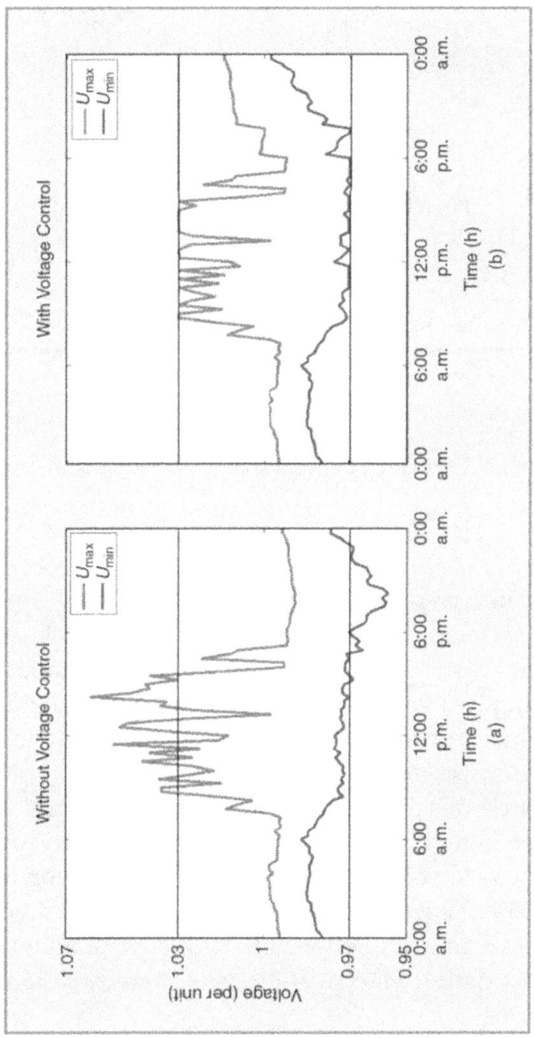

Figure 4-9. Minimum and Maximum Voltage Magnitude (a) Without and (b) With Central Voltage Control (Source: EPRI 1018096)

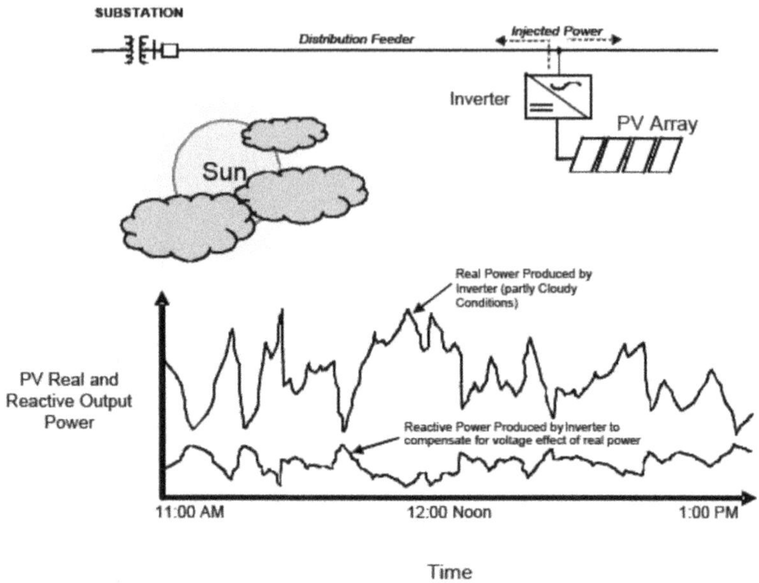

Figure 4-10. Advanced Inverter for Reactive Compensation (Source: EPRI 1018096)

Demand Response

Participating in Demand Response (DR) programs, specifically to provide ancillary services, The Integrated-Grid Approach will provide the connectivity and distribution management systems to facilitate consumers participating in a variety of demand response programs including various customer notification schemes, interruptible tariffs, direct load control, real-time pricing, and critical peak pricing.

Figure 4-11 provides an example of where this control scheme can also enable higher penetration of PV on distribution feeder without necessary reconductoring or reinforcement. The illustration provides an example of how 20% PV penetration can impact primary voltage with and without voltVAR control. Better volt-VAR control allows more real power to flow on existing electrical circuits thereby accommodating more PV installations.

PJM experienced record demand during September 2013. In order to meet demand the system operators deployed their demand response program on September 9, 10 and 11. Figure 4-16 and illustrates the load shedding which occurred on September 11. It illustrates how demand response (refer to Emergency DR Event Period) was deployed on Sep-

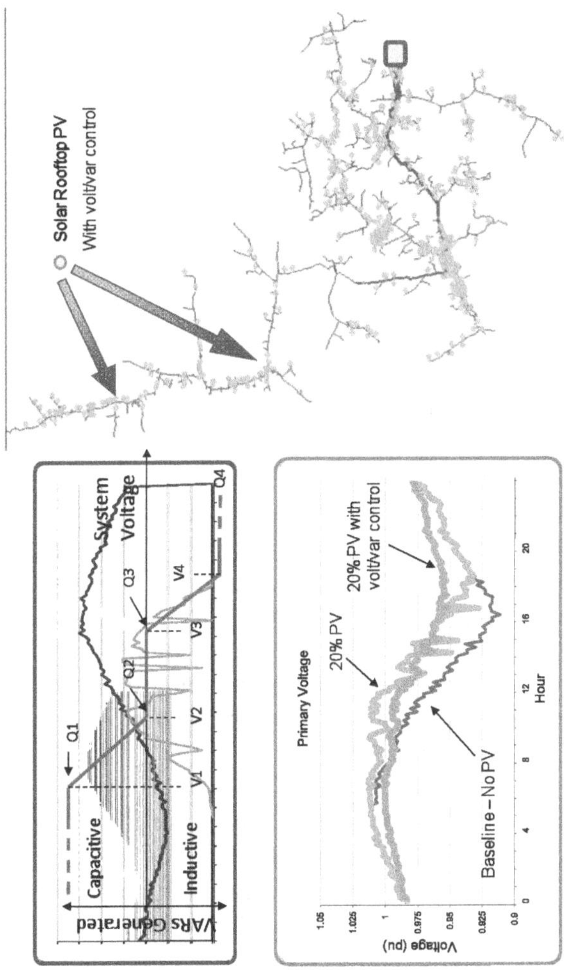

Figure 4-11. Volt-VAR Control with High-Penetration PV (Source: EPRI 1026640)

tember 11. Demand response was an essential, but insufficient resource to PJM during that period. The illustration does not reflect what the load would have been if the DR systems had not been activated. However, this event was reportedly the first time in ten years that load on the PJM had to be shed

However, in reviewing the event several attributes are apparent:

Although overall demand response performed well during this period, restrictions such as the required two hour notification for invoking demand response and the lack of granularity to utilize demand response on a more targeted sub-zonal basis limited the operator's flexibility in using demand response as a tool in real time to prevent some of these events. PJM has indicated that it will examine that issue as well as other issues surrounding generator and transmission system performance through its stakeholder process. The downside is that experience suggests that if they make demand response resources more useful, more dispatch able, then participation drops.

The lack of divisibility of Demand Response resources is usually because by design the stakeholders have never been able to agree on if only some were needed, who would get called. Some want to be called because they get paid on performance. Others get paid up from and never want to be called. Since Vocational Marginal Pricing (LMP) during these periods is set using shortage pricing, that keeps a large load drop from Demand Response from effecting it. There is little enthusiasm for establishing a lot of dispatch granularity. However, this is a complex and controversial issue. DR advocates have prevented DR from being treated like a fungible resource, which reduces its value relative to a generator, yet DR usually gets the same level of compensation.

Feeder Hosting

Figure 4-13 illustrated another example of the value of the Integrated-Grid Approach. In this figure a distribution system was analyzed to determine the feeder hosting capacity or the amount of Local PV generation the feeders could support without substantial upgrading. As shown in the figure, the use of advanced inverters and control could substantially increase the feeders' capacity. In this case, potential overvoltage situations can be mitigated eliminating the need for extensive reconfiguring. The "y" axis represents the voltage in "per unit" or PU values. A PU of 1.0 would translate into a service voltage on a 120 volt system of 120.

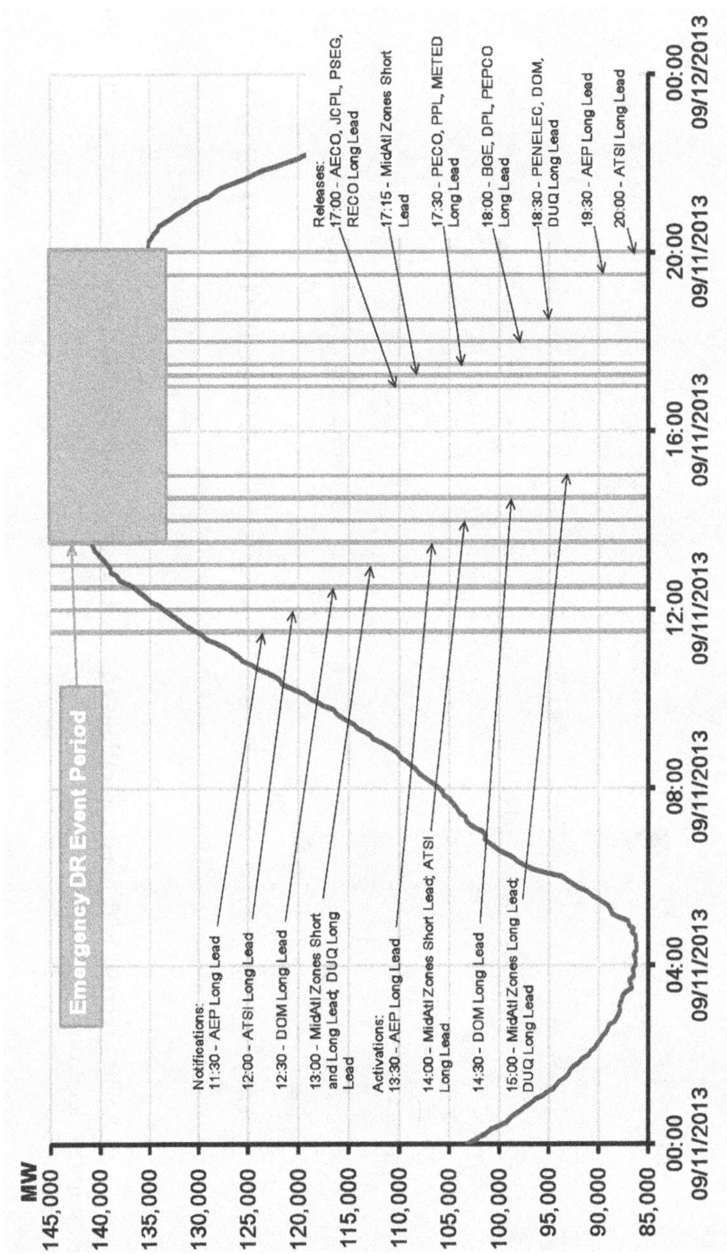

Figure 4-12. PJM Geographic display of load shed events on September 11, 2013 (www.pjm.com)

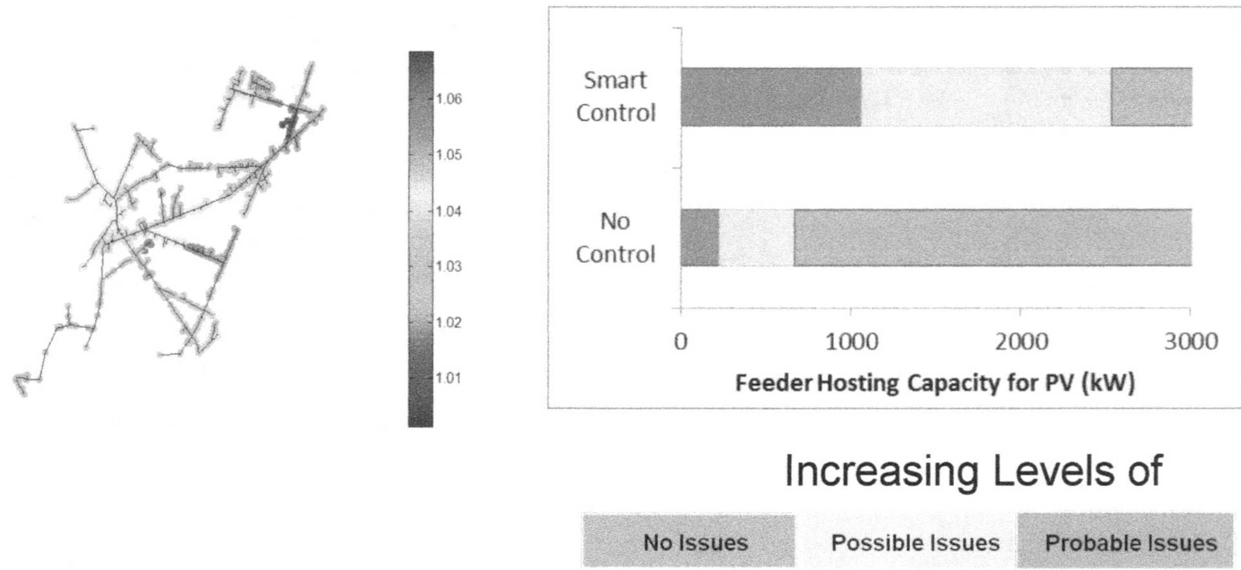

Figure 4-13. Smart Inverters with communications and control can substantially increase the amount of PV a distribution systems can support.

Resiliency

The presence of local generation also has the ability to increase the overall resiliency of the power system. In the event of an outage, local generation can help support portions of the distribution system which may be without power. It can also allow consumers to sustain part of their building services.

The rising interest in the use of local generation has occurred in part due to heightened concerns about resiliency, improvements in generation technologies, power electronics, and the need for new capacity resources on the power system. A key potential benefit of local generation is the ability to improve the reliability of the power system by providing emergency power during interruptions of the bulk system supply. This benefit can only be realized if the local generation is operated in a configuration that facilitates islanded operation. The Integrated-Grid Approach allow for this type of operation while also being able to capture all of the other benefits of local generation such as waste-heat recovery, load reduction on the transmission and distribution system, and power quality improvements.

Low refers to EPRI low estimates of total smart grid costs and high to the estimates of high costs. These costs total between $338 and $476 billion for the U.S. and can yield benefits of between $1,294 and $2,028 billion. (Refer to EPRI report 1022519 for details.) (Residential, commercial and industrial numbers were apportioned based on 2009 sales as reported by EIA (38% residential; 37% commercial; 25% industrial.) Source: EPRI 1022519

This does not include the total cost of reconductoring the distribution system in the event there is substantial penetration of local generation. Based on estimates from the German experience, it is estimated that this could cost $1500 per residential customer or an amortized value of $14/mo. Reconductoring costs can be calculated based on the assumption that potentially 30 customers/mile based on US estimates. This estimate uses German data from Table 4-1.

The resultant Total Cost of Medium Voltage plus Low Voltage = $53K + $36K + $89K ≈ $90K. Discounting by 50% since U.S. has far less low voltage than Germany yielded approximately $1500/customer. The resultant marginal cost to the average customer for a fully modern grid with a high penetration of local generations would be $12 for the cost of the smart grid and $14 for the cost of reconductoring totally $26 per month. In moving forward with the second phase of this study, EPRI intends to carefully conduct a more detailed analysis of costs and benefits while evaluating these assumptions. In particular, it is believed that there are synergies between the grid modernization deployment and the grid reconductoring and distribution upgrades that may well reduce these costs.

It should also be noted that these costs will vary considerably depending on the density of customers per mile of distribution feeder and the actual system. For example, Table 4-2 lists a range of interconnection costs by project size for two California utilities: Southern California Edison (SCE); and Pacific Gas & Electric (PGE). This cost are substantially higher than the German experience and reflects, in part the fact that the California installations are reportedly in more rural parts of their systems while the German system is very dense and their installations are much closer together such that upgrades cover several customers at once. Detailed analysis is needed to further investigate these differences.

Other studies have accessed the macro-level implications of grid modernization. Table 4-2 summarizes these benefits and apportions them to the average residential electricity customer.

The total expected benefit from improved quality, security, reduction in overall capacity needed, increase in reliability and reduction in the overall cost of energy is $39/month. As with the previous estimates in this paper, these are average and likely will vary widely across the U.S. In addition the actual costs incurred to both modernize the grid and to upgrade the distribution system to accommodate substantial penetration of local generation will be made in the short-term and are substantial commitments to be made based on speculation that the widespread use of local generation and storage will indeed evolve. On the other hand, the projected benefits are even less certain and for the most part will not be realized until after the investments needed to enable the Integrated-Grid Approach have been realized. However the $39 benefit does compare favorably with the cost of $26/month

Practices vary as to who pays for elements of interconnection costs. Figure 4-4 illustrates the breakdown for Southern California Edison.

Germany has over one million local generation installations (mostly PV). They have adopted a procedure which requires that a determination be made as to what upgrades are needed from a "system" perspective. This is based on assumptions regarding a certain amount of DER going into certain geographic regions. Then they make plans to upgrade lines/transformers to accommodate expected generation levels. The cost

Table 4-1. Estimates of Requirements for German Upgrades resulting from PV

Requirements for German Upgrades	EC2020	Cost (mil euro)	Cost (mil USD)
Medium-Voltage Lines (km)	55000	3500	4725
Low-Voltage Lines (km)	140000	6000	8100

Source: "E-Bridge Consulting" http://www.e-bridge.de/fileadmin/templates/downloads 20_Baumanagement-BDEW_Verteilnetzstudie_2011_Schreibgeschuetzt_.pdf

	<10 kW	<1 MW	1-3 MW	3-5 MW	5-10 MW	10-20 MW
SCE	26,576	30,225	20,974	10,487	9,580	54,106
PG&E	26,675	159,630	110,772	80,909	43,151	21,576
Overall	26,576	30,225	20,988	14,966	18,187	28,613
Low Estimate	13,288	15,112	10,494	7,483	9,093	14,307
High Estimate	39,864	45,337	31,482	22,449	27,280	42,920

Note the wide range of interconnection costs!

Figure 4-14. Median Interconnection Costs (Levelized $/MW) by Project Size Category

Table 4-2. Benefits to the average residential consumer from implementation of the Integrated Grid Approach (Source: based on EPRI 1022519)

Net Present Worth ($B)		Residential Amortized/mo. ($)	
Reliability	444	11.10	11
Security	152	3.80	4
Quality	86	2.15	2
Cost	475	11.88	12
Capacity	393	9.80	10

of all upgrades is then socialized across all customers. The DER owner is not responsible for distribution upgrades.

Figure 4-15 gives a more simplistic illustration which summarizes a possible framework for evaluating The Integrated-Grid Approach.

The figure shows avoided energy and capacity costs to represent the benefits that inure to all rate payers for PV energy generated. Avoided costs are equivalent to benefits because they reflect costs that would have been incurred except for the PV energy produced. Avoided costs are the standard currency for valuing energy efficiency investments—it is appropriate as a general characterization of benefits. The avoided costs shown reflect estimates of 2016 capacity and energy costs portray in a range of values to characterize the variety of circumstances of utilities across the country.

To the extent that local generation, including PV, reduces the need for capacity to serve consumers, because peak demand is decreased, then it theoretically reduces marginal capacity cost. In some cases, local generation may also impact ancillary service cost, depending on the penetration of local generation. As the penetration of local generation increases then the need for ancillary services will also increase.* Displacing energy reduces ancillary services costs, based on today's system, but to the extent local energy production is highly variable, more ancillary services may be required in the future and the net results may be positive or negative.

---

*The **avoided energy cost** ($/MWH) reflects the CA estimate of 2016 avoided energy costs, with a range EPRI researchers constructed using EIA estimates of natural gas price volatility (because gas is generally assumed to be at the margin). ISO-RTOs do not produce forward energy price estimates, so the CA value was used as it is referenced.

The **avoided capacity cost** ($/kW-year) value is for 2016 PJM average forward capacity market price for 2016 (the low value) and the average CA state avoided capacity (the high value). The difference is likely based on the particular circumstances of Californian (citing, environment compliance, land costs, etc.)

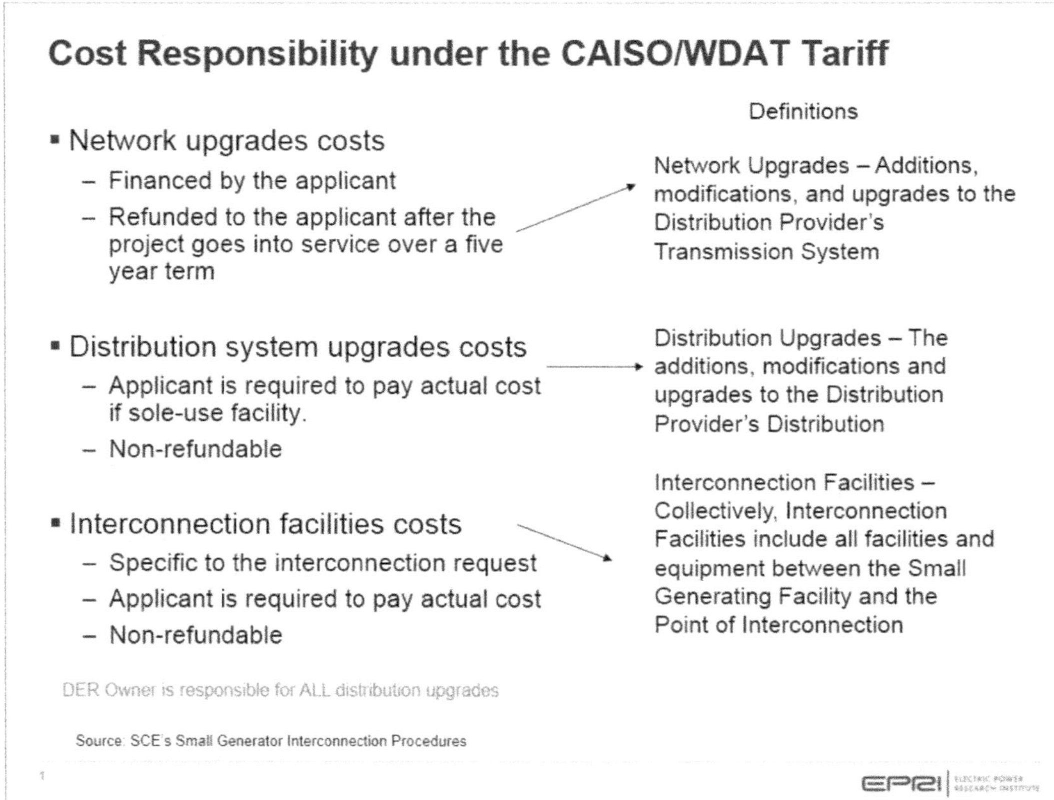

Figure 4-15. Cost Responsibility under the CAISO/WDAT Tariff

Figure 4-16. Potential Value of Local Resources in an Integrated Grid

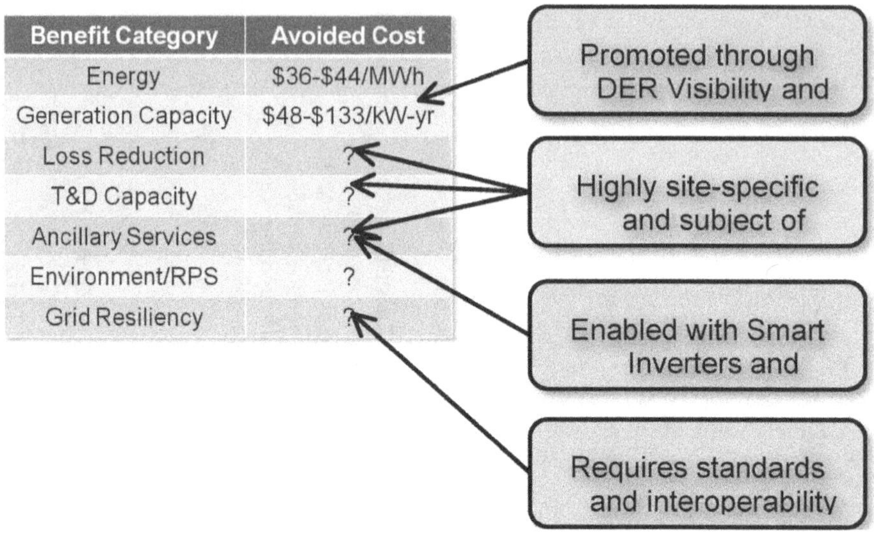

Avoided Costs

On September 26, 2013, Energy and Environmental Economics, Inc. (E3) produced a report for the California Public Utilities Commission's (CPUC) Energy Division which evaluated California's net energy metering programs. In the report, E3 updated an avoided cost framework which has been "developed in numerous proceedings at the CPUC since it was adopted in 2004." The framework is contained in Table 4-3.

Externalities

Externalities are secondary impacts of energy systems which are highly uncertain and not easily monetized. Examples of externalities are emission impacts of local generation and of the central station power, they may offset, economic activity resulting from expansion of local jobs, health impacts of increased or reduced emissions, National Security impacts from shifting energy use, and a few benefits. Often these impacts are assumed to be positive and are the motivation for policy enactment to encourage local generation.

EPRI's framework development focuses on the value of the Integrated-Grid Approach and excludes consideration of specific values of external costs or benefits. It is assumed that there are already imputed in policies such as Renewal Portfolio Standards and Net Energy Metering

Table 4-3. Components of Marginal Energy Cost in California (E3 2013)

Component	Description
Generation Energy	Estimate of hourly marginal wholesale value of energy adjusted for losses between the point of the wholesale transaction and the point of delivery.
System Capacity	The marginal cost of procuring Resource Adequacy resources in the near term. In the longer term, the additional payments (above energy and ancillary service market revenues) that a generation owner would require to build new generation capacity to meet system peak loads.
Ancillary Services	The marginal cost of providing system operations and reserves for electricity grid reliability.
T&D Capacity	The costs of expanding transmission and distribution capacity to meet customer peak loads.
CO_2 Emissions	The cost of carbon dioxide emissions (CO_2) associated with the marginal generating resource.
Avoided RPS	The cost reductions from being able to procure a lesser amount of renewable resources while meeting the Renewable Portfolio Standard (percentage of retail electricity usage).

programs and would require complex analysis to avoid duplication. Individual studies, conducted at more disaggregated levels may elect to recognize and monetize these factors.

The Integrated-Grid Approach as described in this paper would allow both the visibility at a more granular level of DR potential and local generation as well as the ability to offer more flexible programs to engage customers.

The Value of the Grid to Consumers with Local Generation

Unlike the telecom industry where 30% of landline phone customers have decided to forego land line service and rely instead solely on cell service, most all consumers who install local generation remain connected to the grid. This is for very good reasons—cell service gives consumers nearly the same level of service as landline service from an entirely different "grid." In the case of electricity, energy flows through the same electrical grid whether locally or centrally connected and without that grid, a consumer generating locally will not receive nearly the same level of service.

However the actual value of connectivity to the grid for customers which choose to generate electricity locally is not well understood. Without connectivity to the grid, the reliability and capability of the consumer's power system would be diminished unless substantial investments were made in the local power system to achieve equivalence of service quality to what the grid provides. The grid provides electrical support to consumer installations. Connection to the grid, called interconnection or grid connectivity, provides substantial improvements in power quality, reliability and generator operating economics as well as in voltage and frequency regulation, harmonic distortion, efficiency, and operating costs. Figure 4-16 illustrates the primary benefits of grid connectivity. From an operating perspective, this support enables the robustness of the electric infrastructure to provide the following:

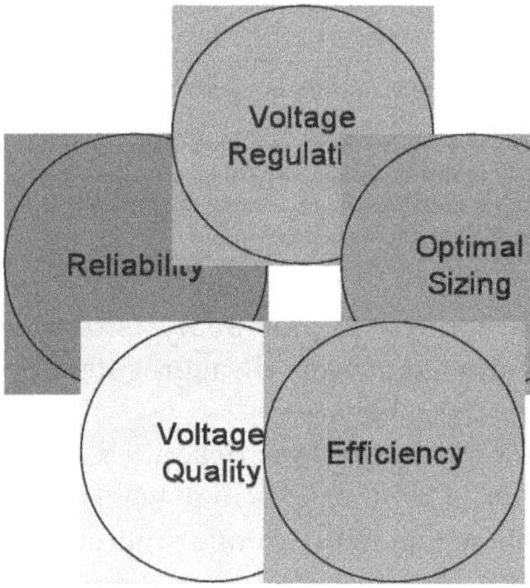

Figure 4-17. Primary Benefits of Grid Connectivity to Consumes with Local Generation

Reliability (Balancing and Backup Resource)

The grid backs up disruptions in the performance of local generation systems. This includes compensation for variable generation output of photovoltaic and wind power generation. When the sun doesn't shine and the wind doesn't blow—the grid provides backup. In this role, the grid is acting as bulk storage. This arrangement may change as storage costs decline potentially allowing for storage as a way for storing excess local generation for us later on when generation drops below demand. This has the grid providing balancing in all timeframes—hours, minutes and seconds—to balance the variability and intermittency of the distributed resources. This resource is based on the combination of the entire power delivery system (transmission and distribution) as well as the generation and storage which are part of it.*

The average annual reliability provided by the grid is over 99.9% including major events such as storms like Sandy and Katrina. Figure 4-18 illustrates the ten-year average of system average interruption duration index (SAIDI) broken into quartiles. SADI excludes those major storms. Without the, grid customers using local generation could experience many days of interruption per year, unless they made substantial investment in redundant back-up generation. This illustrates how the strength of the grid can offer reliability to local power systems.

Variability

Figure 4-20 illustrates the variability of a PV load generation system. These variations are typical in locations anywhere in the US. The

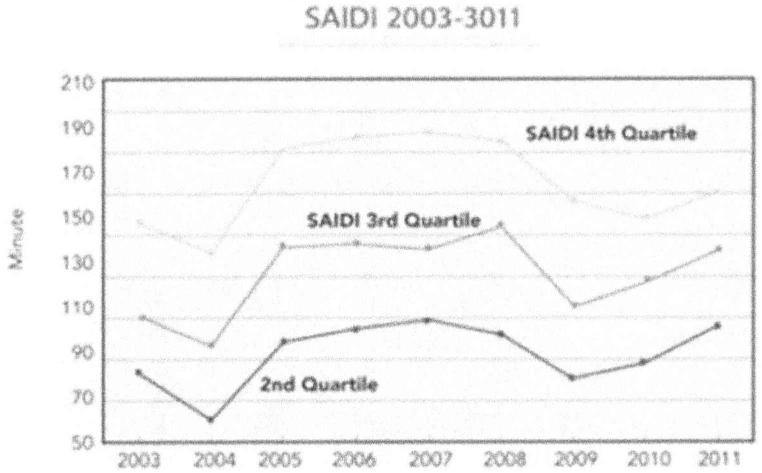

Figure 4-18. Grid Provides Reliability Service (Source: Wilson)

*The conventional measure of reliability is the value of un-served energy or so-called "lost load," which uses customers' value of electricity to set the value of service reliability. In the case of total grid disconnection, it is more logical to ascertain what would be required for a customer to disconnect from the grid but retain the same level of reliability. This will emphasize the implicit value of the grid connection, especially because the disconnect cost is likely to require an annual cost (amortized) of $1000 or more and the embedded distribution cost is probably between $150-250/customer/year (about 15-20% of the residential customers total bill).

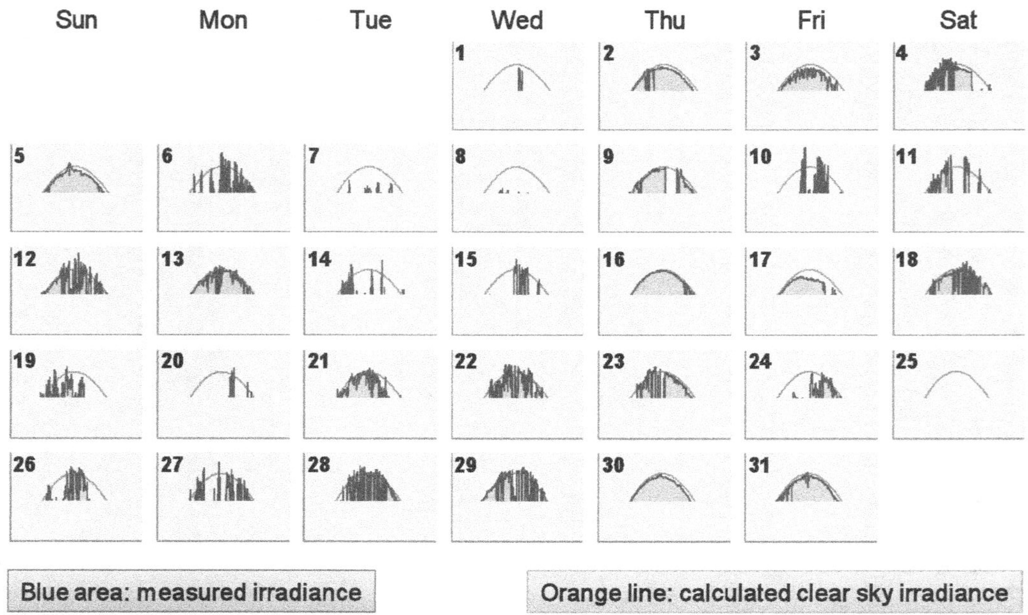

Figure 4-19. Solar System Output—Typical Month for a PV Load Generation System (Source: Trueblood, 2013)

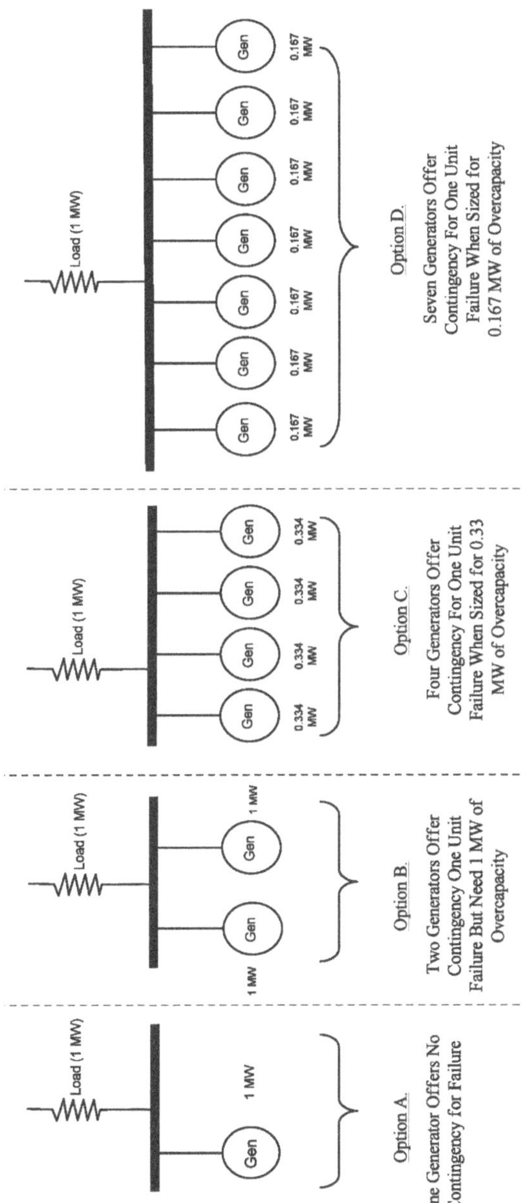

Figure 4-20. Multiple Generation Sources Increases Reliability (1001668)

variability needs to be supplemented by either the grid or by the addition of technologies in order not to support reliability, power quality and availability to supplement local generation.

Net-zero energy buildings are those which generate a sufficient amount of electricity locally so that the average national electricity demand from the grid is a net "zero." The popularity of this concept is growing. Where these net-zero energy buildings do not have on-site electric energy storage, the distribution system is typically the equilibrium mechanism for the building's diurnal variations in electricity demand. For example, the U.S. DOE's National Renewable Energy Laboratory (NREL) in Golden, Colorado, recently commissioned a new 222,000 ft^2 office building, called the Research Support Facility (RSF). It is a net-zero site energy/source electric/energy emissions facility. NREL's RSF is an excellent example of a passive design building, which largely uses passive energy systems for heating, ventilation, air conditioning, and lighting.

To reach net zero, NREL is installing 2.6 MW of PV on the building's roof, over a parking lot, and on the roof of a new parking structure. However, the building does not use electric energy storage. The RSF is electrically interconnected with the local distribution utility. In fact, it depends heavily on that interconnection to support its net-zero energy goals. The extensive solar PV system generates net-positive energy on sunny days and feeds the grid. At night, it relies on the utility for service.

In the case of larger local generation systems the reliability is often a function of the redundancy built into the local power system itself. Figure 4-21 illustrates the reliability inherent in a large power system (such as the grid) as opposed to a local generation system. In order for a local generation system to achieve the same reliability as the grid, additional sources (generators are used in this example) need to be added. In these examples, a one megawatt electrical demand is supplied by one, two, four and seven generators sized to provide maximum reliability with minimal capacity. The more generators that are added, the greater the expected reliability of the power system becomes.

Voltage Regulation (startup resource)

The grid provides startup service or adequate "in rush" current for consumer appliances and devices such as compressors, air conditioners, and welders. This is so they can start reliably without severe voltage dips that can damage equipment. Without grid connectivity or other support technologies, a central air conditioning compressor attached to

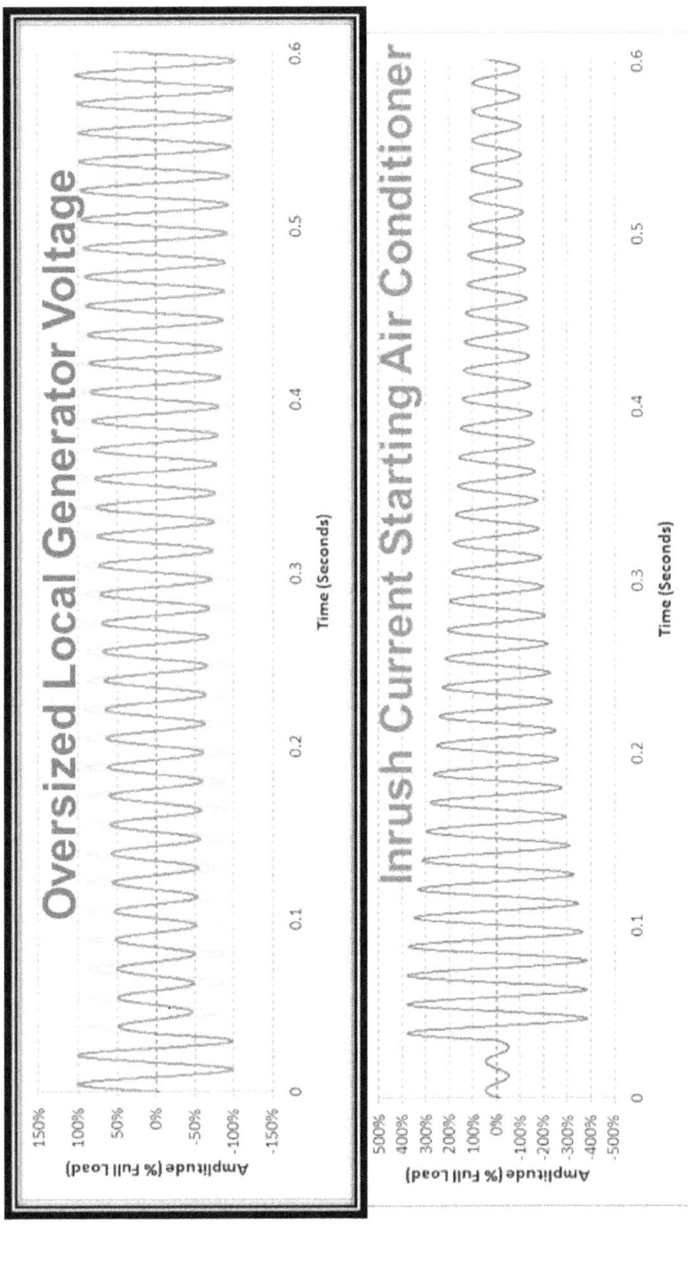

Figure 4-21. Inrush Current Effect on Service Voltage for a Residential Air Conditioner and Voltage Profile for a Local Generator

a PV system may not start at all.

Figure 4-22 illustrates inrush current and effect on service voltage for starting a residential air conditioner. Note that peaks are six to eight times average power draw. A local generation system without batteries or other technologies may not be able provide adequate "energy" to start such motors. The figure also illustrates the potential premise overvoltage generated by a local generator not connected to the grid. Without the grid, certain independent electrical systems can experience overvoltage.* With grid connectivity these impacts are moderated so there is little no impact on the consumers overall electrical service.

Voltage Quality

The robustness of the grid provides mitigation against several types of power quality events. In particular, transients, voltage flickers, sags and swells, as well as harmonic distortion created within customer facilities, are dampened. The electrical stiffness of the grid mitigates the impacts harmonics created from non-linear loads can have as an independent power system. This is illustrated in Figure 4-23.

Figure 4-24 illustrates the difference between voltage quality delivered by the grid as compared to that from a local generation system (referred to here as local generation)

In addition, harmonics are a dimension of voltage quality that can have a deleterious effect on electrical devices—both consumer end-use devices and utility distribution equipment such as transformers. Harmonics cause heating in many electrical components which affects dielectric strength and deteriorates the life of equipment. The economic impacts of reduced life are hard to quantify and not obvious to consumers. Connecting a local generation system to the grid can mitigate harmonics caused by devices in a consumer power system.

In addition, the operation of generation systems in parallel with the utility system allows the local generation system to focus primarily on real power generation and not provide reactive support, the extent of which depends on the utility's interconnection requirements. Without interconnections, each premise has to manage its real power needs with live voltage. Some local generation systems can not only prevent such

*Overvoltage is a condition where the voltage in a circuit or system is above its design limit. When the overvoltage is substantially high or of long duration, it can lead to overheating of electrical components, create fires, or damage equipment. As a minimum, overvoltage will always deteriorate electrical equipment.

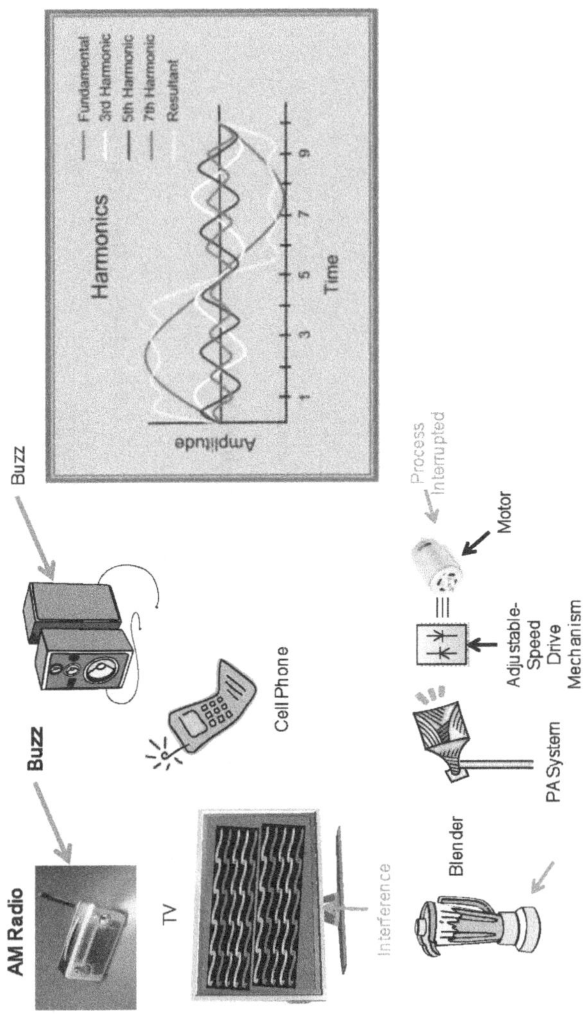

Figure 4-22. Illustration of Harmonic Mitigation Provided by the Grid

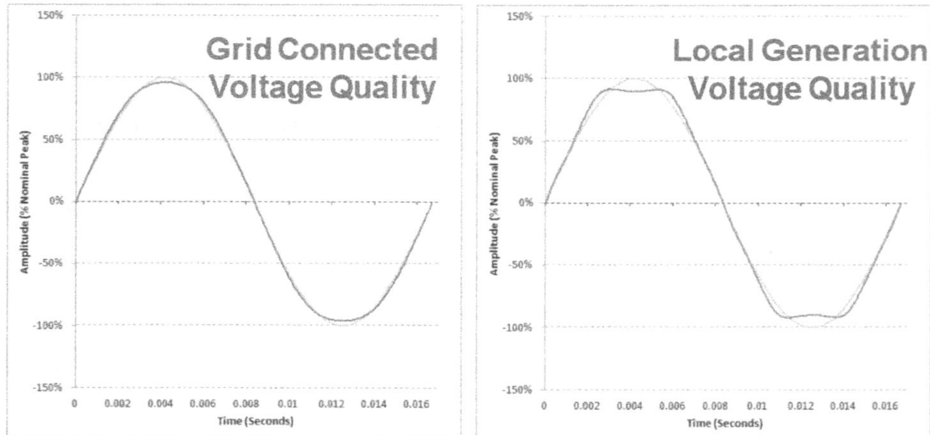

The grey line indicates the baseline for a sinusoidal wave with no harmonics

Figure 4-23. Grid Delivers Higher Quality Power

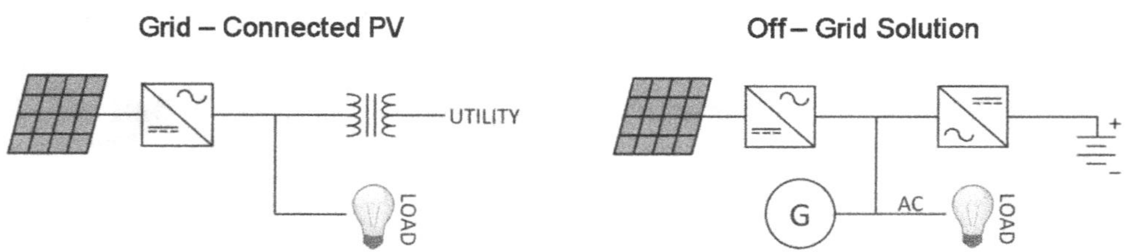

Figure 4-24. Illustration of Harmonic Mitigation Provided by the Grid

adverse effect, they can serve as a system resources to provide real power support. In future system development, local generation systems may be compensated for providing reactive power support on demand.

Efficiency

Operating micro-turbines or combined heat and power (CHP) with the grid will allow internal combustion engine- (ICE-) driven generators sized for local requirements to run 10 to 20% more efficient. In addition, for all local generation systems, there will be a reduction in transmission and distribution losses.

Overall, lower harmonics also reduce hysteresis losses in distribution transformers that ultimately translate to reduced costs for the utility. This reduction is ultimately passed on to consumers in some form, often reduced revenue requirements for the utility. (Kefalas, et al., 2010)

Optimal Sizing

Parallel connections with the grid allow the owner/operator of local generation systems to design the optimally size the local generation system, taking advantage of needs that are better provided by the grid, which allows best use of local generation in assets. For example, an interconnected system could allow for dispatching with the objective of minimizing the impact of power production on the environment.

Table 4-4 depicts the result of one study intended to illustrate some of the potential benefits from grid connectivity. It compares a small building serving a multi-family residential structure or a small commercial or industrial building served by a 100 kVA transformer-fed grid connection with an internal combustion engine-driven (ICE-driven) synchronous generator with a similar rating.

Table 4-4 shows that the utility service is stiffer, more reliable, and better regulated for load-steps than a typical distributed generator operating as a standalone entity. This stiffness results from the network of wires and transformers coupled with a variety of central generators connect to form the grid. In fact, Table 4-4 is conservative in that it compares a 150-kVA generator to a 150-kVA utility service transformer. In many cases, the service transformer is greatly oversized relative to site load, so the utility system is even stiffer than indicated in Table 4-1. The only performance areas where the utility service is worse are with incoming deep voltage sags (due to the exposure to the power system) and in the efficiency when a co-generation unit where waste heat is recovered is

Table 4-4. Performance Characteristics of a Standalone Local Generation and a Utility Service Connection (EPRI 1001668)

Characteristic	Standalone Generator (150-kVA, Three-Phase Unit)	Utility System Service (150-kVA, Three-Phase Transformer and Service Drop
Source capacity	150 kVA	150 kVA
Site peak load	100 kVA	100 kVA
Impedance (percent of source capacity rating, 150 kVA)	20%	4% (includes service drop)
Fault level (per unit of site peak load current, 100 kVA)	7.5	37.5
Frequency dip during 100% step-load	Can be more than 10%.	No change
Voltage drop due to 20-HP motor start (locked-rotor current = 5 per unit)	15-20%	4%
Exposure to fault-related voltage sags coming from utility system	None	One moderate to severe event per week
Impact of load variation on generation efficiency and economic dispatch	Strong impact	No impact
Reliability (annual availability in percent)	97% or less (assumes a single DG with no redundancy)	99-99.999% (average U.S. value is 99.97%; range represents different T&D system types and conditions)
Typical effective power generation efficiency	20-40% (no cogeneration) 55-85% (with cogeneration)	35-55% (depends on mix of utility generation resources; range includes T&D losses)

installed and operated.

Studies have shown that the value of connectivity to the grid exceeds the cost. The cost of connection includes standby service fees, which in most cases range between $25 and 200 per year per kilowatt of expected peak facility demand (gross usage), including interconnection controls and equipment, which can have an incremental cost of $25 to $50 per kilowatt of generation capacity (for larger local generators). This typically results in annual connection costs between $3.75 and $12.50 per kilowatt per year.

In an attempt to further estimate the value the grid provides to consumers with local generation, EPRI evaluated the two systems illustrated in Figure 4-24 and described in Table 4-5.

To put the cost of grid modernization in perspective with the cost of grid services or their equivalent, the authors computed their cost in $ month. Modernization, using the residential consumer as the case, results in an additional utility charge of approximately $12/month.

Table 4-5. Comparative DER Systems

Grid-Connected PV System	Stand-Alone PV System with Storage & Backup Generator
• Location: Santa Maria, CA • Residential Load: 35 kWh/day @ 4.1 kW peak • PV Array Sizing: 8 kW DC (zero net) • $48,000 Initial Investment ($6/watt installed) • Utility arrangement is net-metered (in CA time of day rates sweeten grid connected PV economics) • State/SCE Rebate (CSI): $1.3K • Federal Credits: $16K • Out of pocket $30.7K • LCOE $0.16/kWh	• Location: Santa Maria, CA • Residential Load: 35 kWh/day @ 4.1 kW peak (zero net energy) • PV Array: 11 kW Array @ $66K • Battery Bank: 72 lead-acid batteries, 115.2 kWh/10,000 lbs. $36K installed • Battery Charge Controller: 12 kW, $5.4K • Generator: 14 kW, 92hr/yr. operation, $4K • Out of pocket = $94K ($114-$20) • LCOE $0.98/kWh

The amortized cost of the stand-alone PV system, designed to be self-sufficient and provide equivalent electric service to the premise using storage and back-up generator, would cost the owner a levelized payment of $240/month.

These calculations suggest that it would be far less costly for consumers to employ grid-connected PV and to support grid modernization than to undergo the expense of an equivalent system that offers equivalent grid-like value.

These estimates are for a residential case using technologies that likely will be used for off-grid generation. Costs for other technologies, such as larger campus style grid disconnected microgrid, would be different. In all cases, the cost to reproduce the quality and reliability of electric service available from the grid will be expensive on a premise-specific level. Interconnection ensures grid-level reliability and other services at a far lower costs. From a societal perspective, interconnection ensures the best use of scarce resources so that all consumers benefit from the value of local grid-interconnection distributed generation.

In order to estimate the value of the grid, two houses with different distributed generation configurations can be compared. One is a house which remains grid-connected but has installed photovoltaic power generation such that the consumer is able to operate the home on an annual net-zero energy basis. This house has an 8 kW solar array which produces approximately 11,800 kWh/year. The installation will cost approximately $48,000.

For the second system, it can be assumed that there is no grid connection. The cost of the added technologies which would be required to provide electric service approaching that supplied by the grid includes the following:

• Increasing the PV array size from 8 kW to 11 kW	$18,000
• 72 lead acid batteries	36,000
• Battery charge controller	5,400
• Standby Generator (14 kW)	4,000
Total	**$63,400**

If consumers elect to install these technologies in order to replicate the same level of service as the grid provides, they would incur an additional capital cost of $63,400. Assuming amortization at 6% over 20 years, this translates to a monthly cost of $272.

EPRI has also conducted sensitivity analysis on these estimates to understand how costs vary across locations and implications of projected reductions in cost and improvements to performance of PV systems and of batteries.

HOW TO ENABLE THE INTEGRATED-GRID APPROACH

Today's grid is integrated electrically from central generation to where power is fed into consumers. But, optimization of generation resources is primary done from the perspective of bulk power system that includes the transmission grid. In the Integrated-Grid Approach, the distribution grid is strengthened (by reconductoring), to enable a higher penetration of local generation, and modernized to operate more as an end-to-end system to optimize local generation resources with central generation resources to provide the reliability and quality of power to consumers. An integrated grid overall will provide the lower overall cost to consumers

As described previously, from a societal perspective, the overall lowest-cost approach to meeting future consumer needs for electricity, including use of local or distributed generation, is to enable effective connectivity to the grid. In order to embrace wide-scale connectivity, the industry needs to modernize the grid. This will require additional technologies over and above what would be required to respond to interconnection needs on a case-by-case basis. The power delivery system provides benefits beyond simply delivering energy, and those benefits become more pronounced for consumers who generate some of their own electricity.

Utility distribution systems are essentially designed for one-way flow of power from the bulk power system through substations to distribution systems and ultimately to consumer's premises. The advent of local generation and locally installed electrical storage devices affords an opportunity to leverage those devices to societal advantage. However, the power systems do not now allow the significant two-way flow of power; they do not enable generation to flow upstream efficiently into the power system. As a minimum, this necessitates several changes to the local grid:

- Conductors and some transformers on the distribution system need to be reinforced.
- Depending on the saturation of local generation and storage, sensors and communication, and computational devices need to be installed along with an enhanced distribution management system so that utilities have situational awareness of the new, more dynamic electrical environment.
- As the use of local generation increases, a means to control the output, frequency response, voltage and reactive support from local generation and storage will also be needed. Without this control, the power system may not be able to operate reliably.

The key technologies that are required to enable this functionality are as follows:

- Smart inverters are needed on photovoltaic power generation systems to provide additional functionality such as voltage ride-through, frequency ride-through, and VAR (reactive power) support.
- Enhanced distribution management systems that allow real-time adaptive protection and control of distribution system and all connected local resources.

- Information and communication technology, including a robust telecommunication network with high-speed data processing to enforce interconnection rules overlaid with enhanced cyber security.
- A distributed control architecture which will utilize these technologies for monitoring and control and to communicate instructions to devices.

The Integrated-Grid Approach includes these technologies to provide the ultimate flexibility needed. This would enable full integration of customer resources, end-uses, and electrical and thermal storage with grid components.* This modernized power system would allow the Integrated-Grid Approach and enable a "flexible load shape." The "flexible load shape" concept first envisioned in the initial construct of demand-side management, as is illustrated in Figure 4-25.†

Flexible Load Shape

ElectriNet[SM] enables the Flexible Load Shape Concept which also allows utilities to operate their distribution system in concert with consumer's power systems for a variety of other potential benefits (Farzan, 2013):

- To optimize the energy consumption of the consumer's end-use equipment in balance with enhancing the efficiency of the grid.

---

*The architecture of this modernized power system has been referred to frequently in the literature as the ElectriNet[SM]. The energy management system (EMS) which would be needed to control this system has been referred to as Grid 3.0.

†How did the Flexible Load Shape Concept originate? The Flexible Load Shape Concept originated from the foundations of demand-side management (DSM) as envisioned in the late 1970s and early 1980s.

Demand-side management is the planning and implementation of those utility activities designed to influence customer use of electricity in ways that will produce desired changes in the utility's load shape—i.e., changes in the time pattern and magnitude of a utility's load. Utility programs falling under the umbrella of demand-side management include load management, new uses, energy efficiency, electrification, customer generation, and adjustments in market share.

In DSM planning, six load shapes were envisioned as representative of optimal strategies for given power systems. These include peak clipping or the reduction of system peak loads; valley filling or building off-peak loads; load shifting which involves shifting load from on-peak to off-peak periods; energy efficiency is the load shape change that results from reducing end-use consumption; electrification or strategic load growth is the load shape change that refers to a general increase in sales beyond valley filling; and flexible load shape.

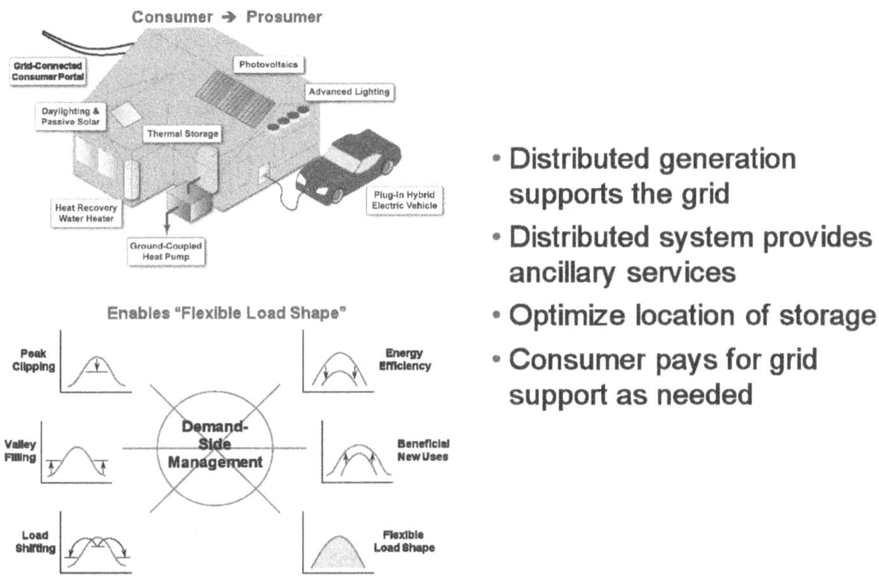

Figure 4-25. Connectivity Enables the Modern Power system to Support the Integrated-Grid Approach (Based on EPRI EA/EM 3597)

- To utilize on-site generation, possibly in conjunction with energy storage, to avoid peak energy costs and even create revenue streams for the consumer by selling energy back to the grid as price signals rationalize.
- To facilitate enrollment in demand response programs which can generate revenue by reducing load on the grid. Demand response can be provided by both self-generation and end-use loads.
- While grid energy transactions and fuel costs dominate the economics of a power system, participation in capacity and ancillary services markets by local generators may also be important incremental revenue drivers to distributed generation economics.
- The reliability improvements obtained through islanding capability can be quite valuable, depending on the mission of the facility and the critical load served during islanded operations.
- Local generation installations can also be viewed as a means of creating zero-net-energy communities and meeting other environmental goals established by states or regulatory agencies.

The ElectriNet^{SM}

EPRI has adopted the term ElectriNet^{SM} to refer to the modernized power system which is used to support the Integrated-Grid Approach. The ElectriNetSM embodies the value of support provided to the local generator by a parallel utility system connection. The support value is substantial and can make local generation viable in applications where if it were standalone (not grid-connected), the performance would be marginal.*

Local Energy Networks

The ElectriNet^{SM} is an aggregation of Local Energy Networks (LEN)†. LENs are defined by interconnected distributed end uses, local generation, storage and other technologies at the building, community, or distribution level that increase the independence, flexibility, and intelligence for optimization of energy use and energy management at the local level. The power delivery (transmission and distribution) system is the enabling infrastructure for Local Energy Networks. Most of today's LENs are fairly simple in design, consisting of a single generator supplying a dedicated load or of several co-located generating units linked electrically to operate much like a single unit of twice the size. Compared to traditional power systems, LENs offer the potential for improvements in energy delivery and efficiency, power quality, reliability, and cost of operation for very concentrated and localized loads. To realize fully these benefits, however, it will be necessary to coordinate the control of complex LENs potentially comprised of several different kinds of local generation and storage systems, which may also be geographically dispersed.

There are several other benefits to aggregating privately interconnected loads and distributed energy resources within clearly defined

---

*Why is the grid so strong? The utility system is a strong source and stable from a frequency perspective. It represents a vast interconnected network of thousands of megawatts of generation capacity, and any single distribution feeder load is tiny in comparison and has no significant impact on the bulk generation dispatch needs or operating efficiency. By comparison, the DER system is weak, and the starting of large motors and other loads can create severe power quality problems. Typically, reliability of utility system power is in the range of 99% to 99.999% (national average is about 99.97%). This compares to a single DER that is on the order of 94 to 97% available.(why are we stating this here when we discussed start up services or inrush current several sections before this.

†This concept of a network of Local Energy Networks was first described by EPRI as the ElectriNet^{SM}.

electrical boundaries, and that act as a single controlled entity (Smith and Ton, 2013), who refer to these groups as microgrids* and posit that they provide benefits that include:

- Enabling grid modernization and the integration of multiple smart grid technologies.
- Enhancing and easing the integration of distributed and renewable energy sources that help reduce peak load and also reduce losses by locating generation near demand.
- Meeting end-user needs by ensuring energy supply for critical loads, controlling power quality and reliability at the local level, and promoting customer participation through demand-side management and community involvement in electricity supply.
- In addition, in some jurisdictions and on some systems local generation can achieve benefits from avoided renewable portfolio standards (RPS) purchases. Depending on the generation mix, utilities may also be able to use local generation resources as a hedge against volatility in natural gas prices.
- Supporting the macrogrid by handling sensitive loads and the variability of renewables locally and supplying ancillary services to the bulk power system.

In the Integrated-Grid Approach, Local Energy Networks, energy sources, and a power distribution infrastructure are first integrated at the local level. This could be an industrial facility, a commercial building, a campus of buildings, or a residential neighborhood (refer to Figures 4-26, 4-27 and 4-28.) Local energy networks are interconnected with different localized systems to take advantage of power generation and storage through the Smart Grid enabling complete integration of the power system across wide areas. Localized energy networks can accommodate increasing consumer demands for independence, convenience, appearance, environmentally friendly service, and cost control.

---

*The reader should note that a collection of microgrids is not a sufficient construct to establish an ElectriNet^{SM}. An ElectriNet^{SM} enables flexible load shape, and as such, an entirely transactive topology where consumer uses, consumer generation and storage devices and utility assets are monitored and potentially controlled in real time.

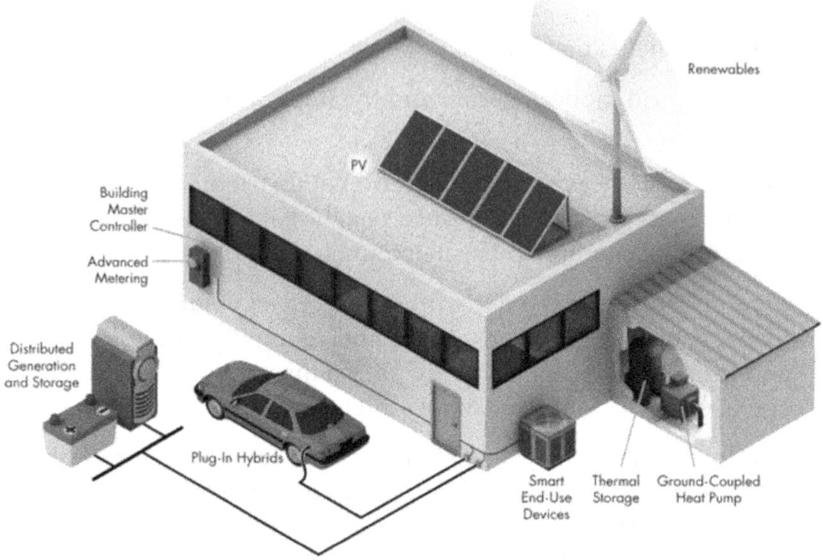

Figure 4-26. Building-Level Local Energy Network (Source: Gellings 2010)

In addition, Local Energy Networks allow for:

- The optimization of energy availability at the local level across a larger variety of energy sources, resulting in improved economics.
- The creation of an infrastructure for a more robust management of a facility's overall energy requirements (heating, cooling, and power).
- The control and management of reliability at the local level to meet higher requirements.

CONCLUSION

Changes in the retail electric business are the ordinary, not the exception. In order to provide society with affordable, reliable and environmentally sustainable electricity, the Integrated-Grid Approach should be adopted because it ensures the best use of scarce societal resources when deployed in combination with the functionally essential technologies (local generation, smart distribution, etc.) and supported

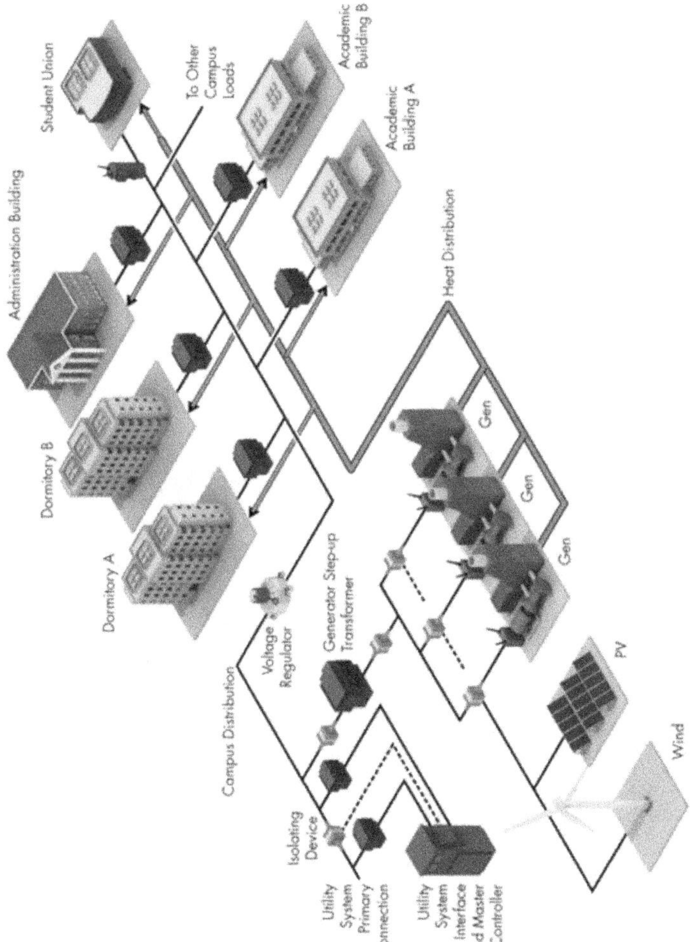

Figure 4-27. Campus-Level Local Energy Network (Source: Gellings 2010)

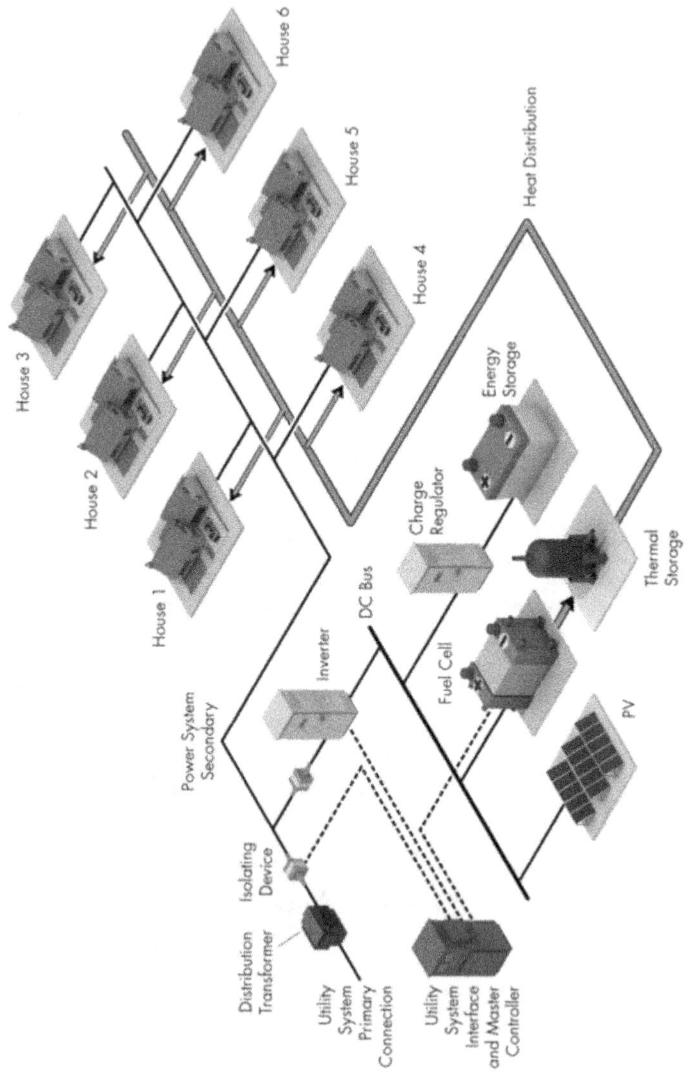

Figure 4-28. Community-Level Local Energy Network (Source: Gellings 2010)

by enabling rate structures and interconnection guidelines. As indicated in the analysis presented, from a societal perspective, it is more sustainable and economic to interconnect local generation systems to the grid rather than to have them undertake the added expense to allow them to operate independently. In addition, integrating local and central resources provides even greater opportunities to enhance the value of all of these resources.

References

- Disruptive Challenges: Financial Implications and Strategic Responses to a Changing Retail Electric Business, P. Kind, Energy Infrastructure Advocates, Edison Electric Institute (EEI), Washington, DC, January 2013.
  German Renewables Fee Rises 47%, Opening Government Rift, S. Nicola, www.bloomberg.com, October 15, 2012.
- Distributed Photovoltaics: Utility Integration Issues and Opportunities, EPRI, Palo Alto, CA: August 2008. 1018096.
- "Finding a Bright Spot: Utility Experiences, Challenges, and Opportunities in Photovoltaic Power," T. Key, IEEE Power & Energy Magazine, May/June 2009.
- "The End of Paying for Electricity by the Kilowatt-Hour: Changes to Utility Rate Designs are Afoot in the U.S. and Europe," K. Twee, Greentech Media, April 24, 2013.
- Advanced Metering Infrastructure (AMI) for Distributed Solar (PV) Integration: Utility Survey Results. EPRI, Palo Alto, CA: 2009. 1019585.
- "The ElectriNetSM," C. Gellings and P. Zhang, ELECTRA, CIGRÉ, June 2010.
- "Report to NIST on the Smart Grid Interoperability Standards Roadmap," (Contract No. SB1341-09-CN-0031), D. Von Dollen, EPRI, June 17, 2009.
- Regulated Utility Solar Business Models: Concepts and Case Studies. EPRI, Palo Alto, CA: 2012. 1024365.
- Benefits Provided to Distributed Generation by a Parallel Utility System Connection. EPRI, Palo Alto, CA: 2003. 1001668.
- Program on Technology Innovations: Tracking the Demand for Electricity from Grid Services. EPRI, Palo Alto, CA: 2013. 3002001497.
- Regulated Utility Solar Business Models: Concepts and Case Studies. EPRI, Palo Alto, CA: 2012. 1024365.
- Distribution Research Area Strategic Plan: September 2010. EPRI, Palo Alto, CA: 2011. 1022335.
- Engineering Guide for Integration of Distributed Storage and Generation. EPRI, Palo Alto, CA: 2012. 1023524.
- Demand-Side Management—Volume 1: Overview of Key Issues. EPRI, Palo Alto, CA: 1984. EA/EM-3597
- "Key Connections," M. Smith & D. Ton, IEEE PES Magazine, July 2013.
- "Microgrids for Fun and Profit," F. Farzan, et al., IEEE PES Magazine, July 2013.
- Program on Technology Innovation: Using Scenario Planning to "Stress Test" EPRI's R&D Portfolio. EPRI, Palo Alto, CA: 2013. 3002001496.
- Ratemaking, Solar Value and Solar Net Energy Metering—A Primer, Solar Electric Power Association, Washington, DC, July 2013.
- "Grid Voltage Harmonics Effect on Distribution Transformer Operation," T.D. Kefalas, et al., Proceedings of IEEE 7th Mediterranean Conference and Exhibition on

Power Generation, Transmission, Distribution and Energy Conversion, 2010.
- Electric Power Migro-grids: Opportunities and Challenges for an Emerging Distributed Energy Architecture, D.E. King, Dissertation, Carnegie Mellon University, May 2006.
- Annual Energy Outlook 2013, Electricity Supply Disposition, Price and Emissions, Reference Case.
- Assigned Commissioner's Scoping Memo and Ruling, Before the Public Utilities Commission (PUC) of the State of California, FER/Gd2, January 18, 2012.
- "Network Tariff Structure for a Smart Energy System," A Eurelectric Paper, May 2013.
- "Are Europe's Utilities Dying or Can they Reinvent Themselves," J. Berst, ________, September 4, 2013.
- "How Electricity Became a Luxury Good," Spiegel Stuff, Der Spiegel, 9/4/2013.
- "Carpe Diem—Why Retail Electricity Pricing Must Change Now," A. Schurr and S. Hauser, in Energy Efficiency, edited by F. Shiushanse, Elsevier, 2013.
- Distributed Photovoltaics: Utility Integration. Issues and Opportunities. EPRI, Palo Alto,CA: 2008. 1018096.
- Market Driven Distributed Energy Storage System Requirements for Load Management Applications. EPRI, Palo Alto, CA: 2007. 1014668.
- Energy Storage Market Opportunities: Application Value Analysis and Technology Gap Assessment. EPRI, Palo Alto, CA: 2009. 1017813.
- Program on Technology Innovation: Distributed Photovoltaic Power Applications for Utilities. EPRI, Palo Alto, CA: 2009. 1018097.
- Requirements for German Upgrades, E Bridge Consulting, 2011, http://www.e-bridge.de
- "European Utilities: How to lose half a trillion euros—Europe's electricity providers face an external threat," The Economist, October 12, 2013.
- "PJM Capacity Auction Attracts Record Amount of New Generation and Improved Capacity," PJM News, Valley Forge, PA, May 24, 2013.
- SMA: PV Inverters Can Be Made Grid-Smart for Little Additional Cost, M. Krause, Greentech Media, October 17, 2013.
- "Germany Breaks its Own Record for Solar Power Generation" K. Kroh, Climate Progress, August 22, 2013.
- Average Size of U.S. Solar Installations: Small, J. Farrell, Institute for Local Self-Reliance (ILSR), January 28, 2011.
- California Net Energy Metering (NEM) Draft Cost-Effectiveness Evaluation, California Public Utilities Commission Energy Division, Energy and Environmental Economics, Inc., September 26, 2013.
- Guidebook for Cost/Benefit Analysis of Smart Grid Demonstration Projects. EPRI, Palo Alto, CA: 2012. 1025734.
- Technical Potential for Local Distributed Photovoltaics in California: Preliminary Assessment, March 2012, E3 report for CPUC, page 55.
- Smart Grid (Made) in Germany, T. Becks, VDE Association for Electrical, Electronic & Information Technologies, Presentation to Cleantech Forum, Munich. April 18, 2012, http://events.cleantech.com/munich
- Communication Systems for Grid Integration of Renewable Energy Resources, F. Yu, P. Zhang, W. Xiao, and P. Choudhury, Network, IEEE Vol 25, Issue 5, September-October 2011.
- "PV Measures Up for Fleet Duty: Data from a Tennessee Plant are Used to Illustrate Metrics That Characterize Plan Performance," Power and Energy Magazine, IEEE, Vol. 11, No 2., pp. 33, 44. C. Trueblood, S. Coley, T. Key, L. Rogers, A. Ellis, C. Han-

sen, E. Philpot, March 2013.

- "Smart inverter volt/var control functions for high penetration of PV on distribution systems," Power Systems Conference and Exposition (PSCE), 2001 IEEE/PES, vol., no, pp. 1, 6, 20-23. J. Smith, W. Sunderman, March 2011.
- "Benefits Provided to Distributed Generation by a Parallel Utility System Connection" EPRI Report No. 1001668.
- Stochastic Analysis to Determine Feeder Hosting Capacity for Distributed Solar PV, EPRI Report No. 1026640, December 2012.
- Estimating the Costs and Benefits of the Smart Grid, EPRI Report No. 1022519, March 2011.
- IEEE Distribution Reliability Working Group 2012 as Reported In: Wireless Connectivity for a Reliable Smart Grid, Electric Light and Power, J. Wilson, April 1, 2013.
- Time to Short Your Electric Utility?, P. Kelly Detwiler, Forbes, July 19, 2013.
- A Review of Solar PV Benefit & Cost Studies, Rocky Mountain Institute, 2013

Chapter 5

Lessons Learned from the World's Smart Grid Demonstrations

PURPOSE OF DEMONSTRATIONS

Smart Grid demonstrations have been underway around the world for a number of years now. Sharing information on these demonstrations is a very effective way to inform utilities, regulators and other policy makers of the successes and failures—literally, the lessons learned. By learning from others we can reduce the costs or uncertainty in our own demonstrations or deployments.

The Smart Grid concept has attracted the attention of the world's power industry. The use of sensors, communications and computational ability facilitating control can yield a substantial improvement in performance and a reduction in cost for utilities and their customers. This has attracted the attention of the industry at large, its many stakeholders and regulators. However, the electric utility industry is among the most conservative in the world and is understandably skeptical about embracing new technologies that have not been extensively vetted. The expression that "utilities are first in line to be second" to adopt new technologies describes the attitude of many of them toward smart grid technologies. The world's response has been to engage in demonstrations. Demonstrations can validate the performance of smart grid technologies on the way toward eventual wide-spread deployment. As of 2015, there are literally hundreds of smart grid demos covering every continent around the globe.

Many smart grid demonstrations are aimed at multiple objectives often involving several technologies and systems. In this chapter the author summarizes select demonstrations from around the world, often by focusing on just one key application in each country or utility.

EUROPE

Europe is currently engaged in 459 smart grid projects covering all 28 European Union (EU) member states and several other European countries. The EU considers smart metering projects an essential part of the smart grid inventory of applications. The EU has maintained records of its projects since 2002 and estimates that 3.2 billion Euro have been expended.

Of the 459 projects, 201 are considered R&D with a budget of 7.5 million Euro having an average duration of 33 months. The EU claims to have a good level of diversity in the smart grid landscape engaging 1670 organizations including universities, Transmission System Operators (TSOs), Distribution Systems Operators (DSOs), manufacturers, I.T. companies and utilities. (European Commission—http://jrc.ec.europa.eu)

Ireland

Ireland is among the leading nations in Europe in terms of deploying smart grid technologies. For example, ESB Networks has a Smart Grid Demonstration Project which consists of our separate but related work streams selected to advance development of ESB's model of a future, integrated network. (EPRI 3002004652) The smart grid demonstration includes new and innovative methods for integrating renewable generation resources and electric vehicles, studies of customer behavior, and optimization of the existing electric distribution networks.

ESB Networks believes that a smart grid is not just about enabling customer responsiveness or a high penetration of renewables, but also about creating "smart green circuits" that enable operational efficiency, monitoring of line conditions, loss reduction, and protection. Smart green circuits are an important part of the Irish strategy, as the size and scale of the country's electricity distribution system presents unique challenges in terms of maintaining continuity of supply to customers within a high standard, and ensuring that network losses are minimized. ESB conducted tests of smart green technologies on four distribution circuits. Three are rural, and one is urban.

A number of systems and applications were evaluated to help understand how best to create efficient and "self-healing" circuits, including:

- A smart fault passage indicator system was added to one circuit. This system locates and analyzes faults and communicates in-

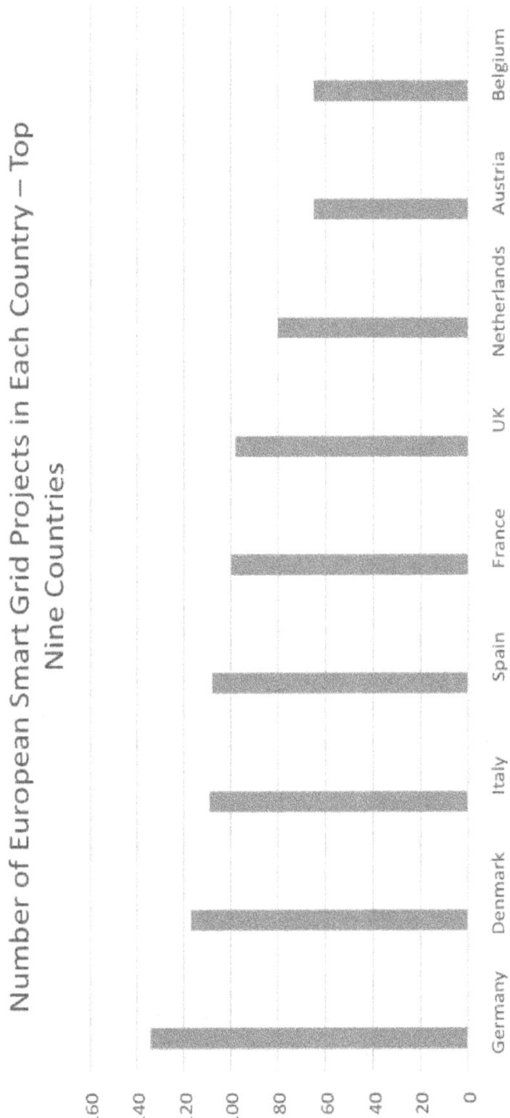

Figure 5-1. Top nine countries engaged in Smart Grid projects in Europe. (Panek 2014)

formation to network operators via GPRS (general packet radio service) communications to facilitate quick response. Fault notification is sent to the iPhones of network technicians.

- An arc suppression coil protection system was installed on ESB's 20kV circuits. This protection system is designed to carry earth faults safely while continuing to supply customers.
- A cost-benefit analysis of replacing conventional transformers with low-loss transformers was conducted. ESP has a large number of lightly loaded transformers in operation in Ireland, and core losses contribute significantly to overall network losses.
- Circuits were converted from a 10kV to a 20kV operation voltage. One network, which serves 2,200 customers, was operating at voltages outside of the standard so conversion to 20kV was done to improve the ability to maintain operation within prescribed voltage levels.
- Simulations and field tests of conservation voltage reduction (CVR) were performed. CVR was done on two rural circuits, as well as on the urban circuit.
- The self-healing circuit has operated successfully on over 12 separate incidents of faults. On all occasions, a faulted section of network was isolated and supply was recovered for remaining customers within seconds. The success of the trial has led to plans to change the worst performing network sections in the country into self-healing circuits. The performance of the arc suppression coil protection system, complemented by a range of earth fault management facilities in operation on ESB's 20kV network, has been proven successful. ESB has achieved cost reductions, fault-finding time has been reduced by 84%, and measured continuity of performance improved by 100%.

The case to continue doubling the operation voltage of the medium voltage network is overwhelming. Conversion of networks from 10kV to 20kV resulted in a 75% reduction in network losses, an improvement in voltage drop by a factor of four, and an increase in network capacity of more than 100%.

Conservation voltage reduction resulted in significant reductions in energy demand, with measured CVR factors ranging from 0.83 to 1.74. Cost-benefit analysis shows that a strong economic case can be made for amorphous core transformers rather than conventional silicon steel units in cases where a transformer needs to be replaced for reasons other than just reducing losses.

Germany

One key smart Grid undertaking in Germany is their "E-Energy: ICT-based Energy System of the Future" project which has strong support and funding. It is undertaken by the Federal Ministry of Economics and Technology (BMWi) as part of the technology policy of the Federal Government. Due to its high importance both in terms of innovation and the national economy, it was declared a beacon project at the National IT Summit of the Federal Chancellor. Just like the terms "E-Commerce" or "E-Government," the abbreviation "E-Energy" stands for the comprehensive digital interconnection and computer based control and monitoring of the entire energy supply system.

The primary goal of E-Energy is to create E-Energy model regions to demonstrate where information and communication technologies (ICT) can be used to achieve optimal efficiency, supply security and environmental compatibility in the power supply, while creating new jobs and developing new markets. E-Energy projects are innovative in that they integrate ICT system concepts with the entire electricity system all along the entire supply chain—from generation and transport to distribution and consumption. Further, the projects are developed and tested in real-time in regional E-Energy model projects.

E-Energy projects are supported and funded in an interdepartmental partnership with the Federal Ministry for the Environment, Nature Conservation and Nuclear Safety (BMU). The BMWi will appropriate up to €40 million for four model regions and the BMU will make up to €20 million in funding available to two additional model regions. Together with the equity capital of the participating companies, some €140 million will be mobilized for the development of six EEnergy model regions.

The E-Energy projects are overseen by Ludwig Karg, Geschäftsführung, B.A.U.M Consult, Munich, Germany. B.A.U.M. Consult is a leading consulting firm whose services range from energy efficiency audits to energy and ecology planning for communities to planning, building, financing and operating biomass and photovoltaic power stations. B.A.U.M. Consult also offers strategic consulting in the areas of sustainability and climate protection.

Three of the projects are summarized below. (EPRI 1020303)

The three projects are:

- eTelligence in the Model Region "Cuxhaven"
- MeRegio—Minimum Emission Regions
- Renewable Model Region Harz

The Cuxhaven eTelligence project shows an efficient system which intelligently integrates electricity generated by decentralized producers into the energy supply system. The ensemble of project components—ranging from the regional marketplace, over the virtual power plant, the intelligent distribution network, controlling for decentralized generators and large-scale consumers, to feedback systems for household consumers—pave the way for the future distribution network. eTelligence shows us one way to achieve greater security of supply and to improve economic efficiency. At the same time, eTelligence enables better use of renewable energies, which will prompt a significant reduction in CO2 emissions. In future, when the wind blows or the sun shines on the North Sea coast—making plenty of electricity from renewable sources available—the needs of Cuxhaven residents and companies will be adjusted accordingly.

After the field test has been completed in 2011, the next step will be to assess the gathered data. Following this, the results will be applied throughout Germany as a whole.

MEREGIO or Minimum Emission Regions are areas possessing energy supply systems that are optimized with respect to their greenhouse gas emission. In particular, electric and thermal energy should be produced, transported, and consumed efficiently using low-emission power plants as well as Smart Grid innovative functions and components.

The RegModHarz-project (Renewable Model Region Harz) is to show the potentials of information and communication technologies (ICT) in energy supply systems. In this project a virtual power plant (VPP) consisting of different renewable generators, controllable loads and storage locations will be planned, simulated and set into operation. Possibilities for the marketing of the electricity produced by this VPP will be analyzed and simulated. Advanced methods and technologies to support grid operations will be developed as well.

Interconnection Standards and Grid Codes: Lessons from Germany

The experience with photovoltaic (PV) power generation in Germany highlights the importance of both establishing interconnection standards and of having grid codes so that as the penetration of PV increases, unintended consequences which threaten power quality, reliability and safety do not result.

Figure 5-2 illustrates the mix of power generation technologies which comprise the German power system. At present, the German power system has an installed capacity totaling 147 W, nearly 30% of which is solar photovoltaic or approximately 32 GW. In addition, Germany has about 31 GW of wind power generation.

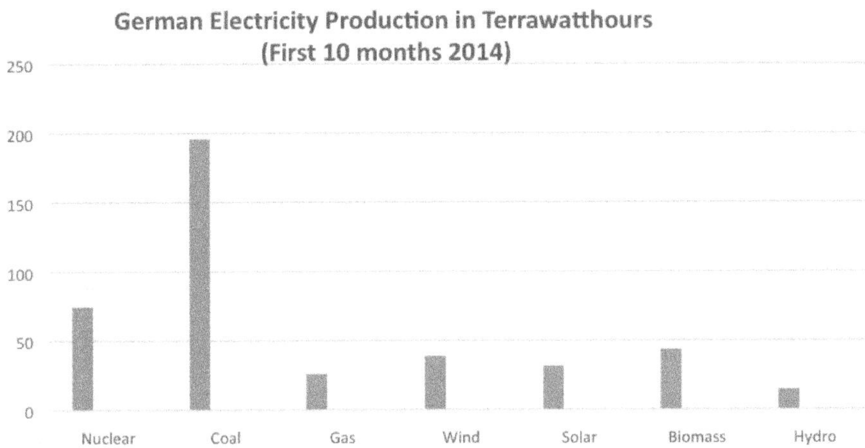

Figure 5-2. The Germany Power system Resource

On the surface, this might appear to be an ideal situation—as peak load varies from 40 to 80 GW. There are times when PV and wind can supply nearly all of Germany's needs. For example, Figure 5-3 illustrates the performance of Germany's renewable portfolio on December 31, 2012. At the time, 45 GW of demand was supplied by those resources.

It is important to note that 90% of German's PV generation is distributed.

Germany made a conscious decision to begin installing increased PV some years ago by instituting a country-wide feed-in tariff initially of 50¢/kWh. At the time, they put standards in place which mandated that inverters be arranged so as to drop the PV system off-line whenever the frequency reaches 50.2 Hz.* At the time this was established, this was a logical choice if one assumes modest PV penetration. However, when demand is low, there can be circumstances where wind power genera-

*The reader should note that the European frequency standard is 50 Hz.

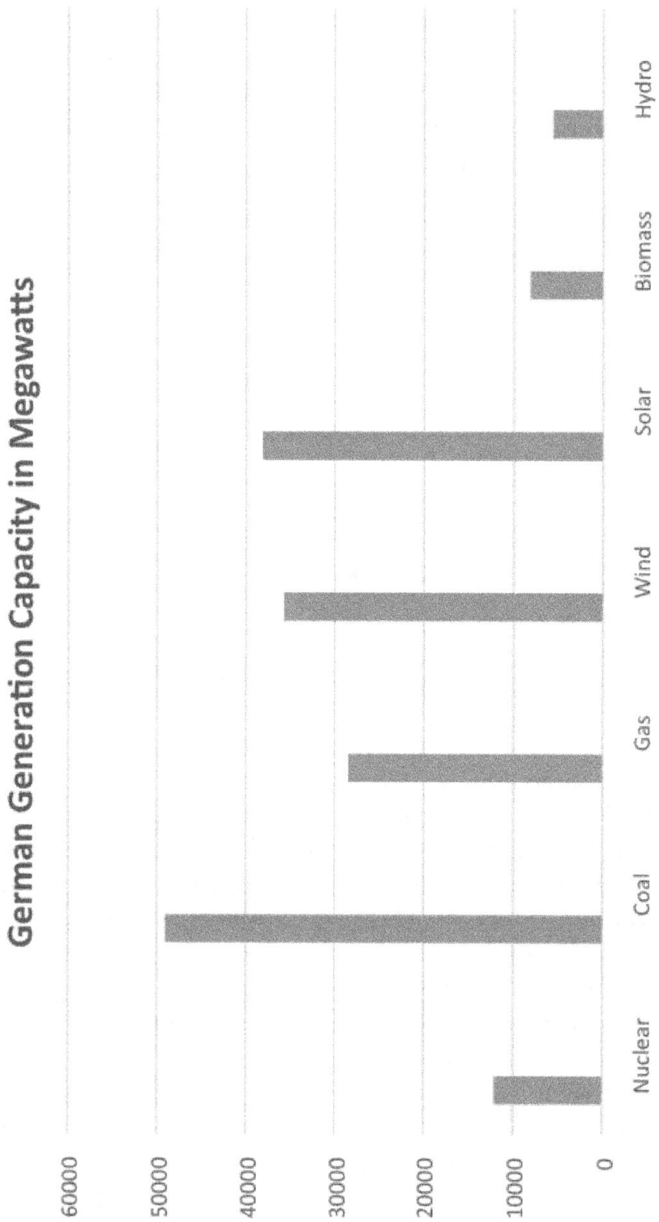

Figure 5-3. German Generation Capacity

tion naturally fluctuates such that frequency exceeds 50.2 Hz, potentially causing *all* of the PV systems to drop off instantaneously. This abrupt change can, as a minimum, cause instability or even fatal collapse of the power system.

Figure 5-4 illustrates the consequences of this on one feeder. As the frequency crossed the 50.2 Hz threshold, PV systems dropped off-line causing instability.

For example, on June 16th something very peculiar happened in Germany's electricity market. The posted wholesale price of electricity fell to minus €100 per megawatt hours (MWh). That is, generating companies were paying grid operators to take their electricity. It was a bright, window Sunday and electricity demand was low. From 2:00 pm and 3:00 pm, solar and wind generators produced 28.9 gigawatts (GW) of power, more than half the total. The grid cannot cope with more than 45 GW without becoming unstable. At the peak, total generation was over 51 GW; so prices went negative to encourage cutbacks and protect the grid. (*The Economist* 2013)

As a result, Germany has revised its rules for interconnection of PV systems. The new rules have two components: (1) PV generation must reduce when frequency increases based on the slope shown in Figure 5-5; (2) PV systems greater than 10 kW must allow operational control by the distributor and systems smaller than 10 kW# must hardwire a limit at 70% of system output.

Each of these rules is intended to prevent abrupt changes in PV output as the frequency varies on a heavily-based renewable power system. For example, on a high-sun/low load-day, this will moderate PV output. This will reduce the risk for blackouts.

At present, Germany is faced with an enormous undertaking to retrofit many of the PV inverters. Retrofit of inverters in Germany is estimated to cost €200M. While PV adoption in the U.S. is modest, at present it is timely to carefully consider interconnection standards.

France

France's leading smart grid project was launched by Electricte de France (EDF). It is referred to as PREMIO (Production Repartie, Enr et MDE, Integrees et Optimisees). It is an open architecture system designed to optimize the simultaneous control of distributed energy resources (DER), such as thermal storage and electric batteries, that can serve as a virtual power plant when aggregated.

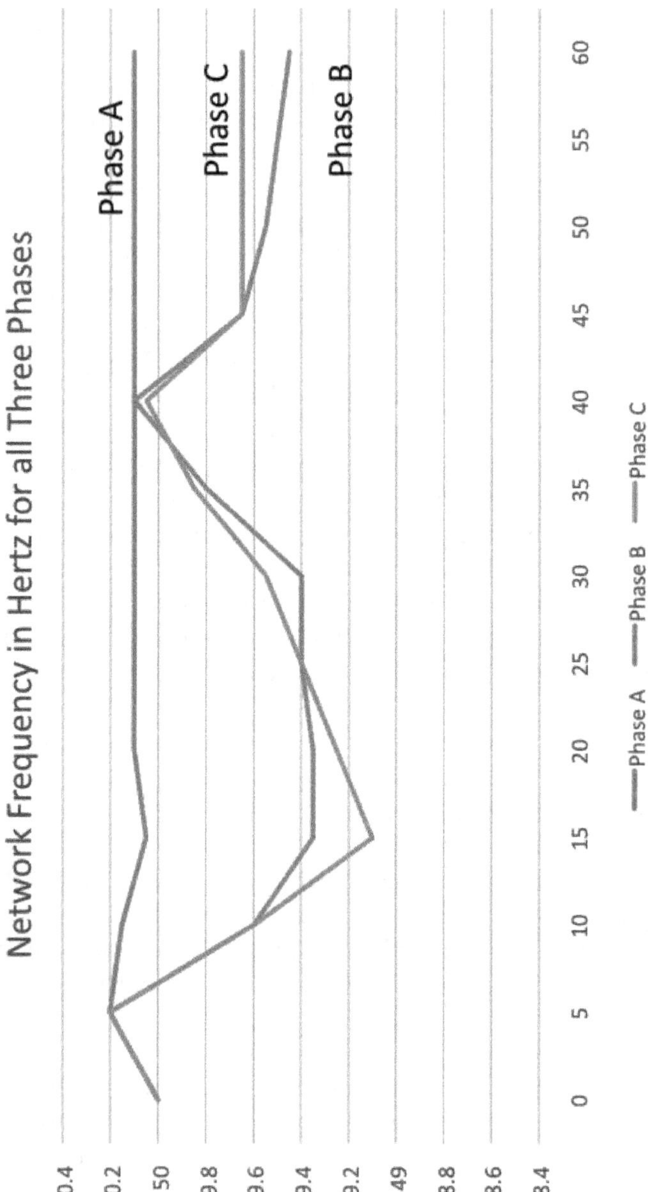

Figure 5-4. Grid Impacts resulting from inverter settings on one feeder in the German power system

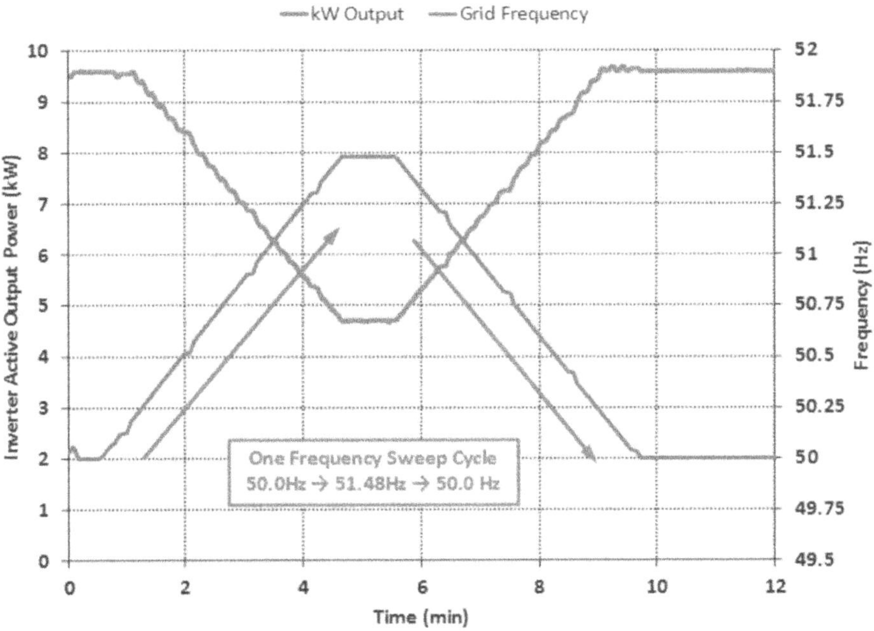

Figure 5-5. Importance of Interconnection Requirements: German Experience

The PREMIO project was designed for peak shaving in southeast France, where the power system faces increasing constraints. The project is exploring reducing loads in response to operator requests via:

- A central control unit and infrastructure located at customer facilities.
- A communication and information system that enables operation, planning, and maintenance tasks.

The control unit simulates an upstream operator by sending an offer for a variable length "critical period" as either a day-ahead request or an intraday request. (EPRI 3002004652)

Tests of the system were conducted during the "cold" season of October 2011 through April 2012. Measuring response precision was a key goal of the tests, which were initiated every two days. A total of 231 load reductions of various amplitudes and durations were measured. The control unit aggregated three different types of DER to achieve the load reductions:

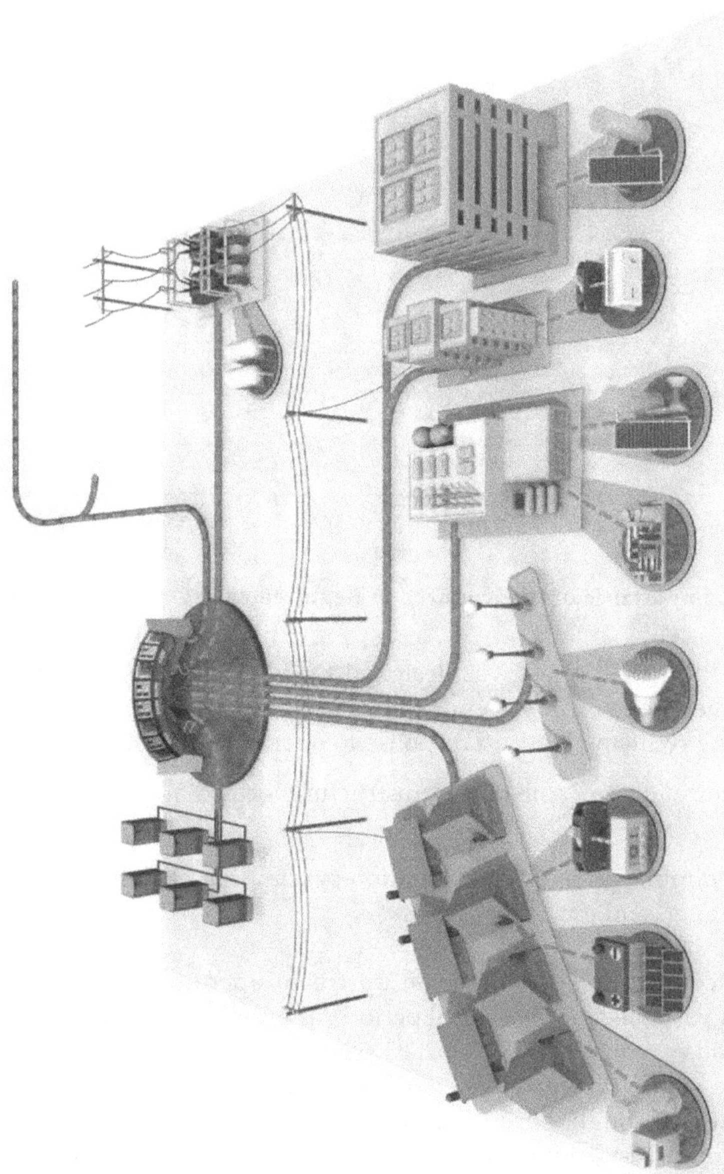

Figure 5-6. An illustration of the PREMIO project in southeast France

- A "smart box" that controls electric space heaters and water heaters (10 included).
- A heat pump coupled to a hot water tank that delivers hot water to the heating circuit when the heat pump is turned off (6 included).
- Lead-acid batteries that inject power into the grid (21 included).
- Control settings for each type of DER included the minimum delay between consecutive load shedding periods and the maximum duration allowed for a load shedding.

The analysis of results included the following observations:

- The system responded very reliably and on time, resulting in a good load reduction profile although the precision of the response profile was irregular.
- The response velocity can reach 60 kW/hour.
- The load curve is sensitive to consumer behavior and outside temperature.
- Technical issues, such as communications losses, may impact the availability of each DER.
- The smart boxes may shed only a part of the load due to optimized device control settings.
- Due to open-loop design, control requests cannot be modified once issued.
- Only four customer overrides of smart box control occurred.

Australia

In 2010, the Australian Government committed approximately $100 million in funding a consortium led by Ausgrid (one of the three New South Wales electricity distribution entities) to fund a program referred to as the Smart Grid, Smart City Program. of approximately $490 million in both cash and in-kind support. (Ausgrid 2014)

The objectives of the Smart Grid, Smart City Program were to:

- Deploy a demonstration and/or commercial scale rollout, as specified in the Guidelines, that informs a business case for key applications and technologies of a smart grid.

- Build public and corporate awareness of the economic and environmental benefits of smart grids and obtain buy-in from industry and customers.
- Gather robust information and data to inform broader industry adoption of smart grid applications across Australia.
- Investigate synergies with other infrastructure (such as gas, water and the National Broadband Network).

As part of awarding the funding, the Australian Government specified a range of activities, outcomes and data priorities for each of the trial work streams. In essence, the Smart Grid, Smart City trials focused on producing results which could be used to determine whether individual or combinations of smart grid technologies could achieve economic benefits for Australian electricity consumers. These activities and data priorities were designed to answer a number of research questions. These work streams were as follows:

Supporting Information and Communication Technology Platforms (Common Platforms)

Investigated the feasibility of various high-speed, reliable and secure data communications network and associated IT systems which integrate with the electrical distribution network. It also examined interoperability with the National Broadband Network.

Grid Applications

Investigated the ability of grid-side monitoring and control technologies to reduce network operating costs and support the future planning and implementation of lower cost networks.

Customer Applications

Focused on residential electricity consumption, reliability, customer behavior and responses to feedback technologies and pricing models. It also included an electric vehicle trial and investigations into the interoperability of electricity metering with gas and water metering.

Distributed Generation and Distributed Storage

Investigated the feasibility and potential benefits of distributed generation and distributed storage within electricity grids.

Electric Vehicles

Investigated the potential impact of wide-scale uptake of electric vehicles in Australia on the electricity distribution network.

Geographic Locations

Six geographic locations were selected for the program. These included the following:

Greater Newcastle area

The greater Newcastle area was selected as the focal point for the trial, due to its good mix of regional and suburban characteristics, representative geography, climate, socioeconomic and demographic factors of the broader Australian population. The customer demographics and socioeconomic indicators of the area closely reflect the demographics of a typical Australian city.

Sydney CBD

Part of the City of Sydney Local Government Area (LGA), this area provided additional high density residential buildings and large scale co-generation.

Ku-Ring-Gai area

Situated on the north shore of Sydney, this area provided additional testing of high income demographics. The area also provided a high number of customers with swimming pools (approximately 36 per cent in some areas of the Local Government Area) for testing new products.

Newington area

Part of the Auburn LGA, this area provided a typical Western Sydney climate zone and suburban environment for broad testing. The suburb added a high multicultural population and contains the highest penetration of customer photovoltaic energy generation, assisting in the trial of renewable generation and storage applications.

Scone Area

Part of the Upper Hunter LGA, this area provided additional rural characteristics and a more extreme climate zone. The area provided a rural network to perform end of feeder trials and much lower levels of internet use representative of more rural geographies.

Nelson Bay Area

An area north of Newcastle in NSW, including two zone transformers with eight feeders exhibiting signs of moderate voltage constraint, supporting around 10,000 customers spread across 210 distribution transformers. This area is typical of older (brownfield) distribution zones with existing constraints which made it amenable to testing the potential benefits of grid application technologies.

JAPAN

Japan is only recently embraced the smart grid concept in earnest. On April 8, 2010, four sites were selected from four cities in Japan to run large-scale, cutting-edge pilot projects on the smart grid and smart community. What the Japanese refer to as a community Energy Management System (EMS) will be achieved based on a combination of the home Energy Management System (EMS), the building EMS, (Figure 4). (Ling 2012) Within the next year two other major projects were added. These are described briefly below:

Yokohama City

There are 900 units of PV systems installed in the city of Yokohama with 2,000 more being added. The project focuses on the development of the EMS, which integrates the home EMS, the building EMS, and EVs. It is expected to generate PVs with a capacity of 27,000 kW.

Kyoto Keihanna District

The Smart Grid Pilot Project in the Kyoto Keihanna District makes use of the 'smart tap', which visualizes energy consumption controlling the home' usage. It is also a pilot project to test 'electric power virtual coloring' technology, which actualizes the overall home EMS. This project calls for the installation of PVs in 1,000 houses and an EV car-sharing system.

Toyota City

The project focuses on the use of heat and unused energy as well as electricity. It has a demand response with more than 70 homes and 3,100 EVs. Through this project, houses that contain an IT network of electrical appliances and other household equipment, solar panels, household storage batteries, onboard automobile storage batteries, and other devices, can develop household power leveling and optimized energy usage.

Kitakyushu City

This project focuses on real-time management in 70 companies and 200 houses. Energy management is controlled by the home EMS and the building EMS. This project implements various demonstrations, including communications, urban planning, a transportation system, and lifestyle, with an emphasis on the demonstration of energy projects such as electric power.

UNITED STATES

Not unlike demonstrations elsewhere, U.S. smart grid demonstrations consist of digitally based sensing, communications, and control technologies and field devices that function to coordinate multiple electric grid processes. These demonstrations include the application of information technology and associated systems to handle new and expanding data and permit more effective and dynamic grid operation. Often, the information available because of these smart grid systems also enable customers to make informed choices about their energy management. (USDOE 2014)

The U.S. electric industry spent about $18 billion for smart grid technology over the four year period 2010 through 2013. (BNEF 2014) Smart Grid investments funded by the U.S. Department of Energy as part of the American Recovery and Reinvestment Act of 2009 (ARRA). (USDOE 2014). The remainder were funded by utilities themselves, a number of those as part of the Electric Power Research Institute (EPRI) Smart Grid Demonstration project.

Annual smart grid spending nationwide reached $5.2 billion in 2011 and declined to $2.5 billion in 2014. (BNEF 2014) The decline was due largely to a slowdown in spending for advanced metering infrastructure (AMI). Industry analysts expect that spending to increase reaching $1.9 billion in 2017.

According to DOE, Progress in smart grid deployment is being made in many areas: (USDOE 2014)

Advanced Metering Infrastructure

Advanced metering infrastructure (AMI), including smart meters, communication networks, and information management systems, is enhancing the operational efficiency of utilities and providing electricity customers with information to more effectively manage their energy use. DOE estimates that 65 million smart meters will be installed by 2015.

Customer Technologies

Customer technologies, such as programmable communicating thermostats and building energy management systems, coupled to smart meters to make energy usage data accessible and useable.

Integrating Technologies

The integration of sensing, communications, and control technologies with field devices is improving reliability and efficiency. Smart grid applications enable utilities to easily locate and isolate faults to reduce outages, dynamically optimize voltage and reactive power levels for more efficient power use, and monitor asset health to guide maintenance.

Advances Sensors

The deployment of advanced sensors and high-speed communications networks on transmission systems is advancing the ability to monitor and control operations at. For example, the placement of phasor measurement units (PMUs) throughout the transmission grid permits grid operators to identify and correct for system instabilities, such as frequency and voltage oscillations, and operate transmission lines at greater capacities.

EPRI SMART GRID DEMONSTRATION INITIATIVE

Over a 7-year period from 2007 to 2014, the Electric Power Research Institute EPRI collaborated with utilities to conduct 48 smart grid demonstrations referred to as the EPRI Smart Grid demonstration Initiative. (EPRI 3002004652) Several of the EPRI projects are also summarized here. The Smart Grid Demonstration Initiative is a collaborative research effort to design, deploy, and evaluate how to integrate distributed energy resources (DER) into utility grid and market operations. The Initiative leverages multi-million dollar investments in the smart grid by the electric utility industry, with the goal of sharing information and research results on a wide range of smart grid technologies and applications. Twenty-four utilities from Australia, Canada, France, Ireland,

Japan and the United States collaborated in the Initiative.

Details on EPRI's demonstrations can be downloaded at www.epri.com and many are publicly available at no cost. A periodically updated list of all of the materials delivered can be seen at www.smartgrid.epri.com/. EPRI has created or been strategically engaged in the creation of resources that can support the electricity industry in making business and technical design decisions about the smart grid and grid modernization.

Cost Benefit Framework

A cost-benefit analysis framework—Co-funded by EPRI and the U.S. Department of Energy (DOE), a Methodological Approach for Estimating the Benefits and Costs of Smart Grid Demonstration Projects was developed as well as a practical guidebook on the step-by-step process for estimating the costs and benefits associated with smart grid demonstration projects. The framework report (product number 1020342) and the guidebook (product number 3002002266) are publicly available at no cost.

Distributed Energy Resources Architecture

A distributed energy resources architecture that served as the foundation for OpenADR—Concepts to Enable Advancement of Distributed Energy Resources (product number 1020432).

Consumer Electronics Associates (CEA)

EPRI was instrumental in the initiation of a modular communications interface to facilitate communications of energy management signals to reach devices that range from appliances and consumer products to energy management gateways and controllers. Open source software, produced for OpenADR by EPRI, contains a reference implementation of CEA-2045 that manufacturers can incorporate into residential devices as a cost effective way for products to be made smart-grid ready. (product number 3002000450)

Smart Inverter Standards

EPRI has engaged inverter manufacturers, system integrators, utilities, universities, and research organizations in a collaboration to provide smart inverter function descriptions to standards development organizations, documented in the EPRI report Common Functions for

Smart Inverters (product number 3002002233).

Use Case Repository

Software and other high-technology companies have employed "use cases" to identify and document project requirements for many years. EPRI's IntelliGrid program developed a use-case methodology that is adapted for the electric power industry, and numerous utilities used this methodology as an initial step in designing smart grid projects. This resource is available to the public at no cost at www.smartgrid.epri.com/.

US DEPARTMENT OF ENERGY SMART GRID INVESTMENT GRANT PROGRAM

Since 2009, the U.S. Department of Energy (DOE) and its electricity industry partners have jointly invested over $7.9 billion in 99 cost-shared Smart Grid Investment Grant (SGIG) projects involving more than 200 participating electric utilities and other organizations to modernize the electric grid, strengthen cyber security, improve interoperability, and collect an unprecedented level of data on smart grid operations.

The SGIG program is a cost-shared initiative that seeks to accelerate the transformation of the US electric power grid through the deployment and use of smart grid technologies and systems. The American Recovery and Reinvestment Act of 2009 (Recovery Act) provided $3.4 billion in federal funding, and project recipients have invested an additional $4.5 billion in private funding, for a total budget of $7.9 billion.

These funds are helping to build a smarter and more modern electric grid that will be needed to accomplish some of the most important economic, energy, and environmental priorities. More intelligent electricity delivery systems use two-way communications and automated controls to provide essential capabilities to meet the needs of a growing digital economy, enable greater levels of clean energy development, and strengthen the electric grid to be more resilient to natural disasters and cyber-attacks. The SGIG program is demonstrating progress in each of these areas and is expected to encourage future investments by providing greater know-how for accomplishing grid modernization.

DOE'S RESULTS

DOE finds that Recovery Act investments in SGIG and SGDP projects before March 2012 (almost $3.0 billion) ultimately generated an estimated $6.8 billion in total economic output. Economic benefits from these investments extend beyond the energy sector, as for every $1.0 million directly spent, the gross domestic product grew by $2.5 million—a higher multiplier than other federal infrastructure investments. In addition, SGIG and SGDP projects supported about 47,000 full-time-equivalent jobs in multiple sectors, including about 12,000 direct jobs among smart grid vendors. (DOE 2013)

The results of the analysis are summarized in Table 5-1 and the major findings include:

Smart grid deployments positively impact employment and labor income. Among smart grid vendors—an ecosystem of manufacturers, information technology companies, and technical services providers—about 12,000 direct jobs were supported, with the remaining jobs being in those companies' respective supply chains and induced by the money spent throughout the broader economy.

Investments in smart grid industries support high-paying jobs. Industrial sectors that benefit directly include computer systems design, technical and scientific services and consulting, and electrical and wireless equipment and component manufacturing. Industrial sectors that experience indirect and induced benefits include real estate, wholesale trade, financial services, restaurants, and health care.

Table 5-1. Total Impact DOE Smart Grid Investment Grant (SGIG) Program (DOE 2013)

All Vendors		**Smart Grid Vendors Only**
Employment (jobs)	47,000	33,000
Labor Income (2010$)	$2.86 Billion	$2.07 Billion
GDP (2010$)	$4.18 Billion	$2.91 Billion
Economic Output (2010$)	$6.83 Billion	$4.79 Billion
State and Local Taxes (2010$)	$0.36 Billion	$0.26 Billion
Federal Taxes (2010$)	$0.66 Billion	$0.49 Billion

CENTRAL MAINE POWER

Central Maine Power (CMP) installed more than 5,000 network devices and 620,000 smart meters as part of its smart grid project. CMP also installed a wireless mesh system so as to provide two way communications between the meters and their communications infrastructure. The results include a 92 percent reduction in physical meter readings along with a reduction of 750,000 fleet miles and 34,000 truck rolls totaling $500k in annual cost. Additional benefits include a more than 15-minute saving in responding to outages. (USDOE 2013)

CONSOLIDATED EDISON COMPANY

Consolidated Edison has numerous projects which can be referred to as smart grid projects. Two key efforts are: Distributed Energy Resource Management System (DERMS); and Demand Response Management System (DRMS).

DERMS provides near real time status of DR/DG resources and utility equipment on one platform it expands Control Capability via secure messages through NOC or directly to buildings. It includes a Load Flow Decision Aid which allows for more efficient dispatch of distributed energy resources (DER). It also makes detailed resource information available to operators. DERMS has the ability to visualize DERs effect on utility system (feeders, transformers.)

DRMS Automates Demand Response (DR) Administrative Processes including: Enrollment; Event notification; Performance; Settlement; Reporting; and Forecasting. It measures near real time DR performance. DRMS has the capability to perform targeted DR reductions by area or substation. It can also enable Peak Load Shifting and enable dispatching economically or to target based on Pricing/Profit.

In ConEd's system, DR MWs and Customer count is rising resulting in high administrative costs for the DR portfolio is very high. In addition the baseline calculation is very complex the lookback window is 30 days with 15-min interval data. The Most similar 10 days are selected and the bottom 5 dropped. DR aggregators are evaluated on portfolio performance with penalties for non-performance. DRMS reduces the Cost/error/time of many DR administrative processes. It also has the ability to forecast and evaluate DR performance in near real time. It an enabling platform allowing for development of more advanced DR programs. (Wellington 2014)

FirstENERGY SERVICE COMPANY

In order to resolve to a critically stressed area, of the New Jersey shore, FirstEnergy launched a smart grid project designed to manage and reduce load. In order to enable load reduction at times of critical system peak, and to participate in the wholesale market for ancillary services, the following technologies were employed:

- Direct Load Control—Residential customer air conditioner load control by BPL Global
- Distribution Line Sensors—21 single-phase Grid Sentry units and 10 three-phase PowerSense units
- Substation Meters—34 existing Satec meters integrated

In addition, permanent peak load shift units were installed in the form of four "Ice Bear" cool storage units which make ice at night to cool buildings during the day. (Gardow 2014)

PNM RESOURCES

PNM's smart grid project had as its primary objective to create a dispatchable renewable resource. PNM installed 50 kW of fixed axis silicon carbide photovoltaic modules coupled with a one MWH advanced Lead Acid battery for "shifting" and a 500 kW "Ultra" battery system for smoothing. This method of "stacking" the battery systems had a number of benefits. Flexible control could be developed in algorithms which included smoothing parameters shifting parameters. The flexibility allows for prioritization and optimization with minimal or zero latency in control signals. (Arellano 2014)

KANSAS CITY POWER AND LIGHT COMPANY

Kansas City Power and Light Company's (KCP&L) smart grid project was designed to combine commercial innovation with a unique

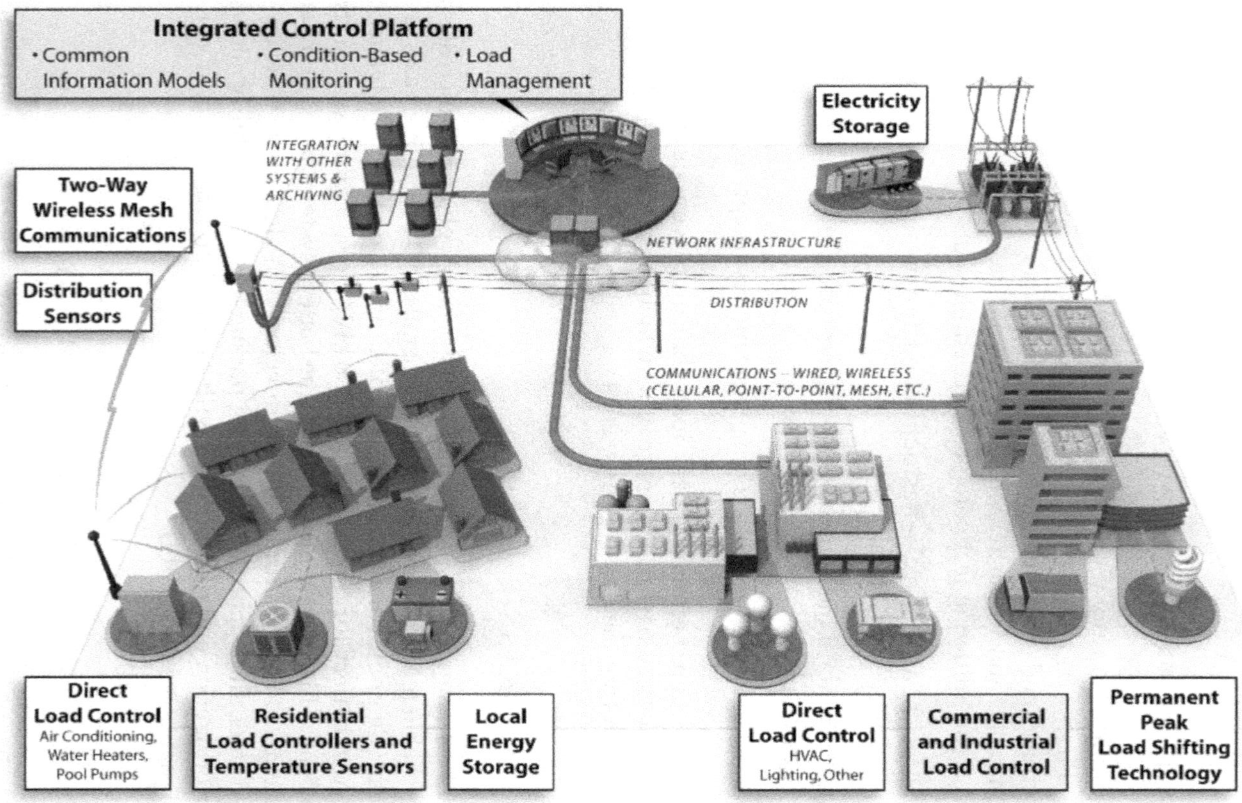

Figure 5-7. Illustrates FirstEnergy's Integrated Energy Resources Management Project (Gardow 2014)

smart grid approach. It created a complete, end-to-end smart grid—from smart generation to smart end-use—that will deliver improved performance focused on a major substation in an urban location. It introduces new technologies, applications, protocols, communications and business models that will be evaluated, demonstrated and refined to achieve improved operations, increase energy efficiency, reduce energy delivery costs and improve environmental performance. It incorporates a best-in-class approach to technology integration, application development and partnership collaboration, allowing KCP&L to advance the progression of complete smart grid solutions—with interoperability standards—rather than singular, packaged applications. It consists of the following components:

- SmartSubstation—IEC 61850 Automation and Protection Network
  - GOOSE Protection Schemes
  - Local DCADA (substation controller) & Operator HMI
- SmartDistribution—Consolidated UI for OMS, D-SCADA, DMS
  - IP WIFI Mesh Network, Reclosers, Capacitors, & Fault Indicators
  - Distributed Substation based '1st Responder' application servers
- SmartMetering—SmartMeters, AMI, & MDM integrated to legacy CIS & OMS
- SmartEnd Use—mySmart Portal—home energy management portal (HEMP)
  - mySmart Display—in-home usage, pricing, and daily bill true up
  - mySmart Thermostat—PCT without Home Area Network (HAN) mySmart Network—Gateway, PCT, 120v & 240v switches
  - mySmart Rate—TOU rate with 4 hour peak period (3-7p.m.) with 6x price differential
  - EV Public Charge Stations with Vehicle Charge Mgmt. System (VCMS)
- SmartGeneration—1 MW/1 MWHR Grid Battery Energy Storage System (BESS)
  - 170 kW Roof-Top Solar Generation
  - DR/DER Management System (DERM)

— Implemented Direct Load Control and Designed Price Based DR Events

KCP&L implemented Demand Response architecture consisting of Centralized, System-Wide DR Event Planning (DERM) which: enables awareness/knowledge of all DR/DR Resources & Programs; analyzes Requests for Demand Response; and schedules and Communicates DR/DER Events to Control Authorities. In addition that architecture allows Distributed DR Management so as to respond to multiple DR Control Authorities (HEMP, DMS, Commercial BMS, and VCMS). Also accommodates Direct Load Control & Pay for Performance Models. Finally the architecture enables Local Event Execution—the device acts on scheduled events and pricing signals. (Hedges 2014)

WISCONSIN POWER AND LIGHT

Wisconsin Power and Light's (WPL) smart grid project focused on the deployment of a new centralized energy management system. The goals of the new system were to improve distribution system efficiency through the deployment of capacitor banks and volt/VAR control systems and to ensure reliability by automating the capacitor banks. The system enables an annual reduction of 150,000 mega-volt ampere—hours amounting to a saving of $300,000/year. (USDOE 2013)

OKLAHOMA GAS AND ELECTRIC

Oklahoma Gas and Electric's (OG&E) Smart Grid program is an initiative involving system-wide deployment of integrated AMI and distribution automation. Customers were randomly assigned various combinations of customer systems, including IHDs, PCTs, and a web-based customer portal. OG&E tested real-time pricing in the study, which included TOU, CPP and VPP rate structures. The study found that there were measurable demand reductions for all of the combinations of rates and customer systems. With well over 40,000 customers enrolled in the program, the program reduced peak power demand by at least 70 megawatts (MW) (an average of 2.02 kilowatts [kW] per home with PCT installed). (USDOE 2013)

PECO, AN EXELON COMPANY

PECO deployed an advanced metering infrastructure (AMI) to improve overall operational efficiency and service for customers and enable two-way communications. This deployment also involves distribution automation systems including 100 new reclosers and communications upgrades for 300 existing reclosers. These devices will help reduce sustained outages and restoration times and improve operational efficiency. As a result, PECO has experienced a 43 percent improvement in the System Average Interruption Frequency SAIFI on the recloser circuits. (DOE 2014)

MIDWEST INDEPENDENT SYSTEM OPERATOR

The Midwest Independent System Operator operates an 11 state 50,000 mile bulk power system. Sixteen of the utilities MISO serves participated in a smart grid project designed to increase the number of Phasor measurement units installed and to make the data available to its real-time system operators. Figure 5-8 illustrates the range of appli-

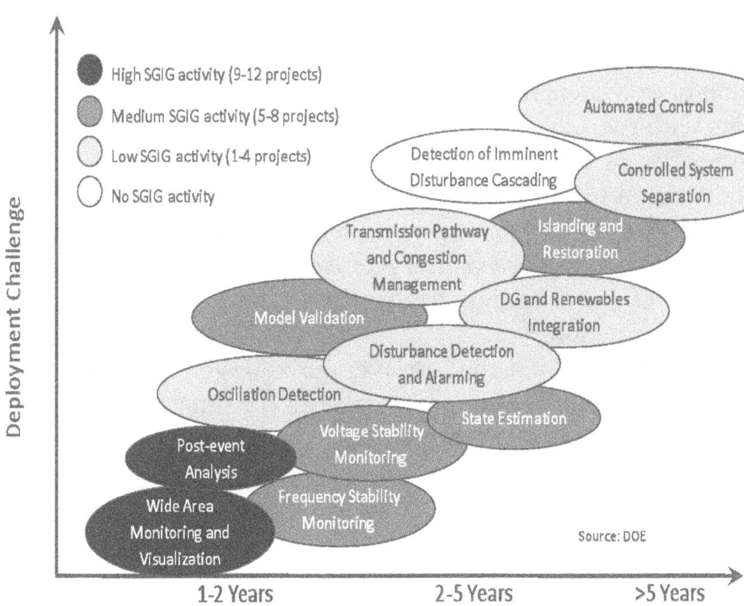

Figure 5-8. Synchrophasor data applications (Paladino 2013)

cations which these Phasor measurement units are enabling the MISO to do. (DOE 2014)

SACRAMENTO MUNICIPAL UTILITY DISTRICT (SMUD)

SMUD's recent Smart Grid projects have targeted Conservation Voltage Reduction. In 2011, SMUD conducted tests to improve distribution system efficiency through a conservation voltage reduction (CVR) experiment and a test of new volt-var optimization software. CVR is a reduction of voltage along the distribution feeder to help reduce electric power demand. By reducing the voltage by a few percentage points, but keeping voltage in the acceptable range +/- 5% of 120V), demand is reduced, and voltage is adequate to deal with voltage drops or sags. CVR is a well-established means of shaving demand, but is being tested on specific SMUD feeders to better understand the performance of new, enabling technologies and to better predict potential demand savings. Volt-var optimization (VVO) is a means of keeping voltage, and reactive power (measured as volt ampere reactive or var) at levels that minimize electricity losses. Advanced software incorporated into the smart grid management system provides better coordination and control of devices in the distribution system to achieve optimal volt-var management.

SMUD's tests entailed the following system enhancements: the installation of new switched capacitor banks, addition of new capacitor controllers with two-way radio communications, implementation of the new VVO software (an enhancement to an existing system called Capcon), modification to the voltage settings and local override control for the load tap changer (LTC) controllers, and implementation of the CVR control capability with the existing Siemens SCADA system. CVR was tested by SMUD on six feeders to determine energy savings as well as peak demand reduction.

The results indicated that an average voltage reduction of 1.7% caused an average daily reduction in energy consumption ranging between 1.3 and 1.5% with an accompanying 1.0% average load reduction.

SOUTHERN COMPANY

Among Southern Company's several Smart Grid projects is an Integrated Distribution Management Systems project. This project includes

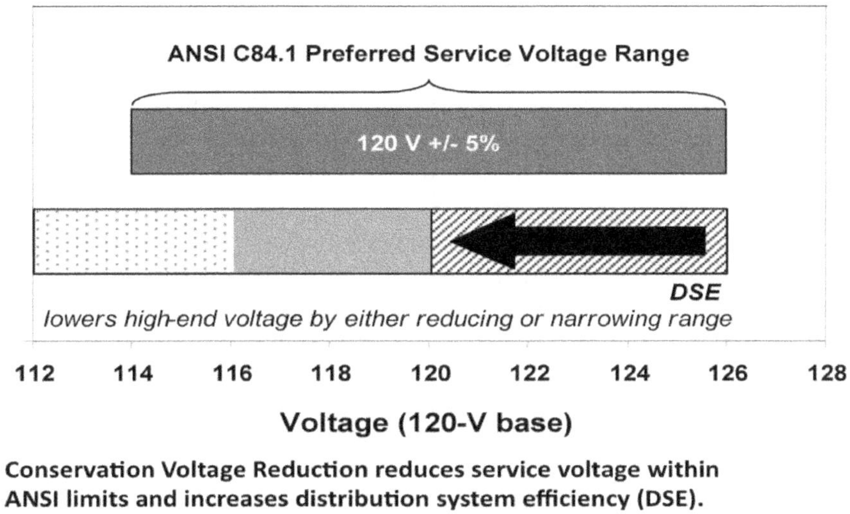

Figure 5-9. Illustration of the Application of Conservation Voltage Reduction (CVR)

installation of a system which has the following attributes:

- Single User Interface
  - Tightly integrate mission critical operations of SCADA, Outage Management and Distribution Management System (DMS) functions into a single user interface
  - Will provide operational efficiency gains
- Intelligent Electrical Model
- Networked topological model from substation to circuit's extremities facilitates the use of advanced network analysis applications to improve operational decisions
- Advanced Network Designs
  - Smart Grid functions for improved:
- Distribution Asset Utilization & Reliability
- Grid Efficiency & Demand Management
- Support of Distributed Generation
  - Networked topological model from substation to circuit's extremities facilitates the use of advanced network analysis

applications to improve operational decisions

- Distribution Training Simulator
  — Networked topological model from substation to circuits extremities facilitates the use of advanced network analysis applications to improve operational decisions

Another Southern Company project under the Smart Grid umbrella is to use the AMI infrastructure to monitor capacitor banks. AMI advanced meter capacitor bank health monitors identified over 650 problems in six months and changed their inspection schedule from once a year to once a day. Figure illustrates the issues they found causing failure of capacitor banks. (EPRI/DOE 2014)

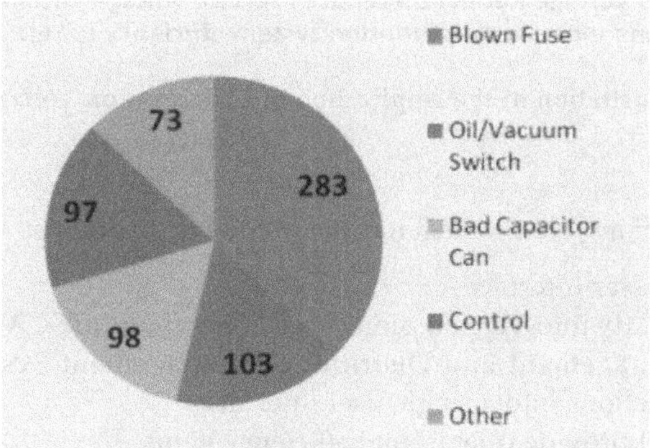

Figure 5-10.

SOUTHERN CALIFORNIA EDISON

Southern California may well be the world's leader in advancing smart grid technology. They are involved in several key projects. The author has chosen one of those, the Irvine Smart Grid Substation to illustrate here. Southern California Edison (SCE) is piloting what some would call the next generation substation automation at its MacArthur Substation as part of its Irvine Smart Grid Demonstration (ISGD). SCE's next generation system is referred to as Substation Automation 3 or SA-3. The SA-3 project team selected IEC 61850 as the standard to use as the foundation of project design. The SA-3 platform enables standards-based communications, automated configuration of substation devices, and an enhanced system design. The MacArthur Substation SA-3 pilot is demonstrating the following (EPRI 3002004652):

- A data-driven Human Machine Interface (HMI)
- Device password management
- Automated device configuration management
- Secure, remote access
- Integration of SCE's Common Cyber security Services (CCS)
- Streamlined data-driven enterprise and substation configuration processes
- Automated HMI and Energy Management System (EMS) test scripts
- Automated fault file collection
- Power systems simulator substation automation testing

The network architecture of the system is shown in Figure 5-11.

The primary objective of the research was to increase understanding of how to use the IEC 61850 configuration capabilities to reduce the levels of human intervention, reduce errors, and improve engineering time. The SA-3 project was broken down into the following three subprojects:

SCE's Substation Engineering Modeling Tool (SEMT) was originally developed for consistent generation of HMI point list for communications to the Energy Management System. The capabilities of SEMT were expanded to serve as an IEC 61850 System Configurator, capable of generating all the configuration files required to configure substation automation devices, including intelligent electronic devices (IEDs), managed switches, substation gateways, and HMI. The SEMT also generates reports such as point lists and test scripts used for automated testing.

HMI Project entailed replacement of SCE's aging legacy HMI system, which required excessive configuration and maintenance support work-hours. HMI was based on the IEC 61850 standard. The new HMI minimized manual keystrokes in development of project files, database, and graphical user interface (GUI).

Substation Gateway 61850 Project resulted in the demonstration

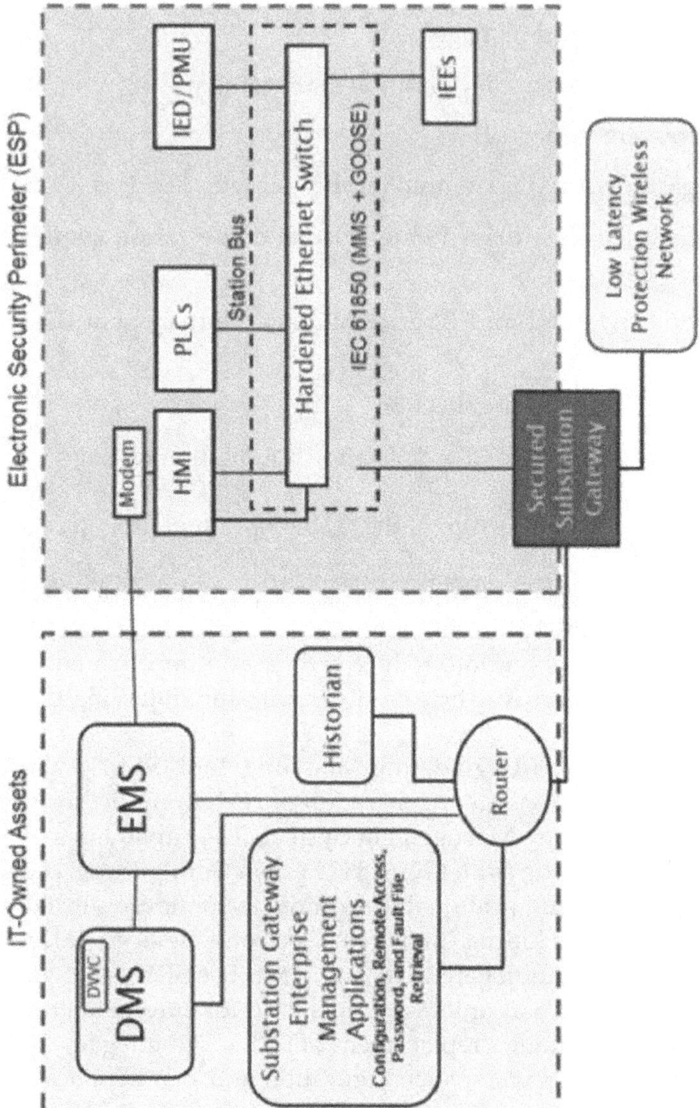

Figure 5-11. Irvine Smart Grid Substation SA-3 Network Architecture (EPRI 3002004652)

of a secure Ethernet communication interface to SCE substations for serving all system Data Beyond SCADA (DBS) to the enterprise server. Development included demonstration of remote, secure access to substation edge devices by authorized Protection and Engineering personnel.

Each of the projects was first developed and tested independently. Once the individual system components were proven, a complete system test was performed. To aid in the system test SCE developed a testing platform using power system simulation where ideas could be implemented and tested efficiently. The setup also served as an ideal training platform for operators of the new HMI. The benefits of power system simulation and scripted testing might prove to be beneficial to Factory Acceptance Testing (FAT) and Site Acceptance Testing (SAT). Key results as of 2014 include:

- HMI configuration: The project has successfully demonstrated an HMI that can be fully configured with the data driven process in a short period of time, providing significant improvement compared with manual processes.
- Security: The SA-3 project is the first substation deployment of SCE's Common Cyber security Services (CCS) platform. The security system was designed to meet NERC cyber security requirements for protection of critical cyber assets.
- Personnel training: Training material on the SA-3 system was developed to provide a reference for test technicians, operators, and maintenance. As part of the training program the test technician and operators in the area were provided hands-on training.

CONCLUSIONS

Overall the demonstrations in Europe, the United States, Japan and Australia highlighted industry stakeholders expectations that the vision of the Smart Grid's ability to: ensure higher levels of security, quality, reliability, and availability (SQRA) of electric power; to improve economic productivity and quality of life; to improve the efficiency of and asset utilization in the power system; to reduce environmental impacts and facilitate the integration of renewable resources and variable net demand; to enhance customer choice and control; and to ensure safety is reasonable and achievable. Several demonstration projects reinforced that achieving this vision will require careful policy formulation, accelerated infrastructure investment, and greater commitment to public/private research, development, and demonstration (RD&D) to overcome barriers and vulnerabilities. It is clear from these demonstrations that utility objectives rely on technology as an enabler of the future grid.

Many of the projects summarized in the chapter have been underway for a number of years. During the relatively short history of the smart grid it has become obvious that utilities and their regulators will be increasingly faced with the challenge of maintaining utility scale electricity production and delivery, while also accommodating policy imperatives that embrace end-user electricity production and a steady move towards deploying carbon free sources of electricity production.

A number of overall conclusions are now apparent as to what it will take to achieve success regarding the smart grid. These are highlighted in the following paragraphs.

Support Active Participation by Consumers

In order to achieve expected value, the smart grid should engage customers and the increasingly intelligent energy technologies in their homes and businesses. The ready availability of information on the Internet of the best, least expensive, most unique, and most customer relevant products and services has produced increasing customer expectations. The power industry is not immune to these trends. The smart grid customer will expect technology to efficiently manage their comfort and energy choices, to be informed regarding and experience a reduced incidence of service interruptions, and to be offered choices for controlling and modifying the way they use and purchase electricity. They increasingly will expect and have choices, incentives, and social media interactions regarding the adoption of intelligent devices and modifying their purchasing patterns and behavior.

Develop New Products and Services

In order to be successful, the smart grid should be able to support a market system that provides cost-benefit tradeoffs to consumers by creating opportunities to bid for competing services. As much as possible, regulators, aggregators and operators, and consumers need to modify the rules of business to create opportunity against market conditions. A flexible, rugged market infrastructure exists to provide different service levels of electricity reliability and quality that are tailored to meet individual customer needs. For some consumers this may entail ensuring continuous electric service and reliability, while also providing revenue or cost reduction opportunities for market participants.

Deploy Intelligent Electronic Devices Throughout the Grid

The grid of the future is likely to contain thousands of intelligent devices. However, the architectures to control or coordinate the operation of these devices consistent with system reliability is not yet certain. Some central station generation and network equipment is likely to continue to be centrally dispatched. Other devices and end uses may transact directly with one another creating a form of distributed coordination through localized markets or so called, master controllers.* These controllers would control local distributed energy resources (DERs) and utilization in buildings, campuses or industrial parks by communicating with one another as well as with the grid.†

Open Standards

These demonstrations highlighted that standards will be key to ensuring the interoperability of the different power and information systems needed to maintain the power grid. The continued ability of this nation to lead and influence standards development will be a major factor in the ability of U.S. companies to participate in and benefit from large scale global initiatives to build and modernize power systems.

The Need to Integrate New and Legacy Technologies

It is clear from the roll out plans which even these highly successful demonstrations intend, that the smart grid will not occur in a few years—it will develop over time. As such, each utility will require a clear roadmap and transition path to the future. At each point along the way next generation systems will have to interoperate with legacy systems.

---

*Master Controllers need not be sophisticated control devices, but will vary considerably depending on the application. In a simple residential photovoltaic installation they may merely be an intelligent inverter. In larger buildings they can be Energy Management Systems operating end-uses, electrical and thermal storage and guiding participation in Demand Response programs.

†Distributed energy resources refer to distributed generation, demand response, energy storage, electric vehicles, and microgrids.

Key Smart Grid Technologies

The demonstrations summarized in this chapter experienced an array of successful applications of smart grid technology. A few of those which had the greatest success are highlighted here.

Synchrophasor Technologies and their Application in the Recovery Act Smart Grid Programs Synchrophasor technologies and systems improved the reliability, operational efficiency, and resiliency of every electric transmission system in which they were installed.

Distribution Automation

DOE published a report that discussed the 48 of the 57 Smart Grid Infrastructure Grants (SGIG) electric distribution projects that they funded which were seeking to improve electric distribution system reliability through the use of distribution automation. Table 5-2 provides a summary of partial results from the period from April, 2011 to March, 2012 covering a total of 1,250 distribution feeders.

According to DOE, these results raise questions about the usefulness of CAIDI as an index for measuring the effects of automated feeder switching on the duration of customer interruptions. Automated feeder switching generally reduces the number of customers experiencing sustained outages (reducing the denominator of the index), relative to the duration of the sustained outages (expressed in the numerator.)

Table 5-2. Smart Grid Distribution Automation Reliability Improvements (DOE 2013)

Reliability Indices	Description	Range of Percent Changes
SAIFI	System Average Interruption Frequency Index (outages)	-11% to -49%
MAIFI	Momentary Average Interruption Frequency Index (interruptions)	-13% to -35%
SAIDI	System Average Interruption Duration Index (minutes)	+4% to -56%
CAIDI	Customer Average Interruption Duration Index (minutes)	+29% to -15%

Voltage and Reactive Power Management

A number of the US demonstration projects implemented advanced voltage and VAR optimization technologies. Most demonstrations witnessed line loss reductions in the range of 0 percent to 5 percent, and a few reported experiencing loss reductions greater than 5 percent on certain feeders. Industry estimates typically indicate that line loss reductions of 5 percent to 10 percent are possible. Feeders with the highest amount of inductive loads showed the greatest reductions in line losses. Results for conservation voltage reductions indicate a potential for peak demand reductions of approximately 1 percent to 2.5 percent. (EPRI 2014 and DOE 2013)

Automated Metering Infrastructure (AMI)

AMI was one of the most widely demonstrated technologies in the US and in Europe. Smart Grid demonstrations observed that the primary benefits associated with AMI deployment can be broadly categorized as: System Operation Benefits; Customer Service Benefits; and Financial Benefits.

System Operation Benefits—primarily associated with reduction in meter reads and associated management and administrative support, increased meter reading accuracy, and improved utility asset management, easier energy theft detection, and easier outage management. DOE segregates these costs into four categories: meter operations costs, vehicle miles driven, vehicle fuel consumption, and vehicle emissions. They reported reductions in meter operations ranging from 13 to 77% and reductions in vehicle use ranging from 12 to 59%.

Customer Service Benefits—primarily associated with early detection of meter failures, billing accuracy improvements, faster service restoration, flexible billing cycles, providing a variety of time-based rate options to customers, and creating customer energy profiles for targeting Energy Efficiency / Demand Response programs.

Financial Benefits—these accrue to the utility from reduced equipment and equipment maintenance costs, reduced support expenses, faster restoration and shorter outages, and improvements in inventory management.

Demand Reductions

Smart grid demonstrations reported demand reductions from: the Application of Advanced Metering Infrastructure; Pricing Programs, and Customer-Based Systems and time-based rates. Results varied by the type of program with the largest demand reductions associated with time-based rates. A significant variable in the results of the demonstrations was the number of customers enrolled. One uncertainty in achieving voluntary participation in customer enrollment in time-based rate programs is the success in recruitment. That is typically measured by recruitment rate, which is defined as the number of recruited customers divided by the number of solicited customers. Differences in recruitment rates reflect the solicitation strategy (e.g., opt-in versus opt-out) but also the amount of time and effort that was devoted to market research and customer education, and the effectiveness of marketing materials and education campaigns.

References

- Smart Grid, Smart City: Shaping Australia's Energy Future Executive Report, Ausgrid, July 2014
- Smart Grids in EU: Integrating Innovation in Energy Supply, Presentation by Jan Panek, Heat of Unit-ENER B3—Retail Markets, European Union, EPRI Innovation Summit, Dublin, 3 June 2014
- American Recovery and Reinvestment Act of 2009: Smart Grid Investment Grant Program—Progress Report II U.S., Department of Energy, October 2013
- Results and Findings from the ARRA Smart Grid Projects, Presentation by J. Paladino, U.S. Department of Energy, 2013 MARC Annual Conference, June 12, 2013
- The Japanese Smart Grid Initiatives, Investments, and Collaborations, A. Ling et al, (IJACSA) International Journal of Advanced Computer Science and Applications, Vol. 3, No. 7, 2012
- EPRI Smart Grid Demonstration Initiatives: Final Update, EPRI 3002004652, December, 2014
- 2014 Smart Grid System Report, U.S. Department of Energy, August 2014
- DERMS DRMS, A. Wellington, Consolidated Edison Company of New York, Presentation to EPRI/DOE Workshop; "The Smart Grid Experience: Applying Results, Reaching Beyond," October 27-29, 2014
- Integrated Distributed Energy Resources Management Project, E. Gardow, FirstEnergy Service Company, Presentation to EPRI/DOE Workshop; "The Smart Grid Experience: Applying Results, Reaching Beyond," October 27-29, 2014
- PNM Prosperity Energy Storage, B. Arellano, PNM, Presentation to EPRI/DOE Workshop; "The Smart Grid Experience: Applying Results, Reaching Beyond," October 27-29, 2014
- A Distributed Resource Management System Architecture for Supporting Distribution Grid Operations, E. Hedges, Kansas City Power & Light Col, Presentation to EPRI/DOE Workshop; "The Smart Grid Experience: Applying Results, Reaching Beyond," October 27-29, 2014
- Bloomberg New Energy Finance (BNEF) 2014, March 6 U.S. Smart Grid Spend Report. Prepared for the U.S. Department of Energy by Brian Warshay and Colin McKerracher.

Chapter 6

Enhancing Smart Grid Resiliency

WHY INCREASE RESILIENCY?

A more flexible, resilient, and interconnected power system will be required to successfully meet the challenges posed by future uncertainties. In turn, the success of the utilities and the companies involved in operating and maintaining the power system will be measured by their flexibility, resilience, and connectivity.

To transition to a more robust power system with these characteristics, it will be necessary to develop and deploy a portfolio of technologies that can address the many issues emerging from the evolving transformation.

Potentially stimulated by climate change, sunspot activity, or other phenomena not now fully understood, there is an increasing occurrence of extreme weather events that impact the electric infrastructure and seemingly causes both widespread blackouts and local power outages. Recent examples include the U.S. hurricane Katrina, the 2011 and 2012 storms in the Northeastern U.S., and the Tohoku earthquake in Japan that led to a subsequent tsunami often referred to as the Fukushima disaster. In each case, the electric infrastructure suffered considerable damage and left citizens without power and often without clean water, food, and fuel for heating or transportation. An additional consequence of the digital age is that the stop-gap communication now often provided by cell phone technology failed as well; cell towers were often powerless and even if in service, citizens could not recharge their phones without solar chargers—which few consumers had. These seemingly random occurrences appear to be increasing in frequency. (EPRI 1026889)

These events add to the vulnerability already present from man-made outages and cyber security incursions. Coupled together, they represent a new reality that the industry's stakeholders must embrace and to which they must respond. Systems now designed for a one-time-in-50-year or one-time-in-100-year event may now need to withstand a one-time-in-200-year event.

WHAT IS POWER SYSTEM RESILIENCY?

Resiliency as it relates to power systems is more than the dictionary definition: "To return to its original state." In the case of power systems, it refers to the ability to harden from or recover quickly from a high-impact, low-frequency event.

The premise of resiliency (as applied to power systems) is that institutions and systems can be designed to "better absorb disruption, operate under a wider variety of conditions, and shift more fluidly from one circumstance to the next." (Zolli 2013) In general, the resilience of any system implies interventions that (1) ensure sufficient reserves; (2) diversify inputs; (3) collect quality, real-time data about operations and performance; (4) enable greater autonomy for constituent parts; and/or (5) utilize firebreaks so that a disturbance in one part does not disrupt the whole.

In broad strokes, the literature on resilience suggests that to expand the array of comfortable positions available, the options include reducing material needs; inventing new technology; liberating from a traditional constraint; and modifying tools so as to enable movement from one topology to another. To create resilience, power system engineers must enhance the ability to resist from being pushed from one steady state and comfortable position to an entirely new position.

Resiliency may include robustness—the hardening of a system—as well as redundancy, which improves the ability to persist when compromised. Keeping redundant, critical parts is costly but can be highly effective. However, resiliency does not always equate with recovery. A resilient system need not necessarily return to its original state. (Zolli 2013)

Distributed systems offer an intriguing method of defense. For example, the internet breaks up data into packages and routes them through a distributed network of routers. So a disturbance in distributed systems can seldom take down the entire information grid. The same concept can be applied to power systems.

Prevention, Recovery and Survivability

The resilience of the power system is based on three elements: prevention, recovery, and survivability:

- Damage prevention refers to the application of engineering designs and advanced technologies that harden the power system to limit damage.
- System recovery refers to the use of tools and techniques to quickly restore service as soon as practical.
- Survivability refers to the use of innovative technologies to aid consumers, communities, and institutions in continuing some level of normal function without complete access to their normal power sources.

Improving the power system's resiliency requires advancement in all three aspects. The most cost-effective approach combines all three. The following sections highlight some opportunities for improvement using existing technology and materials, and some opportunities for accelerating the development and adoption of innovative technologies through collaboration with electric utilities and technology providers worldwide.

Prevention

Damage prevention "hardens" the system to prevent damage through changes in design standards, construction, maintenance and inspection, and in recovery practices. These changes can include:

- Enhanced vegetation management
- Securing power plants against storms and terrorists
- Developing more resilient nuclear fuel
- Selective undergrounding of T&D facilities
- Reinforcing overhead lines
- Deploying cyber security measures and R&D actions to improve pole and line design
- Refining use of hydrophobic coatings
- Enabling dynamic circuit operation

Recovery

Recovery requires prompt damage assessment and repair: the effective use of restoration management tools. R&D efforts to support recovery include:

- Airborne damage assessment
- Development of advanced outage management systems (OMS)
- Effective use of geographical information systems embedded in an asset management system
- Development of field force visualization tools
- Availability of recovery transformers

Survivability

Survivability refers to the ability to maintain some basic level of electricity service to customers or communities. Key to survivability is to facilitate enhanced communications with customers. Additional R&D should focus on the use of plug-in electric vehicles (PEVs) as a power source, enabling photovoltaic (PV) systems to operate as a standby source and enabling urgent service to cell phones, traffic lights, hospitals and prisons.

HIGH-IMPACT, LOW-FREQUENCY EVENTS

Disruptions in electrical service and poor power quality events can occur as a result of a number of factors. Most often so-called "high-impact, low-frequency" or HILF events are cited as those that can threaten electric service reliability and can potentially result in loss of service for extended periods of time. HILF events can result from a variety of natural and manmade disasters, as well as unintended failures of major equipment. Figure 6-1 illustrates a range of HILF events. However, interruptions in electric service can result from less dramatic situations:

- Loss of a substantial quantity of natural gas power generation in a single region due to lack of adequate scheduling of gas pipelines
- Equipment failure in necessary communications systems
- Even snakes causing a fault as a result of crawling between substation bushings in an attempt to reach bird nests.

HILF events can threaten lives, disable communities, and devastate

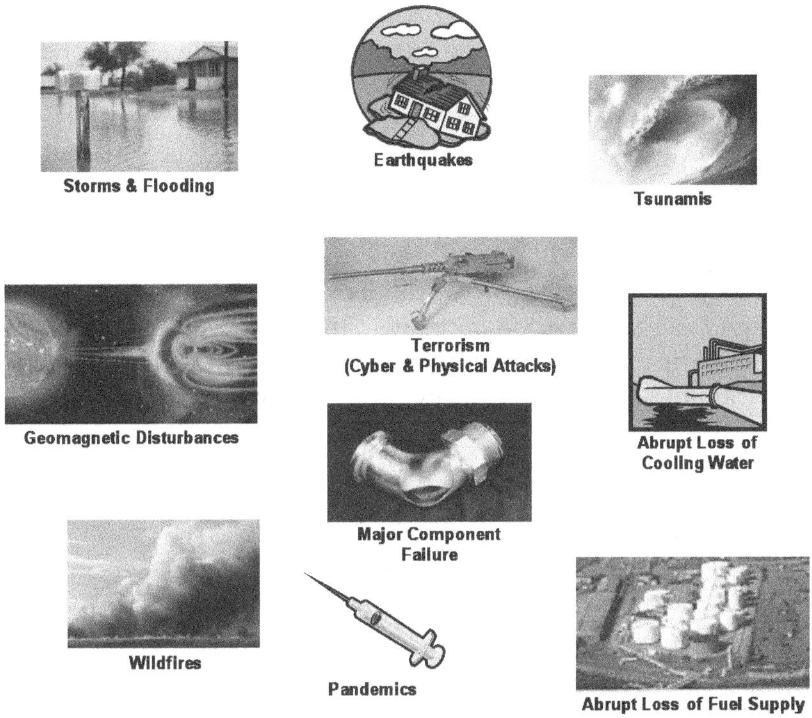

Figure 6-1. High-Impact, Low-Frequency (HILF) Events

generation, transmission, and distribution systems, as well as interdependent systems such as natural gas pipelines and other fuel transport, and telecommunications. While many outages result from damage to the millions of miles of electrical lines, an expanded vision of resiliency requires a broader vision of the consequences of HILF events.

Customers' expectations of service reliability have changed dramatically with the evolution of the 24/7, digitally connected society. Even with enhanced response and heroic efforts by crews, restoration that stretches to days, and in some cases weeks, is no longer acceptable. At the same time, consumers expect electricity to be affordable. The investment required to harden the power system to withstand the worst disasters must be regarded by all concerned as prudent and cost-effective.

A discussion of power system resiliency must also reflect the scope of the power system. A traditional bulk power system is comprised of bulk power generation, a network of transmission lines, and a distribution system that links the grid to consumer buildings and other facilities. However, as technology has evolved and new electrical appliances and devices have entered the electrical marketplace, electricity can now be generated and stored at increasingly diverse points, including near points of end use. Therefore, power systems can be thought of as self-contained devices like a laptop with batteries and a solar charger; a building with photovoltaic power generation, wind or gas-fired microturbine generators; a campus or community "microgrid," or more complex bulk power systems. The ability to interconnect each of these power systems—whether via permanent interconnections or controlled according to operational objectives—can add substantial redundancy, which is another aspect of resiliency (EPRI 1025861).

THREATS TO RESILIENCY

Threats to resiliency include weather-related events, cyber attacks, physical terrorist attacks, geomagnetic disturbances, earthquakes, tsunamis, and flooding.

Weather-related Events

According to a 2008 Edison Electric Institute (EEI) Reliability Report, 67% of electrical outage minutes were weather-related, with contributing factors of lightning (6%), weather (31%), and vegetation (30%). These are typically due to wind, ice, or snow either directly affecting distribution assets or bringing vegetation into contact with utility lines, poles, and transformers (see Figure 8). (EPRI 1025861)

Restoring service after storms is costly. A survey of 14 U.S. electric utilities identified 81 major storms between 1994 and 2004, costing the utilities more than $2.7 billion (Johnson 2005). These direct costs represent only a fraction of the wider economic losses resulting from extended outages. Effects are exacerbated by the interdependence of the grid and other infrastructures such as telecommunications, natural gas and oil, water supply systems, banking and finance, and transportation. Utilities have continued to improve disaster response, focusing on upgraded equipment, advanced communications, rapid deployment of staged resources from their own and others.

For example the storm that struck the northeastern United States on October 29, 2011 carried snowfall exceeding 12 inches which stuck to the still leaf-laden trees, bringing down branches and trees. Damage was widespread to distribution lines and poles. More than 800,000 Connecticut Light & Power customers lost power during the 11-day outage. The duration of the power outage caused inconvenience and frustration among the public and among state and municipal officials (WITT 2011). Figure 6-2 is a photo of the type of damage Connecticut's Power System experienced in 2011.

On average, U.S. electricity consumers can expect to lose power for more than 100 minutes annually due to transmission and distribution outages from major storms. As with any average, individual consumers may never see an outage, while others could see outages of several days—despite the heroic efforts of utility workers in the field day and night.

As with most of us in the Western World, being without heat, hot water, lights and communications is not challenging for a short time but after a few hours your mood changes and the serious nature of what it means to not have electricity sinks in. (Federman 2012)

We have become so accustomed to electricity that even minor changes outages lasting a few hours can anger us along with elected officials, and regulators. Consumer sensitivity is fueled by our increasing dependence on digital devices. To satisfy consumer service demands

Figure 6-2. October, 2011 Northwestern Ice Storm

today, yesterday's standards for restoration is no longer sufficient—consumers want fast restoration.

Table 6-1 lists large US blackouts by their initiating events.

Table 6-1. Causes of Large Blackouts in the United States (Hines 2008)

	% Of Events	Mean Size in MW	Mean size in No. of Customers
Earthquake	0.8	1408	375900
Tornado	2.8	367	115439
Hurricane	4.2	1309	782695
Ice Storm	5.0	1152	343448
Lightning	11.3	270	70944
Wind/Rain	14.8	793	185199
Other cold weather	5.5	542	150255
Fire	5.2	431	111244
Intentional Attack	1.6	340	24572
Supply Shortage	5.3	341	138957
Other external cause	4.8	710	246071
Equipment failure	29.7	379	57140
Operator Error	10.1	489	105322
Voltage reduction	7.7	153	212900
Volunteer reduction	5.9	190	134543

Causes of Large Blackouts in the United States (Hines 2008)

Extreme HILF Weather Events

The nation's electric infrastructure's is increasingly vulnerability to widespread blackouts and power outages resulting from extreme weather and other natural disasters. Recent trends are illustrated in Figures 6-3 and 6-4, and recent examples include hurricane Katrina, the October 2011 storms in the northeastern United States, and the Fukushima Daiichi nuclear disaster in Japan leaving citizens without power and often without clean water, food, and fuel for heating or transportation. In some cases, the stop-gap communication now often provided by cell phone technology can fail as well, because many cell towers were without power, and telephone users are without power to charge their phones.

These seemingly random occurrences appear to be increasing in frequency. As shown in Figure 6-3, the number of natural disasters totaled 247 in 2010.

The impact of extreme weather events is further evidenced by examining power outages in North America. According to the North American Electric Reliability Corporation (NERC), the number of so-called major disturbances increased dramatically from 1992 to 2010 (see Figure 6-4).

According to 2013 National Climate Assessment, "Climate change is already affecting the American people." (NCADAC 2013) Weather events have become more frequent and intense, including heat waves, heavy downpours, and in some regions, floods and droughts. Sea level is rising, oceans are becoming more acidic, and sea ice is melting. These changes are part of the pattern of global climate changes, allegedly driven by human activity.

Climate change means increases in ocean and freshwater temperatures, frost-free days, and heavy downpours. Sea levels have risen, and there have been large reductions in snow cover, glaciers, permafrost, and sea ice. Winter storms have increased in frequency and intensity.

Cyber Attacks

Increased use of computers and wireless communications also means heightened cyber vulnerability—an added complication in the resiliency equation. Power systems are particularly vulnerable to cyber attack with the increased role of automation, as automation is one of the strategies to reduce the impact of both generation unavailability and other outage causes. Attempted breaches are launched by both amateur and professional hackers each day. Fortunately industry efforts have successfully minimized the impact of these incursions. Continued development of cyber defenses will be required on an ongoing basis to enhance power system resiliency. Refer to Chapter 9.

Physical Terrorist Attacks

Terrorists can potentially choose targets and inflict damage in order to impose disruption and make speedy restoration difficult (NAS 2012). Power plants, substations, and transmission lines are the most likely tar-

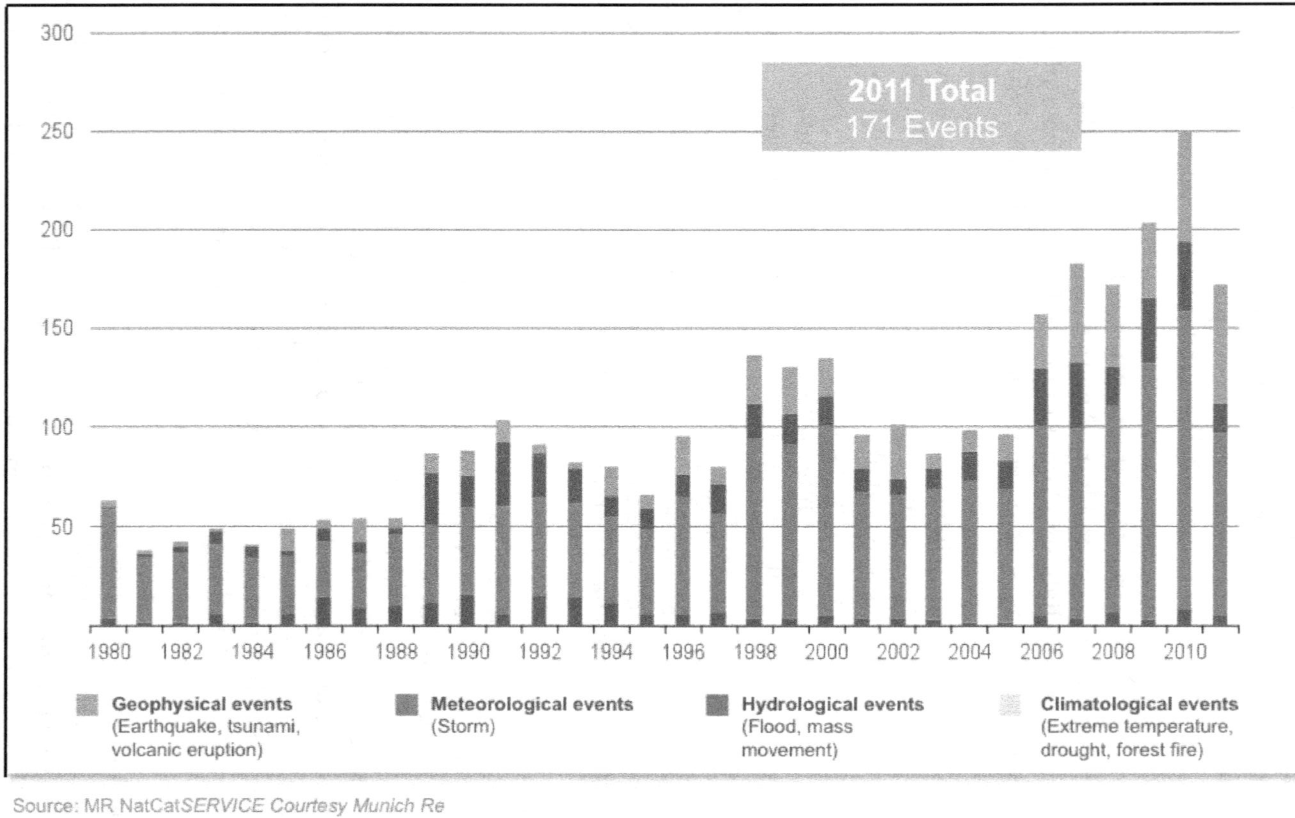

Figure 6-3. Natural Disasters in the United States, 1980-2011, Number of Events, Annual Totals (Source: MRCatSERVICE, Courtesy Munich Re)

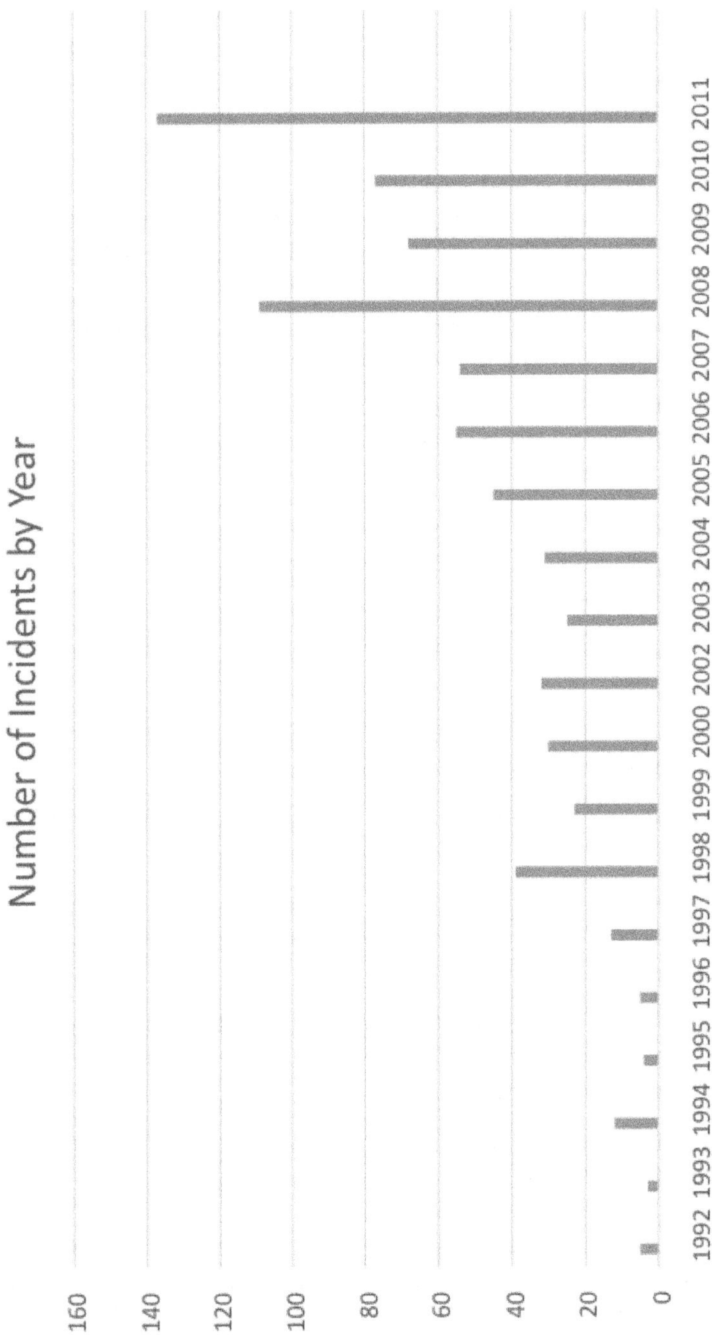

Figure 6-4. North American Electrical Disturbances Caused by Weather (Campbell 2012)

gets. Damage to key substations could be much greater and more extensive than that caused by natural events, requiring replacement of large transformers, circuit breakers, and other equipment. Depending on the availability of spares components, restoration could take weeks, months, or even longer.

Terrorist groups could execute a coordinated planned attack on multiple assets within the bulk power system to effectively disable the system. This risk could assume the form of physical attacks, cyber attacks, or a blend of the two EPRI 3002001935).

Coordinated physical and cyber attacks could incorporate, but not be limited to, the following elements in some combination:

- Distributed denial-of-service attack in which attackers flood network resources to render physical systems unavailable or less responsive.
- Rogue devices used to access the system to manipulate it or provide incorrect data to system operators.
- Reconnaissance attacks that probe a system to provide attackers with information on system capabilities, vulnerabilities, and operation.
- Eavesdropping attacks that violate the confidentiality of communication within a network.
- Collateral damage that occurs as an unplanned side effect of cyber attacks.
- Unauthorized access attacks in which the attacker obtains some control over the system, and accesses and manipulates assets without authorization.
- Unauthorized use of assets, resources, or information that would feed system operators incorrect information presumed to originate from a "trusted" source.
- Malicious code or malware through viruses, worms, and Trojan Horses (EPRI 3002001935).

The distributed nature and diversity of the system also presents some security challenges. Limited on-site supplies and the challenges of securing replacement system components would hinder a smooth recovery from these types of attacks. Attacks on the bulk power system workforce also present a risk.

Economic Impact of Terrorist Attacks

The Federal Energy Management Administration (FEMA) commissioned a report in 2005 to estimate the economic costs of terrorist attacks. The study identified three potential outcomes from successful terrorist attacks on electric power systems: the loss of human life and health, business losses, and declines in property value. The study concluded that the loss of human life can range between \$20,000 and \$7 million per capita. The impact on GDP can be \$113 per person per day, with individual business establishments suffering losses of between \$3400 and \$88,000 per event.

GEOMAGNETIC DISTURBANCES

Geomagnetic disturbances are the result of either electromagnetic pulses or geomagnetic disturbances themselves. Electromagnetic pulses (EMPs) can cause substantial damage to electrical devices including power systems and electric end-use devices. Electromagnetic pulses can be created by high-altitude nuclear detonation yielding EMPs (HEMPS) and geomagnetic disturbances (GMDs).

HEMP is an EMP detonated at high altitude to produce more widespread effects (an EMP, as opposed to a HEMP, can cause more localized but still significant impacts). The HEMP from a high-yield gamma ray weapon can in principle impact the functionality of power grids, communication infrastructures, computing and electronic processing systems, and ground transportation systems dependent on microprocessors or embedded electrical systems that are susceptible to the disruptive effects of large EM perturbations (EPRI 3002001936).

For example, in July of 1962, as a part of its nuclear testing program, the U.S. military detonated a nuclear device 1445 km (900 miles) from Hawaii at an altitude of 400 km (250 miles) over the Pacific Ocean. The blast, dubbed Starfish Prime, besides its bright visibility in Honolulu, knocked out streetlights, damaged a telephone company's microwave link, set off numerous burglar alarms, and reportedly disabled automobiles on the islands (see Figure 6-5). Despite the long distance from the test site, the damage in Hawaii was directly attributable to the

EMP that is a natural by-product of any nuclear explosion (DOD 2005 and Bainbridge 1976).

Figure 6-5. Starfish Prime as seen from Honolulu, July 9, 1962 (DOE 1994)

Earthquakes, Tsunamis and Flooding

Earthquakes

Without proper planning, earthquakes can cause significant damage to power generation and transmission facilities. Earthquake effects on structures, lifelines, and equipment installations range considerably.

In 2006, the Hawaiian Islands experienced the strongest earthquake in the islands in 23 years. Significant power disruptions occurred not only on the big island of Hawaii, but also on Oahu and Maui, as a result of the earthquake. With the exception of switchyard equipment and poorly anchored components, engineered electrical and mechanical equipment relevant to the electric power industry suffered little damage even at sites that experienced high ground motions (EPRI 1021087).

Blackouts were caused by operational issues without major earthquake damage to the respective power system components, and a combination of operating power plants that were either automatically shut down by generator/turbine vibration sensors or manually shut down by operators thinking that the earthquake vibrations were a precursor to turbine failure.

Blackouts in Hawaii were determined to be the result of the direct trip of 12 of the 69 kV transmission circuits by earthquake-caused relay mis-operation. Another contributing factor was the lockout of eight substation and generation step-up transformers caused by earthquake activation of the sudden pressure scheme used to protect the transformers from internal electrical faults. Fifteen substations had primary transformer fuses shaken from the fuse holders by the earthquake. Vibration sensors also shut down the largest power plant on the island of Hawaii, but the effect of this loss of generation was offset by the loss of electrical load caused by the trip of the transmission lines.

Between 1971 and 1995, California had 12 earthquakes that resulted in seismic damage to substation equipment. The majority of the damage relates to equipment operating at 220/230 kV and 500 kV. In a few cases, damage to 60-kV equipment was noted. In California, the Loma Prieta Earthquake of 1989 is most often cited. It is the most costly natural disaster in the United States. Pacific Gas & Electric reported damage to two power plants: Moss Landing and Hunters Point (Dames and Moore 1994).

Tsunamis

Tsunamis result from disturbances in the earth's crust from earthquakes, slides or volcanic activity. Tsunamis generate large waves that can damage power plants, substations, as well as transmission lines and distribution systems.

In the case of power plants, tsunamis can impact cooling water systems, result in loss of cooling water (due to dry intakes during drawdown caused by receding tsunami waves), and damage structures and foundations from hydrodynamic forces (NRC 2009). This can include loss of fuel supply tanks located on-site intended to support standby generations.

Flooding

Power systems can be subject to flooding from sea water resulting from storm surges or tsunamis and flooding from fresh water. Primary risks from flooding are damage to substations (if design criteria are exceeded), substation transformers, poles and towers directly from erosion or from vegetation uprooting, and failed underground cables, switches, network protectors, or transformers.

POWER SYSTEM COMPONENTS

For the purpose of this chapter, the power system has been divided into the major components as illustrated in Figure 6-6. Each of these components will naturally approach the technology solutions for the key elements of prevention, recovery, and survivability differently.

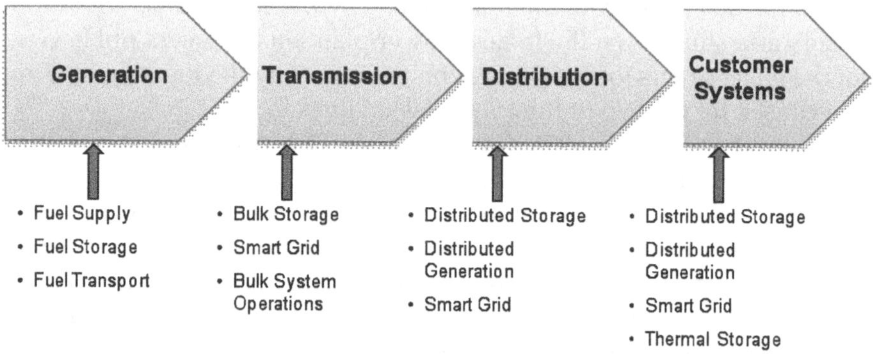

Figure 6-6. Components of the Power System

Prevention and recovery apply primarily to generation, transmission, and distribution, while survivability applies primarily to customer systems.

GENERATION

On the generation system, actions can be taken to enhance resiliency in the areas of emergency management for fossil power plants, nuclear response to external events, assessment of nuclear power plant seismic margin, nuclear power plant load following, severe nuclear accident management, and seismic protection.

Fossil Power Plants

Fossil plants experience most of the damage from HILF events due to flooding and interruptions in fuel supplies, steep ramp rates, fast starts and shutdowns, and extended periods of shutdown. This is simply based on cycling from cold to hot metal in a short time period. The mechanisms of damage they experience are:

- High-temperature, high-cycle creep fatigue.
- Vibrations—increases in valve body and small-bore pipe cracking.
- Stress corrosion cracking (SCC) due to steam quality changes and chemistry (turbine balding).
- Flow-accelerated corrosion due to changes in mass flow rates.
- Wear and tear on pumps and motors due to non-optimal operating points and changing loads.

These impacts are expected to increase due to various state actions. For example, in California, system operators are being pressured to accept energy from renewables at all costs. In Texas, the large wind farms in west Texas are forcing fossil plant to routinely cycle. Likewise, in the upper Midwest and Canada, renewables are challenging the economics of both fossil and nuclear plant operations. EPRI is modifying its PRISM Model to divide the U.S. into 15 regions to determine the sensitivity in each region to increases in renewables and retirement of older fossil units.

In addition, the power industry is expected to face increased competition for rights to both water withdrawals and consumption in the near future. Researchers at the U.S. DOE's National Energy Technology Laboratory (NETL) have estimated that water consumption in the coal-fired power industry will increase 21% (from 2.4 to 2.9 billion gallons/day) from 2005 to 2030, while consumption for all water users will increase only 7%. Competition for water withdrawals and consumption from agricultural, municipal, and industrial users are projected to increase greatly during this period, especially in water-limited areas of the United States including the West, Southwest and, periodically, certain areas of the Southeast.

Emergency Management for Fossil Power Plants

Severe weather events, earthquakes, explosions, fires, and chemical releases are some of the emergency situations that can occur at electric power plants. The individual events may be unpredictable, but the fact that emergencies will occur is inevitable. The degree to which a plant will be adversely affected by an emergency is influenced to a considerable extent by the quality of the plant's emergency management program. EPRI has established emergency management guidelines to support utility companies in developing appropriate programs for their fossil generating stations.

These guidelines provide the necessary structure and guidance for electric generating companies to create site-specific strategic plans and procedurize planned actions necessary to evaluate, prepare for, respond to, and recover from various emergency situations (EPRI 101499). Additional guidelines are also available in EPRI's report on Operational Experience with Lessons Learned on Hurricanes (EPRI 613363).

There was heavy damage to a number of U.S. Gulf Coast power plants caused by several hurricanes in 2004 and 2005. Because several years typically pass between hurricane impacts in a specific area, complacency and failure to capture or implement the lessons learned from the last storm is inevitable. There is sometimes a tendency to bind the potential damage from an approaching storm with that experienced in a previous storm. Some plants have learned the hard way that the destructive potential of any two storms is difficult to project. The size, speed, and direction of a storm can significantly affect the amount of damage it inflicts. A large, slow moving storm can include considerably more rain and lead to flooding, and high sustained winds can cause much more wind damage than a short burst of extremely high winds.

Nuclear Power Plants

The principle actions that govern the resiliency of existing nuclear power plants are response to external events, assessment of seismic margins, changes in operation an accident management.

Nuclear Response to External Events

Nuclear structures themselves are robust, but issues do arise when there is a loss of off-site power. In those cases, on-site emergency generators and batteries provide temporary power for a period of time. Increasingly, nuclear plant operators are interested in advanced battery capabilities. Efforts are underway to engage in battery testing and load profiles. Plant operators are also reviewing power demand to optimize power management using technologies such as LED lighting and low- and zero-power instrumentation.

A key issue regarding nuclear plant resiliency is severe accident management. These actions entail engaging emergency operating procedures to prevent core damage. These procedures are being reviewed to capture insights from the Fukushima Daiichi disaster. The key findings address spent fuel pool implications, multiple-unit failures, containment integrity, and hydrogen generation leading to explosions.

Fundamental to overall resiliency of nuclear plants is the prevention and management of accidents in the first place. Barriers to fission product release must be considered. This may involve changing the fuel cladding that is used, redesigning the reactor coolant system, and revising containment designs. Containment must be managed during a severe event. A performance-based approach needs to be developed. Installing a vent filter is not always the answer.

Each nuclear plant needs to develop strategies to mitigate radiological releases. This reflects a change in conventional thinking on management of containment and releases. Implementation will require a plant-specific analysis resulting in plant modifications, changes in procedures, and training to implement those changes. This work may result in the development of other containment types with varying results.

Assessment of Nuclear Power Plant Seismic Margin

Nuclear power plant owners and operators can assess a nuclear power plant's ability to withstand an earthquake greater than design and still safely shut down for at least 72 hours. This methodology uses a generic screening of systems and component seismic ruggedness and does not require probabilistic calculations (EPRI NP-604-SL).

It is a widely held belief in the technical community that engineering methods and design requirements are substantially conservative and that the resulting seismic margin can accommodate ground motions well above safe shutdown earthquake (SSE) levels. The utility industry and the Nuclear Regulatory Commission (NRC) have searched for ways to quantify the margin without requiring traditional reanalysis of the entire plant or a seismic probabilistic risk assessment. Data from actual earthquakes and shake-table tests form the basis for conducting cost-effective evaluations.

These guidelines include systems- and experience-based seismic ruggedness screening guidelines, detailed plant walkdown procedures to verify component screening, and analytic methods for evaluating unscreened components. These methods are used to determine the seismic capacity for which there is a high confidence of a low probability of failure (HCLPF).

In particular, the Fukushima accident highlighted the importance of considering the quality of water available for use in the event of a severe accident so that proper and timely decisions can be made during the management of the accident. There is substantial value in injecting water, even if it is sea water or from another raw water source, to re-establish cooling of a damaged core or of core debris. The Fukushima accident also focused attention on the potential for a severe accident involving the spent fuel pool. EPRI's reports address this issue as well as the issue where external events pose a special challenge because they can constitute a common cause of failure of plant systems. Although the nature of the actions that might be taken to manage a severe accident initiated by an external event is not fundamentally different, such an event might lead to loss of a broader range of plant systems. These guidelines also address considerations related to accidents that might challenge more than one unit at a multiple-unit site. This includes the ability to identify unique challenges to containment integrity, and in particular, possible pathways involving failure of containment isolation. In addition, the Fukushima accident also focused further attention on the management of hydrogen generated during a severe accident.

EPRI has also prepared detailed documents and procedures for investigation of strategies for mitigating radiological releases in severe accidents. The work evaluated strategies that maintain or enhance the containment function in scenarios involving long-term loss of electric power. The strategies evaluated include water injection (by flooding or spraying), alternative containment heat removal, venting, controlled venting, filtered venting, and combinations. Key insights from the recommendations include the following:

- No single strategy is effective.
- Active core debris cooling is required.
- Existing severe accident management guidelines (SAMGs) strategies provide substantial benefit.
- Spraying the containment atmosphere is beneficial.
- Venting prevents uncontrolled release and manages hydrogen.
- Control of the vent provides benefit.
- Low-efficiency filters can further reduce radionuclide releases.

Seismic Protection—Generation seismic protection is similar to substation protection and generally focuses on:

- Installation of new concrete flood walls isolating tunnel openings.

- Temporarily seal tunnel openings with steel plates and gaskets.
- Seal all manhole covers with steel plates and gaskets.

Nuclear Power Plant Load Following

Nuclear power plants are typically base-loaded because this is the most economical and technically least challenging mode of operation. This picture is changing—there is experience with load-following operations in countries that have a large portion of their power supplied by nuclear power plants (France and Germany) and an increasing demand in other countries due to renewable energy sources. Using nuclear power plants in load following will require a careful review of fuel performance as well as the consequences of increased aging of components due to pressure/temperature transients (fatigue) and gradients as well as changes in mechanical loading resulting in localized stresses and vibration. Before nuclear plants can be used for load following, the capability of existing instrumentation and control (I&C) systems to monitor and control core performance, turbine/generator controls, pressure temperature transients, and vibrations needs to be assessed. There may also be challenges to plant safety and safety margins resulting from a reduction in margins due to equipment aging and fatigue and the limitations of existing probabilistic risk assessment PRA models to quantify affects of load following. Load following may also result in potential increases in plant trips, challenges to operators, and a reduction in fuel burn-up.

Countries that have a large share of their electrical power supplied by nuclear power plants have experience with load following—both the pre-scheduled variable mode of operation and primary or secondary frequency control. Operating experience exists in France, Germany, Slovakia (Russia), and Belgium with load-following operations. In these countries, load following is a part of their operating procedures, and the regulatory body has established safety limits included in their licensing basis. Typically, the ramp rate is maintained at 1 to 1.5% per minute but can be as high as 5% per minute. In France, the operating band is 30 to 100% power with the plants typically kept above 50% power, and the grid operator is allowed to maneuver the plant +/- 50 MWe.

Severe Nuclear Accident Management

EPRI has developed extensive guidance for severe accident management at nuclear power plants. Guidance for responding to severe accidents was first developed as a result of insights gained from examining the Three Mile Unit 2 accident that occurred in 1979. To support effective severe accident management guidance, EPRI assembled an extensive description of the phenomena that could be relevant for a core damage event and, based on those descriptions, identified potential actions that could provide effective response to an accident. The actions and the technical information that supported their definition were then used as the basis for severe accident management guidelines (SAMGs) that were eventually put into place at operating nuclear power plants. These guidelines were updated as a result of the March 2011 accident at the Fukushima Daiichi plant (EPRI 1025295).

Silicon Carbine Cladding and Channels

The Fukushima accident has renewed interest in potentially safer nuclear fuel designs. One option being considered is the use of silicon carbide (SiC) as a cladding material, which has been under evaluation for many years by various organizations. The high-temperature properties of SiC offer many advantages over conventional metal-based fuel materials, especially under loss-of-coolant-accident (LOCA) conditions. The structural and chemical stability of SiC at elevated temperatures offers several benefits, such as the elimination of hydrogen production and maintenance of fuel integrity for longer durations in a LOCA event.

One possible nearer-term application for SiC is as a channel material in boiling water reactors (BWRs). Nuclear fuel in BWRs is typically placed inside a long square channel typically made from a zirconium-based alloy. The channel functions to hydraulically isolate the fuel and to provide support to control rods. The use of SiC in this application could eliminate a phenomenon called "channel bow" that is currently experienced by metal-based BWR channel boxes. If not detected and addressed, channel bow can potentially lead to control rod insertion issues and reduced shutdown margin. This application may be more promising than SiC cladding because a hermetic seal is not required for proper functioning of the component. EPRI is fabricating several test articles to evaluate fragmentation resistance and properties under irradiation conditions. The irradiation testing is to be conducted in cooperation with the Idaho National Laboratory.

Use of Advanced Water Technologies

There are a number of advanced water technologies that have the potential to reduce water needs for power generation. Water management technologies are being developed that have the potential to reduce water withdrawal and consumption at minimal impact on plant heat rate, as well as to reduce pollutant discharges at lower cost and complexity than current options. These technologies can be categorized into the following areas:

- Power plant wastewater treatment
- Advanced cooling
- Water conservation and recycling

This assumes the availability of water for thermoelectric power generation is important to secure the resilience of power generation; a variety of technologies have the potential to improve the water use at power plants. One such technology is a Thermosyphon Cooler (TSC) Hybrid Cooling System. The TSC unit itself is a dry-heat-rejection technology. It is composed of a shell-and-tube evaporator, air-cooled condenser, and interconnecting piping. This technology places a natural convection-driven refrigerant loop between a shell-and-tube evaporator and an air-cooled condenser. The TSC can be utilized either in series, upstream of a conventional cooling tower, or possibly even in lieu of a conventional cooling tower at a power plant (see Figure 6-7) (EPRI 1026528).

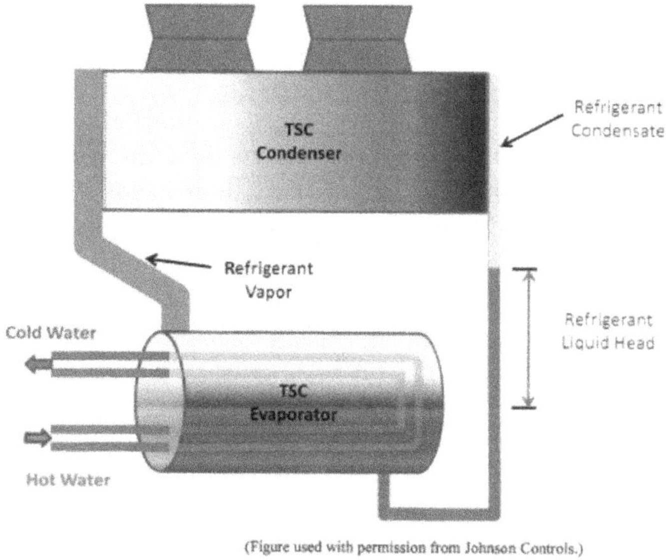

(Figure used with permission from Johnson Controls.)

Figure 6-7. Flow Diagram for the TSC (figure used with permission from Johnson Controls)

Research is currently underway to:

- Determine how much water can be saved by operating a Thermosyphon Cooler in series with a conventional wet cooling tower to reduce the temperature of the power plant's hot return water.
- Determine the energy penalty incurred.
- Compare the TSC Hybrid Cooling System concept to that of a conventional cooling tower system.

To accomplish these objectives, the performance of a TSC pilot unit in series with a pilot- induced draft-cooling tower is being evaluated. The pilot TSC/cooling tower water discharge temperature is being controlled (speed control on the TSC and pilot cooling tower fans) to match that of the host Plant Bowen Unit 4 full-scale natural draft cooling tower. The annualized reduction in water consumption and energy penalty will be quantified under these controlled conditions.

Managing the Interdependency of Natural Gas and Electricity Systems

Historically low natural gas prices, continued demand growth, and increased use of variable resources such as wind and solar using responsive, load-following gas generation have increased the power sector's use of natural gas for generation. And this trend is expected to continue for some time. Differing operational philosophies, regulatory regimes, and market structures have already led to incidents where scheduling, coordination, and resource availability threatened electric system reliability.

Innovation is needed in resolving differences in market timing between gas and electric system operations. Electric generation is bid without fuel cost certainty. From a gas perspective, there is gas system nomination and quantity uncertainty. The gas day is 10:00 AM to 10:00 AM, and the electric day is 12:00 AM to 12:00 AM.

Day-ahead assumptions can be wrong due to:

- More electrical demand realized than expected
- Unforeseen transmission line outages
- Generator trips

All scheduling assumes fuel availability. Day-to-day communication is needed that should define:

- How much gas is nominated?
- How much gas was already burned?
- How much alternative fuel is available?
- How much alternative fuel is on site and how much time is needed to refuel.

Transmission

In general, monitoring of transmission assets is more cost-effective and beneficial than any other asset class (EPRI 1026528). Although transmission lines are one of the critical core backbone elements of the power grid, thousands of miles are unattended and not monitored in any way. Transmission lines have seasonal ratings that need to be considered by operations and planning. For the most part, there is little if any real-time monitoring other than at substations that provide operators with loading information.

On the transmission system, actions can be taken to enhance resiliency in the areas of underground transmission systems, energization of transmission systems, protection from flooding, relays, protection from tsunamis, standby and backup power generation, stored and shared spare components, the recovery transformer, system restoration, optimal black-start capability, substation seismic protection, equipment shielding, and grid hardening and EMP mitigation.

Underground Transmission Systems

In particular, there is an increasing need and public pressure to place a greater portion of new transmission lines underground. Applications include:

- Dense urban areas
- Congested substation exits
- Expanding airport runways
- Sensitive environmental areas
- Long water crossings
- Public objections

An increasing number of power transmission lines are planned to meet energy supply and demand and to improve reliability of the power grids. While overhead transmission lines represent more than 98% of the total transmission circuit length worldwide, there are significant instances where lines must be placed underground (for example, for long water crossings and in dense urban areas) in spite of the increased cost of doing so in most cases. In a growing number of instances, electric utilities face the challenge of evaluating the relative merits of and choosing between overhead and underground transmission line alternatives.

A variety of factors must be taken into account to arrive at a sound engineering-based decision in planning for and building a new transmission line. However, there is no industry-established framework for identifying the relevant factors to assess or quantify their impact and to guide the planner.

Dynamic Thermal Circuit Rating

Beginning in 1991, EPRI began development of the Dynamic Thermal Circuit Rating (DTCR) software to provide utilities with a tool that could be used to optimize the capacity of major transmission equipment such as overhead transmission lines, power transformers, and underground cables. The focus of DTCR is to help utilities realize more power transfer capability through existing transmission equipment with the intent of deferring or, in some cases, avoiding altogether the capital costs for new circuits or upgraded equipment. DTCR proved to be a valuable tool and was applied by a dozen or so utilities, often with detailed engineering support from EPRI (1025003).

Energizing Transmission Systems

Transmission lines have to be energized in the correct order based on:

- Testing
- Critical loads
- Maintaining balance between load and generation
- System stability
- Material inventory
- Restoration drills for major storms is a key to improvement

Additional prevention can be accomplished by clearing more uphill right of ways (ROW) in highly sloping terrain to avoid tree fall-ins; to use more steel instead of wood structures for new/replacement work; and to employ greater sectionalizing of lines to reduce exposure and help restoration. Additional recommended actions include planning for no communications service for 48 hours, and assessing the design flood elevation for each generating station and establish new design levels, as appropriate.

Protection from Flooding

Protection from flooding of transmission assets can be accomplished in a variety of ways including reinforced design of towers, substations and underground systems and equipment. Specific actions may include increasing flood walls, adding to spare parts inventory, increasing the use of submergible transformers and switches, and use of weatherproof enclosures. The following are included:

- Raise moat walls and concrete flood walls at substations to isolate tunnel openings.
- Install new concrete moats around critical equipment.
- Seal all conduit and cable penetrations from trenches or raceways into critical areas.
- Install submersible pumps for dewatering purposes.
- Seal all manhole covers with steel plates and gaskets.
- Raise existing moat walls.
- Install new concrete moats around critical equipment.
- Install removable flood gates or barriers at the station perimeter.
- Seal all conduit/cable penetrations from trenches into critical areas.
- Install submersible pumps for dewatering purposes.
- Seal all conduits, panels and trenches. Assure gaskets on panels are intact.
- Safely store and maintain flood pumps.
- Shrink-wrap all cabinets in flood zones with boat wrap material.
- Store emergency supplies remotely
  - Electric sump pumps
  - Electric sump pump accessories
  - Mobil diesel pump and accessories
  - Industrial shrink-wrap fabric
- Deploy materials before an event
- Establish a corporate emergency response center.

Figure 6-8. Depicts three examples of moats and barrier installed at power system facilities after superstorm Sandy. (Courtesy Consolidated Edison Company of New York)

Relays

Replace electromagnetic (EM) relays with microprocessor-based relays. Standardize relay protection schemes. Incorporate an alternative setting group with microprocessor-based relays. Develop a means to automate the switch over to the alternate "storm" settings group.

Protection from Tsunamis

Actions to protect power plants and substations from tsunamis can include belts of trees and mangroves to provide barriers to wave "run-up;" offshore location of cooling water intake structures; use of design principles that inhibit the scouring or erosion created by waves, and prevent damage from the debris left behind.

Standby, Backup Power Generation

The design and manufacture of storable and easily transportable backup power generators is required for restoring HILF-damaged facilities. Government and power industry stakeholders could warehouse the hardware in hardened and geographically dispersed locations for rapid transport to affected areas to restore power quickly to any section of the grid affected by HILFs. (EPRI 3002001936)

Stored and Shared Spare Components

Another essential component in the recovery process is the availability of components needed in the event of an emergency. These components could be shared among all stakeholders and geographically dispersed for quick access. The industry can assess the vulnerability to HILF exposure of various components and systems and the inventory of available equipment to plan sufficient quantities of backup components where needed. Transformers, generation equipment, Supervisory Control and Data Acquisition (SCADA) systems, aboveground relays and controls, and other embedded processors are examples of potential spare part candidates for stockpiling.

The goal of a NERC Spare Equipment Database (SED) program is to provide a means to securely connect entities that need replacement transformers with entities that have such spares available in the event of simultaneous loss of multiple transformers (NERC 2011). The Edison Electric Institute's (EEI) Spare Transformer Equipment Program (STEP Program), launched in 2006 in response to the 9/11 terrorist attacks, addresses the need to manage resources in response to a terrorist attack. About 50 transmission providers that represent about 70% of the

U.S. transmission grid currently participate in this program (FAMA 2010). Defined stakeholders could take responsibility for the inventory, research, and securing of hardened facilities for these parts depots, including a coordinated and clear command and distribution network for part deliveries and repairs. Realizing this vision would require cost and feasibility studies, as well as research that identifies the most vulnerable components under different HILF scenarios. The fact that many components are now manufactured overseas should also be taken into account as a further complication.

Recovery Transformer

A consortium of the U.S. Department of Homeland Security (DHS) Science & Technology Directorate (S&T), transformer manufacturing company ABB Inc., CenterPoint Energy Inc., and EPRI has developed a prototype transformer that could be deployed to replace a damaged or destroyed transformer in about a week instead of several months to prevent sustained power outages (see Figure 6-14). Known as a "Recovery Transformer," the prototype is being tested at a CenterPoint Energy substation. A large distribution substation transformer can be taken out by disasters such as earthquakes or tornados, geomagnetic disturbances, internal failure, or by terrorist acts. Success so far demonstrates the concept of modular transformer design and accessible substation design to enable rapid deployment of replacements necessary for recovery op-

Figure 6-9. Recovery Transformer—(Photo Courtesy of DHS S&T) System Restoration

erations. Future research could extend this work to other distribution components.

To prepare for the possible need to mount a restoration of service, the following need to be considered. (NAS 2012)

- Black-start capability that supports the ability to supply limited amounts of power to generators and other power-system equipment before they can be brought back on line.
- Line/cable charging strategies and other means of voltage and reactor power control.
- Need to disable or adjust certain protective systems, such as those for under-voltage, under-frequency, and synchronization checks.
- Development of restoration policies including islanding requirements and monitoring of voltages, frequencies, and phase angles.

Setting aside severe weather events, in recent years, several catastrophic and cascading failures of power systems have been seen throughout the world (e.g., the August 14, 2003 blackout in USA and Canada; the August 28, 2003 blackout in London, UK; the September 23, 2003 blackout in Sweden and Denmark; the September 28, 2003 blackout in Italy; and the May 24 and 25, 2005 blackout in Moscow, Russia). Power system restoration following a complete or partial outage is a highly complex decision and control problem. System operators in the control center, working with field crews, re-establish the generation and transmission systems, pick up loads and restore the services, and return the system to a normal operating condition. A general environment of power system restoration is illustrated in Figure 6-10.

System restoration is largely a manual task at present, and few decision support tools are available for system restoration planners and system operations. System planners develop and update restoration plans based on simulation results and accumulated knowledge. These plans developed off-line are used as guidelines for operators in an on-line environment. Operators need to adapt to the actual outage scenario and available resources, and be able to develop the strategy based on their experience. On the other hand, restoration strategies are closely related to the specific system characteristics. Since one system's situation does not conform readily to that of another, restoration strategies cannot be generalized without some degree of customization.

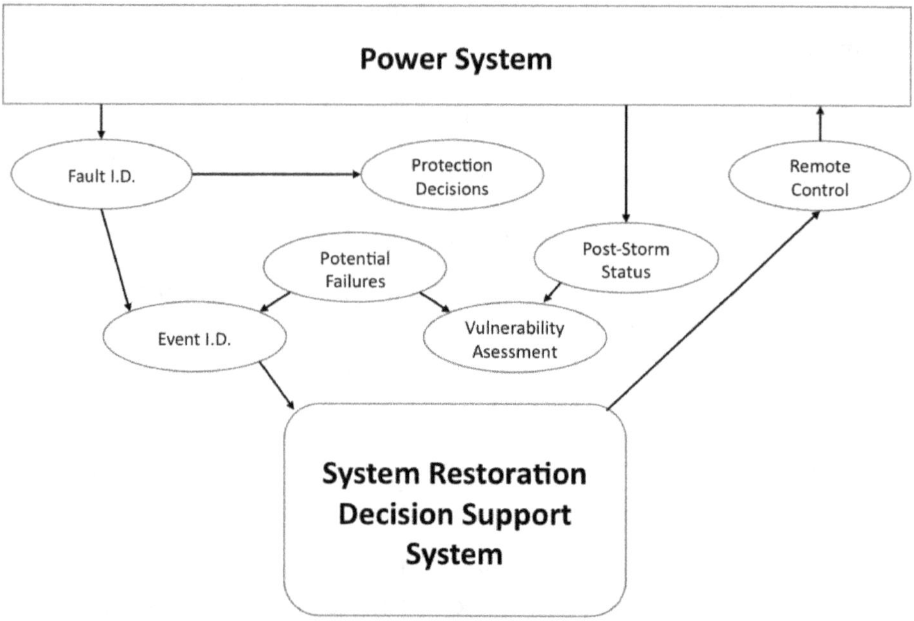

Figure 6-10. System Restoration Environment

Generic Restoration

A decision support tool is needed to assist power grids in system restoration planning, operator training, and ultimately, in an on-line restoration environment. EPRI has developed the Generic Restoration Milestones (GRMs) concept and a comprehensive methodology for power system restoration based on GRMs through multi-year project efforts. EPRI has been working on prototyping a decision support tool for evaluating system restoration strategies. Continuing efforts have been carried on through this project to further develop the tool, demonstrate the capability, and investigate other restoration related issues (EPRI 1020055).

System planners can use this interactive tool to develop and evaluate system restoration strategies. Enhancement of the tool will be performed to incorporate realistic system constraints and user feedback. By utilizing the developed tool, preferred level and location of blackstart capability can be studied. In the future, this tool can be integrated with Operator Training Simulators to help participants in system restoration training gain knowledge and experience in safe and effective power system restoration procedures.

EPRI has developed a computational algorithm to estimate the total restoration time and the system generation capability during the restoration horizon. The optimal installation of blackstart capability is formulated as a Mixed Integer Bi-level Linear Programming problem. Optimization modules have been developed and tested for generator startup sequencing and transmission path search, taking into consideration the time to take Generic Restoration Actions. The optimization tool to find the optimal locations and capacity levels for new blackstart units is a systematic solution to ensure that the additional blackstart units will help to maximize the benefits and minimize the costs. The objective of optimization is for the existing and additional blackstart units to reduce the total system restoration time and increase the total system generation capability during the system restoration period.

Optimal Black-start Capability

The Optimal Black-start Capability (OBC) tool has the following intended functions:

- OBC is a decision support tool to evaluate the blackstart capability based on currently available blackstart capable units.
- OBC is a decision support tool to evaluate optimal locations and capacity levels for installation of new blackstart units.
- Through its optimization capabilities, OBC can be used to achieve a reduced total restoration time and increased total system generation capability after new blackstart units are installed at the optimized locations and amounts.
- OBC's graphical user interface allows the user to input system data easily, displays the results graphically, and provides a text summary.
- The OBC user can save a system restoration plan in a series of *.dat files.
- All restoration actions and steps are saved as logged data.

In identifying the optimal location for installation of a new blackstart capability, the OBC tool considers only the power system needs and does not consider associated environmental or siting needs and/or constraints. As such, the OBC might identify a benefit to locating new blackstart generation at electrical bus for which there would be siting or environmental restrictions that would prevent such an installation. The generators that are typically utilized for blackstart service are 10 to 50 MW small hydro or gas turbine units and, in a few cases, 200 to 400 MW steam units.

Black-start units can be categorized into three typical types of generators (e.g., hydro, diesel and gas turbine units):

- **Hydroelectric generating units**: These generators need little initial power to open the intake gates and have fast response characteristics to provide power to start fossil-fueled or nuclear stations.
- **Diesel generating units**: Diesel generators usually require only battery power and can be started quickly to supply the power to start up larger generating units. They are small in size, and generally cannot be used to pick up any major transmission system elements.
- **Gas turbine generating units**: Aero-derivative gas turbine generators can be started remotely with the help of local battery power. Large gas turbine generators are coupled with on-site diesel generator sets that are started and used to energize plant auxiliary buses and start either the gas turbine or steam turbine. Gas turbine generators can be started and pick up load in a short time. The time required to restart a gas turbine and the subsequent available ramping capability are functions of the duration the unit is off-line.

Substation Seismic Protection

Utilities use the Institute of Electrical and Electronic Engineers (IEEE) Standard 694, Recommended Practice for Seismic Design of Substations to qualify substation equipment for seismic movements. Deficiencies exist in the present standard, and information is unavailable for dynamic response that may be used to better analyze equipment and permit their evaluation in case of limited configuration changes, such as insulator substitution.

EPRI is launching a project to perform tests to assess seismic issues. The project intends to address deficiencies of the present standard, especially those related to details left unspecified, by performing tests in the laboratory.

Equipment Shielding

In one example of shielding, the U.S. military, specifically under the MIL-STD-188-125 standard, deploys technology for protecting fixed and mobile installations from EMP effects. Ground-based shielding detailed in MIL-STD-188-125 refers to "subscriber terminals and data processing centers, transmitting and receiving communications stations and relay facilities for both new construction and retrofit of existing facilities." This standard calls for EMP mitigation for all military command, control, communications, and computer systems, including the electromagnetic energy point of entry to these systems. Adhering to this standard would greatly enhance the protection of North American power grid reliance on its current and future SCADA operations, including smart grid components and systems, but would be extremely costly (DoD 1998).

Grid Hardening and EMPs Mitigation

Protecting the entire power grid system or even all high-value components from HEMP pulse attacks is impractical because the costs, time and effort in proportion to the likelihood of a HEMP-enhanced nuclear strike make it prohibitively expensive. Three approaches are possible in addressing this threat: shielding, standby and backup equipment, and stored and shared components.

- Although infrequent, physical attacks do occur on the power system. They range from intentional collapsing of transmission towers to use of guns to damage insulators and substation equipment. It is impossible and likely uneconomic to completely protect a power system from a determined terrorist. However, there are several actions that could be undertaken including (NAS 2012):
- Engineering design of power systems in a way so as to withstand simultaneous multiple outages.
- Use of monitoring and protection systems to mitigate any impact of an attack.
- Mechanisms to enhance the "graceful degradation" of the system in the event of an attack.
- Mechanisms to add redundancy to the power systems such as using loop flow in distribution systems and employing increasing installations of distributed generation.

On-Line Monitoring of Load Tap Changers

Development of on-line monitors can increase the resiliency of transformers with load tap changers (LTCs). One of the first signs of a problem with an LTC is an increase in the concentration of ethylene in oil. Chemically, that usually indicates overheated oil and is often due to high resistance across load-bearing contacts caused by the accumulation of carbonaceous deposits (coking) between the contact surfaces. A corresponding drop in the acetylene concentration frequently precedes this, which may be due to the acetylene being consumed in the chemical formation of the coking deposits (EPRI 1025003).

Hydrophobic Coatings—Hydrophobic coatings are now being applied to various components in the transmission and distribution system (see Figure 6-11). By helping components shed precipitation, these coatings mitigate water damage on non-ceramic insulators and can facilitate ice removal. Projects are under way at EPRI to assess the performance, reliability, and application of these coatings. (EPRI Program 35)

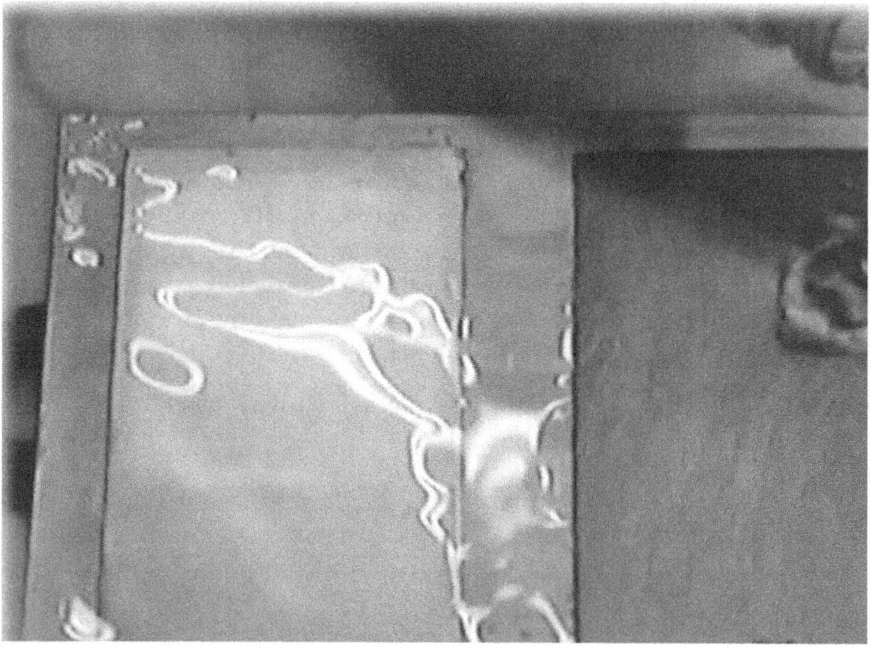

Figure 6-11. Hydrophobic Coatings

For information on this project, click see http://portfolio.epri.com/ProgramTab.aspx?sId=PDU&rId=233&pId=7298&suId=7318

Hydrophobic Coatings

Hydrophobicity is the ability of a surface to repel water so that when it is wetted, small droplets form rather than sheets of water. The hydrophobic properties of materials have been used for decades on transmission line and substation insulation to improve contamination flashover performance. For example, silicone rubber composite insulators or room temperature vulcanized silicone coatings have been used. High hydrophobicity ensures that a continuous film of water does not form on the surface, breaking the leakage path and consequently reducing, or preventing, currents flowing on the insulator surfaces. This, in turn, reduces dry band arcing and subsequent flashovers.

Icephobicity is the ability of a surface to facilitate the removal of ice or slow down its formation. When water comes in contact with a surface temperature below freezing, ice may form. In the case of overhead conductors, if enough ice forms, the mechanical load from the ice and prevailing wind may result in a mechanical failure. With insulators, the ice may bridge all the sheds, and although ice is a relatively good insulator, flashovers can result when it melts.

The application of superhydrophobic and icephobic coatings may have several advantages in improving transmission system reliability. They could potentially be applied to:

- Insulators for two reasons: (1) to mitigate insulator contamination—this could be particularly valuable in coastal and industrial areas; (2) to mitigate insulator icing.
- Conductors and structures to reduce ice buildup and to facilitate ice removal.
- Mitigate conductor corona.

Energy Storage

Energy storage has the potential to transform the electric power infrastructure by enhancing resiliency. This can be accomplished by (1) using storage technology that facilitates the integration of variable energy resources such as wind and solar and 2) improving the capacity factor or utilization of the transmission, distribution, and conventional generation. The power system has traditionally operated on the basis of nearly instantaneous adjustment of the production of electricity to match demand. The industry has constructed infrastructure to meet peak demand with adequate safety margin. In recent decades, the "peakiness" of electric demand has increased such that on average the utilization of the grid infrastructure has decreased. The growth of variable resources in the generation mix adds to the utilization problem. Much conventional generation would still be required for times when wind and solar generation is not available, but would be used on average less and less, resulting in growing inefficiency of the overall grid.

In order for energy storage to be available at a cost and performance to enable wide-scale deployment, the following innovation in research, development and demonstration is needed:

- Ongoing basic electrochemical research aimed at exploring the potential for energy storage based on new electrochemistries and their practical realization.
- Detailed studies of the effects of higher penetration of renewable sources on grid operations and the permanent retirement of a large percentage of traditional generation.
- Energy demonstration storage projects of new technologies to technology readiness level (TRL) 7-8 that expand the use of storage for grid performance enhancement.

Distribution Systems

While a small percentage of electricity customers are served directly from the transmission system, the vast majority of the 65 million customers in the U.S. are served by the distribution system, which is comprised of a complex network of substations, lines, poles, metering, billing, and related systems to support the retail side of electricity delivery.

Utility distribution systems are challenged by an aging infrastructure, conventional designs, and increased demands for digital-quality power. Compared to the transmission system, the greater complexity, exposure, and geographic reach of the distribution system results in inherently lower reliability, reduced power quality, and greater vulnerability to disruption of any kind. Using a reliability measure of average total duration of the interruptions experienced by a customer in a year, over 90% of the minutes lost by consumers are attributable to distribution events. In 2004, EPRI estimated that a fully automated distribution system could improve reliability levels by 40%. Advances in smart grid technologies may prove that estimate to be overly conservative. From a reliability standpoint, feeder automation by itself only deals with outage symptoms, not root causes. For example, it is not a substitute for good maintenance practices, inspection, and vegetation management.

An urban utility may have less than 50 feet of distribution circuit per customer, while a rural utility can have more than 300 feet of primary distribution circuit per customer. Assuming a rough average 100 feet of line for each of the 165 million U.S. customers indicates the U.S. has an installed base of more than 3 million miles of distribution line.

Hardening the distribution system to prevent damage will require changes in design standards, construction guidelines, maintenance routines, inspection procedures, and recovery practices, and will include the use of innovative technologies. A utility's approach to all of these will be determined more specifically by its distribution system and work environment.

System planners have always had the option of configuring the system to maximize distribution service during major outages. Historically, however, the associated costs have not been viewed as prudent.

Several actions can be taken to prevent damage to the distribution system. These actions include vegetation management, targeted undergrounding, overhead distribution reinforcement, enhanced cyber security, load reduction, and restoration management.

Vegetation Management

Vegetation management by tree trimming and use of herbicides is a fundamental management Tree trimming is a fundamental practice for mitigating distribution outages. Vegetation management often faces strong public opposition at least until major storm-related outages occur (see Figure 6-12). Utilities have found it effective to use the results of storm damage and subsequent restoration as part of their assessment of their vegetation management and tree trimming programs. The application of storm data allows a critical review of trim specifications, damage, and clearance standards, while facilitating regulatory enforcement of tree trimming rights (EPRI Program 51).

Targeted Undergrounding

Installing distribution lines underground protects them from trees, collisions with cars, and lightning strikes. However, the cost can be prohibitive—10 times more than overhead distribution—and is not 100% reliable. Severe storms may also bring flooding, and many underground devices, such as network protectors, are not designed to operate un-

Figure 6-12. Vegetation Management can prevent extensive damage

derwater. Once the water recedes, troubleshooting and restoration may actually take longer for underground than with overhead distribution systems.

An effective option may be to bury selected portions of the system as an effective hardening investment strategy. Selective undergrounding would address the common issue that, in many cases, tree damage in remote areas of a feeder can prevent energizing the entire feeder. Selective undergrounding goes against the grain of public perception of undergrounding main trunk lines that are mostly visible. But, in storm conditions, if crews are able to concentrate on the three-phase lines along roadways, large blocks of customers can be restored quickly. (EPRI Project 180C)

Reinforcing Overhead Distribution

Some of the most effective actions are relatively simple and straightforward, such as adding structural reinforcement to existing distribution lines. Examples include adding guy wires or using steel poles to increase the strength of the lines to withstand higher wind loading. Some utilities now routinely make these improvements during system restoration. Another effective action is to reinforce some lines leading to a population center. An example would be to design two of four lines serving a population center to withstand higher winds (EPRI Project 180I).

Load Management

During outages, an established industry practice is to use load reduction programs to reduce demand on the system from customers who still have service. This practice may be necessary when transmission outages occur and when reconfiguration of distribution circuits is needed. Although this practice may reduce demand only 5%-15%, it can be useful in some situations to aid recovery. Options include:

- Media appeals using both conventional and social media
- Voluntary load-reduction programs
- Interruptible tariffs
- Demand-response programs such as direct load control, time-of-use rates, etc.
- Conservation voltage reduction through distribution automation

Restoration Management

Utility restoration management practices include procedures and systems to shift from centralized to decentralized restoration management. These practices must be tailored to events of different magnitudes, and to differing types of disruption (wind, flood, tsunami, earthquake, and terrorist or cyber attacks etc.).

Pole and Line Design

Certain pole and line design configurations are less susceptible to damage from trees and falling limbs (see Figure 6-13). Although hardening of the system is the intuitive solution, significant interest exists in better understanding the way that overhead systems fail and innovating new technologies to ensure that the systems fail in a manner that minimizes the restoration effort. Research is under way to assess overhead components and crossarms, pole treatment options, and the effect of third-party components (telephone, CATV, etc.) (EPRI Project 180I).

Dynamic Circuit Reconfiguration

This operational design offers the opportunity to combine advances in information technology, communications, and sensors with innovations in restoration practices (see Figure 6-13).

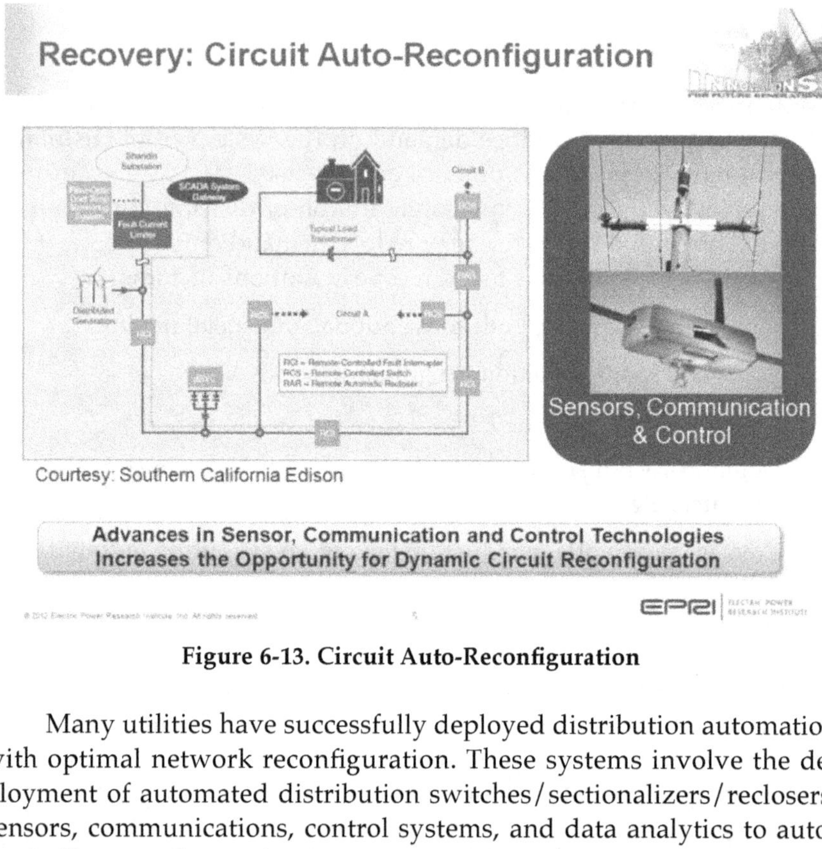

Figure 6-13. Circuit Auto-Reconfiguration

Many utilities have successfully deployed distribution automation with optimal network reconfiguration. These systems involve the deployment of automated distribution switches/sectionalizers/reclosers, sensors, communications, control systems, and data analytics to automatically reconfigure circuit connections in order to maximize service restoration. Once deployed, these systems serve two roles. First, they limit the number of customers affected by faults on the feeder mains by directly tripping (reclosers) or sectionalizing (switches or sectionalizers). Second, once the fault is isolated to one section of the feeder, the systems enable restoration of service to unaffected sections from adjacent feeders. Additional research is needed to ensure compatibility with new technologies such as distributed resources (PV, storage, microgrids, etc.) (EPRI Project 180A).

Damage Assessment and Crew Deployment

Proper resiliency planning must provide for rapid damage assessment and crew deployment. To enhance the ability to respond and recover quickly, the U.S. electric industry has developed effective mutual assistance programs, in which transmission and distribution utilities call in crews from across a region to help restore downed lines, poles, and transformers. The host utility needs to provide logistical support in the form of lodging, food, and fuel; must have on hand the necessary spare parts; and must have the capability to efficiently dispatch crews to the affected areas.

Recovery requires prompt damage assessment, access to damaged assets, and readily available replacement components. In the recent storms in the northeastern United States, pinpointing affected areas was problematic, as was routing crews through streets that were not blocked by fallen trees. As a result, crews were sometimes idled, because they could not reach affected areas.

Airborne Damage Assessment

EPRI recently completed preliminary tests showing that both small piloted aircraft and unmanned aerial vehicles (UAV) or drones equipped with high-resolution cameras, global positioning systems, and sensors can be valuable tools for damage assessment. UAVs equipped with EPRI's Airborne Damage Assessment Module (ADAM) can be small and light enough to be handled by a technician, and can quickly survey devastated areas that are difficult to reach by roads blocked with downed trees or other obstacles. The use of ADAM-equipped aircraft could substantially reduce costs and cut the response time by hours, if not days. It could also aid in assessing system conditions in normal situations.

Implementing ADAM requires integration with three key distribution operations backend systems: the outage management system (OMS), the Geographical Information System (GIS), and the asset management system. These systems are important for enabling incident command operations to understand and access situational information in real time. They also support overall distribution management.

Drones can be used effectively to assess storm damage on utility distribution and transmission systems. Drones can also be used in routine surveys of electric power equipment and support of construction and repair.

However, the Federal Aviation Administration (FAA) and federal aviation rules have failed to keep pace with this new technology. Except for a few hundred waivers and exemptions granted to some commercial operators and governmental jurisdictions, the FAA has imposed a virtual lockdown on commercial drone use. San Diego Gas & Electric is the only electric power utility in the nation to have received a waiver from FAA to test drone use in responding to emergencies and routine inspection of electric and gas lines in remote areas.

Outage Management System (OMS)

At the heart of any storm response and damage assessment is the outage management system, which gathers outage information from a variety of sources and helps to predict outage locations and direct restoration efforts. An OMS enables a focused response, allowing utilities to prioritize restoration efforts, re-energize restored feeders, give customers and the public estimates of restoration time, and coordinate with governmental organizations.

Geographical Information System (GIS)

GIS is rapidly becoming the foundation of distribution system documentation, because the map-based database is suited perfectly to tracking assets and monitoring the state of geographic dispersed assets.

Asset Management System

As with any industry, keeping track of assets is crucial for utilities, because this management provides an interface between the engineering and the accounting sides of the business. An asset register allows utilities to maintain a database of what assets they own, their predicted lifecycle, and their technical specifications. [35, 36].

Field Force Data Visualization—Data visualization technology for utility engineering and field operations are under development. Using computer tablet and smart phone technologies, these systems would allow real-time data retrieval when a user points the mobile device at a distribution pole or at transmission and distribution conductors. As an example of this application, a field technician would aim the hand-held device camera at a pole structure, while the screen displays the camera image, with the one-line circuit drawing overlaid, and all of the pertinent asset data available from the GIS (see Figure 6-14). Recovery crews

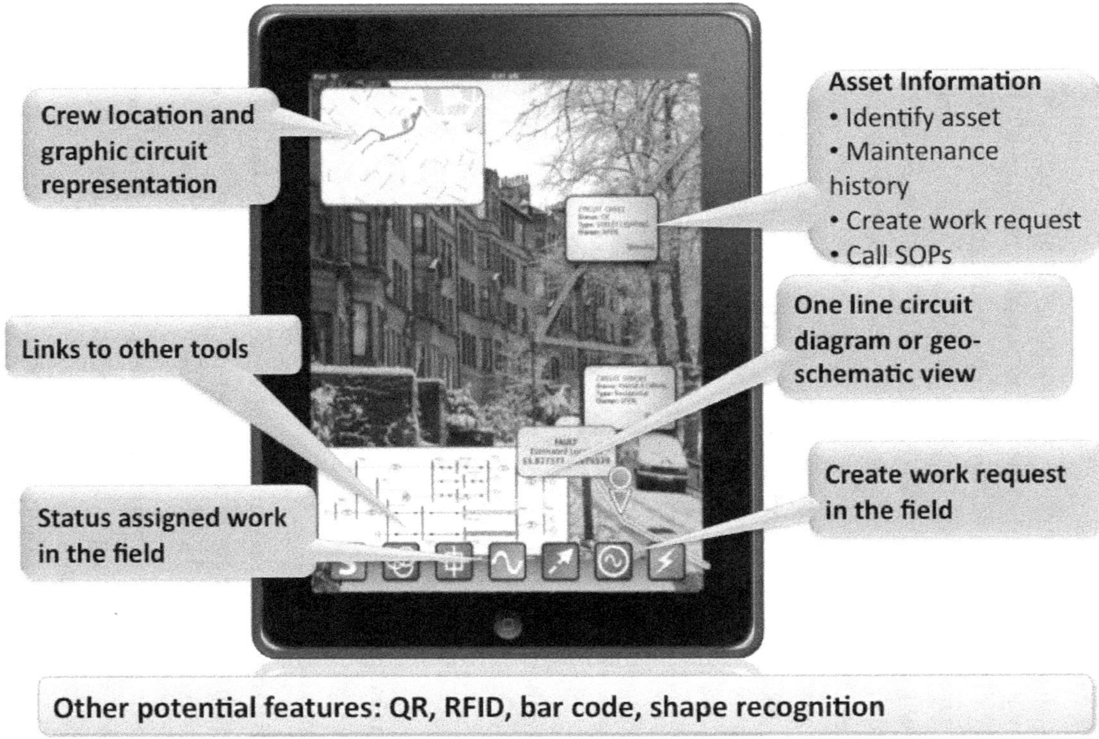

Figure 6-14. Enabling the Field Workforce (EPRI)

can use such tools to retrieve detailed information on assets in the field, and to determine quickly which components should be dispatched for repair and replacement (EPRI 1023304).

Customer Systems

The smart grid allows the electricity customer to become fully integrated with the traditional electricity supply system. Such integration began decades ago with commercial and industrial accounts, but with more advanced and lower cost IT and communications technology, it is beginning to gain traction with residential customers. In 2008, FERC estimated that 4.7% of U.S. customers had advanced meters and that 8% of U.S. customers were engaged in some form of demand response (DR) program. These programs increased the nation's demand-response potential to 5.8% of peak demand by 2008, or more than 40,000 MW. DR potential is expected to grow rapidly over the next 20 years as AMI fully saturates the market.

There are more than 142 million customers in the U.S., of which 13% represent commercial and industrial accounts. The customer base is expected to grow 16% over the next 20 years to more than 165 million. However, since most consumer appliances will by then be DR-ready, the actual number of individual communication-connected end nodes will more than double.

Over the next 20 years, integrating the customer into the smart grid will:

- **Increase DR capabilities** and enable time varying rates to reduce costs, improve load factors, and optimize the economic performance of the grid.
- **Facilitate integration of distributed energy resources (DER)**, including a variety of customer-owned systems, such as rooftop photovoltaic (PV) systems.
- **Integrate the transportation system into the grid** via electric vehicles (EVs) and plug-in hybrid electric vehicles (PHEVs), which can provide the distributed means for large-scale electricity storage for the grid, balance daily load cycles, and substantially reduce the nation's oil dependency.
- **Improve energy management** in the home and commercial buildings, reducing peak demand and improving energy efficiency.

- **Optimize grid performance** by using the demand side as a resource to stabilize the grid, shave peak demand, and reduce the capital investment requirements for peaking generation.

Even as the industry maintains its focus on providing affordable, reliable, and environmentally responsible electricity, the industry must now also broaden that focus to include facilitating the survival of consumers and communities when the lights go out. The basics of this area of activity may include the need to inform and engage stakeholders and regulators in endorsing the use of resources to prepare consumers and communities to survive long-term outages.

In the recent blackouts in India, the use of local generation and uninterruptible power supplies prevalent in that developing country allowed it to recover more easily from outages than many Western countries.

Survivability: Helping Customers Cope

Helping customers cope or survive refers to the ability to maintain some basic level of electrical functionality to individual consumers or communities in the event of a complete loss of electrical service from the distribution system. The key elements of survivability include communicating with customers; using resilient technologies to supply critical infrastructures such as traffic signals, prisons, hospitals, and cell phones; and equipping and enabling consumers to use distributed generation. For distribution utilities, survivability is a new function—and one that will require new business models and innovation.

The concept of assisting customers with survivability features is relatively new to the electric industry. Historically, many customers such as hospitals, banks and data centers have assumed responsibility for their own survivability, relying on generators or uninterruptible power supplies (UPS), and occasionally alternative distribution feeds (custom service coordinated with the electric utility). Given the relatively rapid evolution of customer expectations, now is an opportune time for utilities to reassess their role in maintaining power to essential services as traffic lights, prisons, and hospitals, and in providing customers uninterrupted access to communications and modern conveniences.

As highlighted in reports regarding the October 2011 storms in Connecticut, customer frustration was exacerbated by not knowing an estimated time of restoration. (Customers consistently cite the importance of knowing the estimated time of restoration.) Utilities likewise were frustrated by a lack of accurate information as to which customers or groups of customers were without service.

Communicating with Customers

Utilities are beginning to use both the Internet and smart phones to enhance the targeting and speed of their communications. For example, Commonwealth Edison Company (ComEd) is among a growing number of utilities that facilitate externally facing outage reporting and communication. ComEd offers an interactive outage map on ComEd.com that customers can use to determine the location and scope of outages and to obtain estimated times of restoration. ComEd also offers a mobile application for iPhone® and AndroidTM devices that enable customers to:

- Report an outage and check restoration status
- View their account summary and history
- Manage payments, including budget billing, automatic payments, and one-time payments
- Report a meter reading
- Find a location to make an in-person payment

Community Energy Storage

In the future, the electricity enterprise may benefit from cost-effective and reliable bulk energy storage to help balance and optimize supply and demand of bulk power resources. Distribution-level energy storage offers potential advantages, such as enhanced reliability and support of self-sufficiency. EPRI is examining the functional requirements and storage technology capabilities for mitigating the effects of variable renewable generation, improving reliability, and facilitating service when the distribution system is damaged. Certain battery technologies may be approaching a stage of development at which they can be considered for distributed storage applications at a community or even individual customer scale (EPRI 1025118).

Using PEVs as a Power Source

Plug-in electric vehicles (PEVs), both all-electric and hybrid, could be used to supply energy to a home during an outage. Hybrid electric vehicles also could operate as a gasoline-fueled generator to provide additional standby power. Automakers are interested in the concept, but the technologies require further development.

Nissan Motor Co., Ltd. recently unveiled a system that enables the Nissan Leaf to connect with a residential distribution panel to supply residences with electricity from its lithium-ion batteries. The batteries can provide up to 24 kWh of electricity, sufficient to power a household's critical needs for up to two days.

Using Photovoltaics (PV) Systems as a Backup

Increasingly, consumers are installing rooftop PV systems to augment grid-supplied electricity. Usually limited by roof area and sized to meet an economically viable portion of the building's electrical needs, these systems cannot supply 100% of a residence's typical demand, nor do the systems, as currently configured, allow for operation as independent microgrids to supply part of a residence's needs. EPRI assessments have identified inverter and control designs that could convert PV systems into self-sufficiency technologies, but few inverter manufacturers have stepped forward to serve this need.

In 2011, two U.S. Department of Energy (DOE) projects demonstrated inverter products that can operate while tied to the grid or "islanded." Princeton Power, Inc. (Princeton) demonstrated a 100-kW demand response inverter with two DC terminals and two AC terminals. Princeton is also introducing a 10-kW version specifically for the residential and small commercial markets. Petra Solar demonstrated a micro-inverter-based PV system capable of operating in grid-tie or islanded mode. Both companies report that they have demonstrated seamless transition between these modes. Other companies report development of similar products. With a proper transfer switch (perhaps controlled by the utility), these products could provide backup power to residences or a community during an outage.

Matching Consumer Load to PV Capabilities

The existing controls associated with PV arrays are not able to match the electrical demand of a residence without presence of grid supply or local storage. Companies are developing residential circuit breaker panels that allow the control of individual circuits and appliances. Control devices could be developed to weave these breaker panels into the PV system, so that when grid power is lost, load is automatically curtailed to balance supply and load for the residential microgrid. These systems also could manage the ramps that occur as the sun rises and sets, or as clouds block sunlight.

Urgent Services

An opportunity exists to identify innovative technologies that can provide limited services or urgent services to critical aspects of community infrastructure. Only a few of these technologies have been identified, and fewer still have been researched. EPRI proposes to launch research to address this. Examples include:

- **Cell Phones**—During a prolonged power outage, even if cell phone towers are energized, the cell phones themselves will eventually run dry. Cell phone users whose automobiles have gas or electric storage and patch cords can charge their phones. A few consumers have solar chargers (see Figure 6-15). A resiliency plan would encourage and instruct citizens to have solar chargers. Utilities could consider making them available through various channels and at various prices, or at no cost.
- **Use of Conventional Vehicles**—If appropriate cables and connectors were available, conventional internal combustion engine (ICE)-powered vehicles could be used to energize devices such as cell phone chargers, flashlight battery chargers, and critical medical devices, among others.
- **Traffic Lights**—During rolling blackouts in South Africa in 2007, a major cause of traffic accidents was the loss of functional signals. A self-sufficient traffic light system might be designed with LEDs (see Figure 6-16) and battery backup (or solar PV backup) to operate during an outage.

Technologies described in this section are only a few of those available now or have the potential to be developed, demonstrated and assessed. EPRI is initiating a collaborative project to identify and assess potential solutions to the survivability issue. The preliminary scope includes:

- Identify areas to improve survivability through urgent services such as chargers for cell phones, storage for traffic lights, home generators, microgrids, etc.
- Scan and identify currently available technologies that can improve customer survivability in these areas.
- Screen the state of technology/cost/benefit/performance, application, and maturity of these technologies.

Figure 6-15. Solar Cell Phone Charger (Source: Carnegie Mellon University)

Figure 6-16. LED Street Lights (Source: Carnegie Mellon University)

- Create a knowledge base of what consumers can do now and what requires additional R&D to help accelerate the development and deployment of these technologies.

The Electric Ecosystem

By viewing the electric system as an electric ecosystem, its stakeholders can approach reliability in terms of the central mission of an electric system, not just in terms of preventing blackouts (Talukdar 2003). This concept infers increased use of the power system to bolster reliability, without necessarily relying on the extra-high-voltage and high-voltage transmission grid.

A system geared to survivability is built on a system of sub-units (Farrell 2003). One example involves traffic signals. Traffic signals are typically supplied from the same distribution circuit as other customers in an area. Outages can render traffic lights inoperative, causing traffic accidents. Power system engineers have known for some time that systems using large central-station power plants tied together with a grid is inherently vulnerable. The use of smaller, distributed generators could markedly increase reliability, and if these generators are installed as part of a resiliency plan, they could make the power system more resilient to severe weather or even terrorist attacks. Such a system could configure microgrids with distributed generation, electric and thermal energy storage, and an array of energy management and control systems.

To reconfigure the power system for microgrids and survivability, and to prepare for changes in consumer demand and related technology, utilities need to assess the barriers and incompatibilities associated with integrating distributed resources (e.g., local storage, demand response, distributed generation, and renewable resource) and fully apply them in practice (EPRI 1018945).

Such assessments can support a variety of power system improvements:

- Developing smart grid project plans.
- Evaluating smart grid proposals.
- Reviewing and assessing smart grid projects.
- Assisting in the final assessment of smart grid projects.

Research is under way to develop a methodology and quantitative metrics for evaluating smart grid projects related to integrating distributed energy resources into the grid and market operations, including distributed generation, storage, demand response, and renewables. This methodology includes a smart grid project assessment spreadsheet that identifies the important characteristics of smart grid projects to achieve integration of distributed resources (i.e., those characteristics that indicate the extent to which these projects can fulfill smart grid objectives).

The microgrid is a natural evolution of distributed resources used in areas where conventional power systems could not reliably serve customers. Microgrids may also provide support to conventional power systems that are constrained in meeting demand. When used to enhance resiliency, microgrids can be configured to operate in tandem with the bulk supply system during normal conditions, but disconnect and operate as an independent island in the event of a bulk supply failure or emergency (EPRI 1003973).

In this configuration, not only must the generators all work together to control voltage, frequency, VARs, and other parameters, they must also do so in an environment in which the microgrid may be periodically reconfiguring itself by attaching itself to adjacent microgrids or breaking apart in smaller microgrids. This capability requires the highest level of communication among generation devices and the adaptive grid command and control that can shift from one controller to the next as the system changes state. The generator protection schemes and operating set-points will have to adapt each time that the system changes state. To be implemented effectively, this approach needs considerable research. Such systems could be augmented by the installation of energy storage by the customer.

The concept of allowing electric energy storage to receive benefits from several value streams is still unresolved. A debate is ongoing regarding the best ways to account for value and benefits accruing from energy storage. If consumers could own and operate (automatically) local storage and receive compensation for its use both by distribution utilities and wholesale market participants, while having the asset available for responding to pricing signals, then the storage asset could be available to ensure survivability as well.

References

- EPRI, Enhancing Distribution Resiliency: Opportunities for Applying Innovative Technologies, M. McGranaghan, M. Olearczyk, and C. Gellings, EPRI, Palo Alto, CA: 2013, 1026889.
- A Zolli and A.M. Healy, Resilience: Why Things Bounce Back, Free Press, 2012, ISBN 978-1-4516-8380-6, 2013
- EPRI, Needed: A Grid Operating System to Facilitate Grid Transformation, EPRI, Palo Alto, CA: 2011, 1025861.
- B.W. Johnson, Utility Restoration Cost Recovery, Edison Electric Institute, February 2005, https://www.efis.psc.mo.gov/mpsc/commoncomponents/viewdocument.asp?DocId=4048185.
- Witt Associates, Connecticut October 2011 Snowstorm Power Restoration Report, December 1, 2011, http://nuwnotes1.nu.com/apps/financial/nuinvest.nsf/0/5091F0B5C3FFDAB58525795A00585F4F/$FILE/Witt%20report%20on%20CL%26P%20power%20restoration.pdf.
- B. Federman, "Electricity makes life livable," Times Union, Albany, NY, August 21, 2012, http://www.timesunion.com/opinion/article/Electricity-makes-life-liv-

able-3805478.php.
- NERC, NERC Event Analysis: System Disturbance Reports, http://www.nerc.com/page.php?cid=5166.
- National Climate Assessment and Development Advisory Committee (NCADAC), National Oceanic and Atmospheric Administration (NOAA), Climate Assessment Report (Draft), January 2013, http://www.nesdis.noaa.gov/NCADAC/.
- National Academy of Sciences (NAS), Terrorism and the Electric Power Delivery System, 2012, http://www.wiresgroup.com/docs/WPF_Terrorism%20and%20The%20Electric%20Power%20Delivery%20System.pdf
- EPRI, High-Impact, Low-Frequency (HILF) Events in the Electric Power Industry: Potential Impacts, Mitigation, and Risk Management, EPRI, Palo Alto, CA: 2013, 3002001935.
- EPRI, Electromagnetic Pulses (EMPs) and the Power Grid, EPRI, Palo Alto, CA: 2013, 3002001936.
- Defense Science Board Task Force, Nuclear Weapon Effects Test, Evaluation and Simulation, page 15, Office of the Under Secretary of Defense, Department of Defense (DoD) Washington, D.C. (2005), http://www.fas.org/irp/agency/dod/dsb/nweffects.pdf.
- K.T. Bainbridge, Trinity (Report LA-6300-H), Los Alamos Scientific Laboratory, May 1976, http://www.fas.org/sgp/othergov/doe/lanl/docs1/00317133.pdf.
- U.S. Department of Energy (DOE), United States Nuclear Tests, July 1945 through September 1992, December 1994, http://www.fas.org/nuke/guide/usa/nuclear/usnuctests.htm.
- EPRI, Magnitude 6.7 Hawaii Earthquake of October 15, 2006: Seismic Qualification Utility Group Earthquake Investigation, EPRI, Palo Alto, CA: 2010, 1021087.
- Dames and Moore's Earthquake Engineering Group, "The Loma Prieta Earthquake: Impact on Lifeline Systems," Disaster Recovery Journal, http://www.drj.com/drworld/content/w1_113.htm.
- U.S. Nuclear Regulatory Commission (NRC), Tsunami Hazard Assessment at Nuclear Power Plant Sites in the United States of America—Final Report, NUREG/CR-6966, March 2009, http://www.nrc.gov/reading-rm/doc-collections/nuregs/contract/cr6966/
- EPRI, Emergency Management Guideline for Fossil Generating Stations, EPRI, Palo Alto, CA: 2008, 1014199.
- EPRI, Operational Experience with Lessons Learned on Hurricanes, EPRI, Palo Alto, CA: 2007, 1013363.
- EPRI, A Methodology for Assessment of Nuclear Power Plant Seismic Margin, EPRI, Palo Alto CA: 1991, NP-6041-SL.
- EPRI, Severe Accident Management Guidance Technical Basis Report, Volume 1: Candidate High-Level Actions and Their Effects, EPRI, Palo Alto, CA: 2012, 1025295; and Severe Accident Management Guidance Technical Basis Report, Volume 2, The Physics of Accident Progression, EPRI, Palo Alto, CA: 2012, 1025295.
- EPRI, Thermosyphon Cooler Hybrid System for Water Savings in Power Plants, EPRI, Palo Alto, CA: 2012, 1026528.
- EPRI, Technical Architecture for Transmission Operations and Protection, EPRI, Palo Alto, CA: 2008, 1016055.
- EPRI, Transmission and Substations Research and Development News, EPRI, Palo Alto, CA: 2012, 1025003.
- NERC, Task Force Whitepaper Report: Spare Equipment Database: Interim Report to Planning Committee—June 7-8, 2011 Meeting, June 2011, http://www.nerc.com/docs/pc/sedtf/DRAFT_SEDTF_Interim_Whitepaper_June_2011.pdf.

- J.P. Fama, Edison Electric Institute, "It's Time to Get in STEP," Transmission and Distribution World, April 1, 2010, http://tdworld.com/substations/power_time_step/.
- EPRI, Development of Power System Restoration Tool Based on Generic Restoration Milestones, EPRI, Palo Alto, CA: 2010, 1020055.
- U.S. Department of Defense (DoD) Interface Standard, High-Altitude Electromagnetic Pulse (HEMP) Protection for Ground-Based C41 Facilities Performing Critical, Time-Urgent Missions, MIL STD 188-125-1, July 17, 1998, http://www.wbdg.org/ccb/FEDMIL/std188_125_1.pdf.
- EPRI, Program 35, Supplemental Project, Application of Nano Technology Coatings for Transmission Components (Phase 2), http://portfolio.epri.com/ProgramTab.aspx?sId=PDU&rId=263&pId=7705, 2014
- EPRI, T&D and ROW Environmental Issues—Program 51, 2014 Research Portfolio, http://portfolio.epri.com/ProgramTab.aspx?sId=ENV&rId=254&pId=8036.
- EPRI, Project Set180C: Cable System Management, 2014 Research Portfolio, http://mydocs.epri.com/docs/Portfolio/P2014/2014_P180.pdf.
- EPRI, Project Set 180I: Distribution Systems Practices, 2014 Research Portfolio, http://mydocs.epri.com/docs/Portfolio/P2014/2014_P180.pdf.
- EPRI, Program 183: Cyber Security and Privacy, 2014 Research Portfolio, http://mydocs.epri.com/docs/Portfolio/P2014/2014_P183.pdf.
- EPRI, Project Set180A: Distribution Planning, Design, and Analysis, 2014 Research Portfolio, http://mydocs.epri.com/docs/Portfolio/P2014/2014_P183.pdf.
- EPRI, Project Set 180B: Distribution Inspection, Maintenance, Asset Planning, 2014 Research Portfolio, http://mydocs.epri.com/docs/Portfolio/P2014/2014_P183.pdf.
- EPRI, Project Set 37.111: Risk Based Substation Equipment Asset Management, 2014 Research Portfolio, http://mydocs.epri.com/docs/Portfolio/P2014/2014_P037.pdf.
- EPRI, Supplemental Project 1023304: Airborne Damage Assessment Module, http://www.epri.com/abstracts/Pages/ProductAbstract.aspx?ProductId=000000000001023304
- EPRI, Supplemental Project 1024667: Field Force Data Visualization, http://www.epri.com/abstracts/Pages/ProductAbstract.aspx?ProductId=000000000001024667.
- EPRI, Supplemental Project 1025704: Distribution Grid Resiliency, http://www.epri.com/abstracts/Pages/ProductAbstract.aspx?ProductId=000000000001025704.
- EPRI, Supplemental Project 1025118: Battery Storage Systems and Applications Demonstration, http://www.epri.com/abstracts/Pages/ProductAbstract.aspx?ProductId=000000000001025118.
- S.N. Talukdar et al., "Cascading Failures: Survival versus Prevention," The Electricity Journal, November 2003, pp. 25-31, https://www.ece.cmu.edu/cascading-failures/Talukdar_CascadingFailuresSurvivalvsPrevention.pdf.
- A.E. Farrell et al., "Bolstering the Security of the Electric Power System," Issues in Science and Technology, Spring 2003, pp. 49-56, http://www.issues.org/18.3/farrell.html.
- EPRI, Smart Grid Distributed Energy Resources (DER) Project Assessment, EPRI, Palo Alto, CA: 2009, 1018945.
- EPRI, Investigation of the Technical and Economic Feasibility of Microgrid-Based Power Systems, EPRI, Palo Alto, CA: 2001, 1003973.
- EPRI Smart Grid Resource Center, http://smartgrid.epri.com/
- The Smart Grid Implementation Strategy team, National Energy Technology Laboratory, http://www.netl.doe.gov/smartgrid/.

- GridWise Architectural Council, www.gridwiseac.org.
- The Smart Grid Implementation Strategy team, National Energy Technology Laboratory, "What is the Smart Grid," March 2009, http://www.netl.doe.gov/smartgrid/referenceshelf/factsheets/OE_DER-008_APPROVED_2009_04_15.pdf.
- Weather-Related Power Outages and Electric System Resiliency, R.J. Campbell, Congressional Research Service, R42696, August 28, 2012.
- Trends in the History of Large Blackouts in the United States, http://www.uvm.edu/~phines/publications/2008/Hines_2008_blackouts.pdf
- High-Impact, Low Frequency (HILF) Events in the Electric Power Industry: Potential Impacts, Mitigation, and Risk Management. EPRI, Palo Alto, CA: 2013. 3002001935.

Chapter 7

A Grid Operating System to Facilitate the Smart Grid

ENERGY MANAGEMENT SYSTEMS

Energy management systems (EMS) are computer-based systems used today to operate the complex electric power systems around the world.* These systems assure that the power system or "grid" operates properly and that consumers enjoy reliable electricity supply at the lowest possible cost. These systems operate by balancing the demand for electricity with generation resources (see sidebar). However, the rapid growth of renewable power generation, the increased use of electric vehicles, and increased need to integrate customers with the power system are rendering the current generation of EMS systems obsolete. This paper examines the evolution of today's grid operating system and outlines the development of a new grid operating system which will facilitate the transformation of the grid. Without the development of an advanced grid operating system, the full value of a lot of individual technologies like electric vehicles, electric energy storage, demand response, distributed resources, and large central station renewables such as wind and solar will not be fully realized.

What Does a Grid Operating System Do?

The basic function which a gird operating system performs (see illustration) has not changed since Thomas Edison operated the first power system at Pearl Street in New York City, which is to balance the demand for electricity with the energy available as supplied by generation. The demand for electricity is comprised of the demand from consumers

*The term Energy Management System has also been used to describe building automation systems.

plus exports to other systems and system losses. The components of generation available consist of power generated plus discharge from storage and imports. As alternating current (AC) systems proliferated, a type of generator called synchronous was used. Synchronous generators operate synchronously with each other. The grid operating system's job is to facilitate balancing demand with supply by keeping the frequency of the U.S. grids at or very close to 60 Hertz. If the system deviates from 60 Hertz, it becomes unstable and can collapse. Synchronous generators need to be controlled to stabilize that frequency. When there is too much generation on the system the generators speed up raising the frequency above 60 Hertz. Conversely, where there is inadequate generation on the system, generators slow down and the frequency falls below 60 Hertz. In some countries, 50 Hertz is the standard.

What does a Grid Operating System Do?

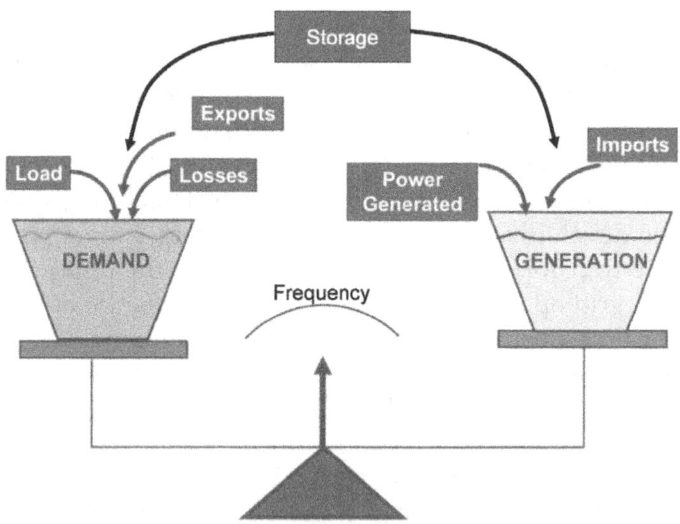

Figure 7-1. Energy Balance

GRID OPERATING SYSTEM 1.0

One of the first challenges which both Edison and Westinghouse faced in the operation of either the first direct current (DC) power systems or alternating current (AC) power systems was to enable reliable operation through control such that in any instant the total generation in a power system was "balanced against" total load. Balancing ensures the generation which is running at any point in time be equal to the total load or demand for electricity at that same moment (see sidebar). When generation is not balanced with load, the system becomes unstable and can collapse. In the 1800s when the first of the Pearl Street Generators was placed in service, this balancing act was a relatively simple endeavor primarily done through control systems in generators. This was a primitive form of a grid operating system which can be labeled as Version 1.0 or Grid Operating System 1.0.

GRID OPERATING SYSTEM 2.0

As power systems evolved, they became larger, more complex, and increasingly more difficult to control. Balancing multiple generators with a network of loads located throughout a city or across town and into the countryside could not be facilitated by generation control alone. By the mid 1950s, some of the first Supervisory Control and Data Acquisition (SCADA) systems were being deployed within power delivery systems. Grid Operating System 2.0 grew out of the understanding of these SCADA systems.

SCADA Systems

The first SCADA systems utilized data acquisition by means of panels of meters, indicator lights, and strip chart recorders. The power system operator manually controlled the power system by turning various knobs or activating switches which in turn sent signals to open or close circuit breakers or to start up generators. These SCADA systems are still used today to do supervisory control and data acquisition in some small utilities, older power plants, and industrial facilities. These primitive SCADA systems were technically simple; they did not require computers or digital sensors. However, the quantity and type of data were minimal and rudimentary. SCADA systems became popular for two principal reasons: they minimized blackouts, and they could significantly increase the utility bottom line through effective dispatch of generation and the marketing of excess generating capacity.

By 1962, there were several large interconnected power grids in the Midwestern, southern, and eastern portions of the U.S. These were the largest synchronized systems in the world. A large blackout in 1965 prompted the U.S. Federal Power Commission to recommend coordination between these systems. As a result, the Electric Power Reliability Act of 1967 was passed, and subsequently, the National Electric Reliability Council was formed.

Federal Energy Regulatory Commission

Late in the last century, the Federal Energy Regulatory Commission (FERC) passed a series of laws which enabled wholesale electricity markets (for example, FERC Order 888 in 1996). Accordingly, many of the grid operating systems in the nation were modified so as to incorporate connection to an increased number of market participants. For some systems, this meant increasing from a few dozen transactions per day to hundreds.

These actions, coupled with an overall concern within the industry that more needed to be done to assure reliable electricity supply, led to the invention of Grid Operating System 2.0. This system was a computer-based operating system and, although enhanced repeatedly since then, is the basic system in use today. During this period, the grid operating system came to be called energy management system or EMS. The major difference between Grid Operating System 1.0 and 2.0 is the capability to balance supply and demand between many market participants, multiple bulk power central generation resources, and a wide ranging control with numerous interconnections area using today's high-voltage grid. In addition, Grid Operating System 2.0 used a much more sophisticated mathematical technique to estimate the condition of the system.

System 2.0 allowed the system operator to estimate the condition of the system at any point in time. Often referred to as real-time monitoring and control, the grid operating system operates in near-real time using data coupled with input from sensors to estimate the condition of the grid 20 to 30 seconds after the fact. A unique development in Version 2.0 was the ability to "fill in" or estimate data from missing sensory inputs. System 2.0 is complex, uses rather sophisticated applications software, and has a large number of input/output (I/O) points and a substantial number of remote terminal units (RTUs). The most prevalent type of intelligent electronic devices (IEDs) which communicate with System 2.0 are reclosers, protective relays, substation controllers, and phasor measurement units.

ABB's EMS System

There are a number of well-developed Grid operating systems on the market and in use today. For example, ABB has a combination SCADA/EMS/GMS (supervisory control and data acquisition/Energy Management System/Generation Management System) which supervises, controls, optimizes and manages generation and transmission systems. ABB also has a companion SCADA/DMS (Distribution Management System) system which performs the same functions for power distribution networks.

According to ABB, both systems enable utilities to collect, store and analyze data from hundreds of thousands of data points in national or regional networks, perform network modeling, simulate power operation, pinpoint faults, preempt outages, and participate in energy trading markets. EMS systems are a vital part of modern power networks and are enabling the development of smart grids. Smart grids will need to handle large quantities of renewable power from both large and small-scale generators. To maintain grid stability in spite of these potentially disruptive source of power and the two-way flow of power in what is currently a one-way system, advanced monitoring systems such as ABB's will be very important.

ABB's system is designed to meet the differing needs of power system operators: Generation companies needed an interface with the energy trading markets, and the capability to plan and optimize supply to meet spot market demand; transmission companies required advanced systems to manage their high voltage networks and prevent a fault in one part of the system from cascading across the entire network; and distribution operators needed to take network management down to the level of individual customer connection points (often numbering several million) to minimize customer outage times. Like its competitors, ABB's network management and utility communication systems enable data, voice, video and protection signaling and other types of critical information to be communicated quickly, reliably and securely.

Changes in the Electricity Grid

The power system is revolutionizing at an exponential pace into a highly interconnected, complex, and interactive network of power systems, telecommunications, the Internet, and electronic commerce applications. Virtually every element of the power system will incorporate sensors, communications and computational ability. No longer will society depend primarily on central station power and what is essentially one-way flow on the grid, since the use of distributed generation and distributed energy storage will proliferate. At the same time, the move towards more competitive electricity markets requires a much more sophisticated infrastructure for supporting myriad informational, financial, and physical transactions between the several members of the electricity value chain that supplements or replaces the vertically integrated utility. Although reliability is now and will remain critical, this complexity will naturally create more opportunities for nodes of failure and potentially allow for increased risk from cyber attacks and coordinated physical and cyber attacks.

The grid has matured steadily since the first high-voltage transmission line was built on overhead wood pole structures to bring hydroelectric power from Niagara Falls to New York City. Today's system is highly interconnected and has benefited from a steady evolution of technologies. Central station coal and nuclear power coupled with the use of gas in combustion turbines have been integrated throughout the nation and evolved in a way that has made the grid what it is today.

Figure 7-2 illustrates today's power system. It is largely comprised of large central station power generation connected by a high-voltage network or grid-to-local distributions systems which serve homes, businesses and industry. Electricity flows predominantly in one direction using mechanical controls. The Grid Operating System 2.0 supports this power system.

Tomorrow's Power System

Figure 7-3 illustrates the elements which will be part of tomorrow's power system. Tomorrow's power system will still depend on the support of large central station generation, but it will increasingly include renewable generation and electric energy storage both at the bulk power system level and in the local distribution system. In addition, tomorrow's power system will have greatly enhanced sensory and control capability configured to accommodate electric vehicles and allow direct engagement with consumers and their appliances and devices. Tomorrow's power system will require a grid operating system which is secure against cyber intrusion and can assure reliable long-term operation of an extremely complex systems comprised of an array of distributed components.

These changes include profound impacts on the generation, deliv-

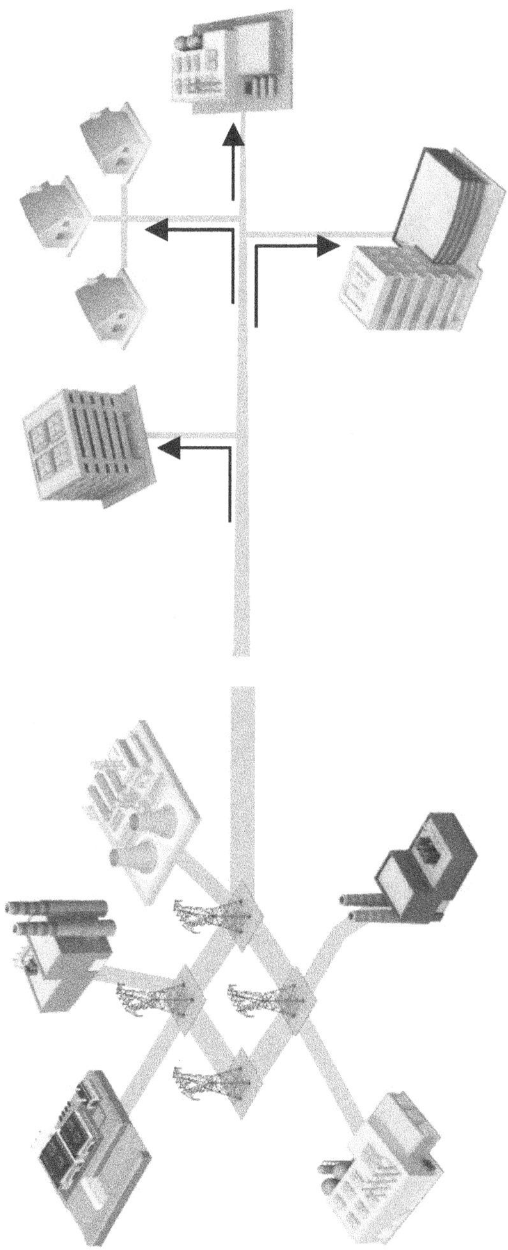

Figure 7-2. Today's Power System

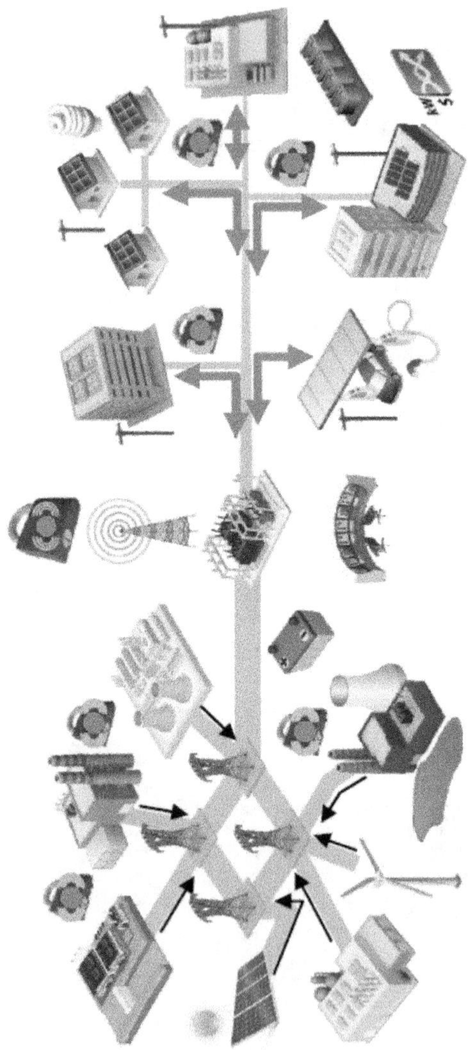

Figure 7-3. Tomorrow's Power System

ery, and use of electricity that require engineers to rethink the way the grid is currently operated. For example:

1. Variable Generation—The North American Electric Reliability Corporation (NERC) predicts that 200 gigawatts of variable generation will be added by 2020. These variable sources, primarily wind turbine generators and photovoltaic systems, present unique challenges to operators.
2. Demand Response—Capacity markets in the Northeast are now opened to demand response bids. In a recent auction, PJM received 17% of their requested capacity in the form of demand response and energy efficiency. Managing demand response in conjunction with generation provides additional challenges.
3. Electric Vehicles—By 2030, it is anticipated that there will be more than 20 million electric vehicles connected to the grid at various times. These vehicles all have battery capacity on board which may enable their use as a fast-acting balancing resource.
4. Smart Meters—There are currently more than 20 million communicating nodes within smart meters. There is a potential for as many as 80% of the approximately 140 million meters in the U.S. to be converted by 2030. Each of these nodes provides opportunities for enhanced energy management and condition monitoring.
5. Distributed Generation—Photovoltaic power generation systems are now in place on over 70,000 roof tops in the U.S., and those installations are growing at over 10% per year. Those distributed sources must be monitored and controlled effectively.
6. PMUs—The increasing deployment of phasor measurement units (PMUs) and the growth in the use of smart substations and sensors will require much faster processing and intelligence to act upon than is present in today's grid operating systems.
7. Communications—The phenomenal proliferation of electronic communications and information technology in all spheres of life and sectors of industry are extending into the electric power system. Sensors, communication and computational ability will transform the power system.

GRID OPERATING SYSTEM 3.0

The present grid operating system was not designed to meet the increasing demands of a digital society or the increased use of renewable power production. There is a national imperative to modernize and enhance the power delivery system. That modernization must include a new grid operating system. Tomorrow's grid operating system must facilitate high levels of security, quality, reliability, and availability (SQRA) of electric power; improve economic productivity and quality of life; and minimize environmental impact while maximizing safety.

The grid operating system must monitor, protect and automatically optimize the operation of its interconnected elements—from the central and distributed generator through the high-voltage network and distribution system, to industrial users and building automation systems, to energy storage installations, and to end-use consumers including their thermostats, electric vehicles, appliances, and other household devices.

Tomorrow's grid operating system must manage a two-way flow of electricity and information to create an automated, widely distributed energy delivery network. It must incorporate into the grid the benefits of distributed computing and communications to deliver real-time information and to enable the near-instantaneous balance of supply and demand at the device level.

A next-generation power delivery operating system, referred to here as Grid Operating System 3.0 or Grid 3.0 must be developed to provide for seamless integration and interoperability of the many disparate systems and components, as well as enable the ability to manage competitive transactions resulting from competitive service offerings that emerge in the restructured and distributed utility environment. Figure 3 illustrates the challenges Grid Operating System 3.0 faces.

Realizing Grid 3.0 depends on developing the architecture—namely the functional requirements and the design requirements based on an open-source design, which can facilitate the informational, financial, and physical transactions necessary to assure adequate security, quality, reliability and availability of power systems operating in complex and continually evolving electricity markets. In addition, the architectural requirements will be designed to support multiple operational criteria including analysis and response to electrical grid contingencies, pricing, and other market and system conditions. The goals of the architecture are to allow for interoperability and flexibility of the power system

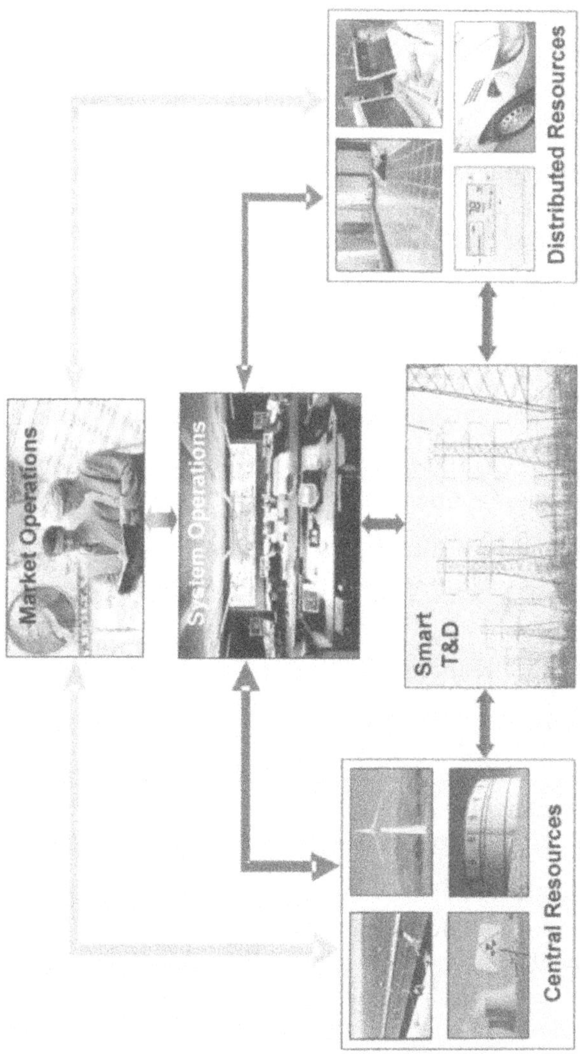

Figure 7-4. Grid Operating System 3.0 Challenges: Integrate Systems and Components

operations while at the same time facilitate and enable competitive transactions to occur. Interoperability can be enabled by the use of open communication protocols developed in EPRI's IntelliGrid Program. Flexibility can be provided by the specification of user-defined business rules which capture the unique needs of various service offerings.

Initial Steps to Grid 3.0

Today's grid operating systems already have begun to expand capabilities leading to EMS 3.0. These changes are a necessary part of sustaining core capabilities even before the nation began to evolve the concept of a "Smart Grid." For example, the development of techniques for real-time simulation and enhanced visualization have been under development since the 1990s. Today, they are considered part of the Smart Grid, but would have simply been viewed as necessary improvements a decade ago.

There are three functions embedded in EMSs used by ISOs and RTOs. They include:

1. Enhancing the visibility of the grid (transparency), increasing reliability, and energy efficiency.
2. Enable the effective integration of increasing amounts of distributed resources including renewables, energy storage, and demand response.
3. Enable effective response to increasingly sophisticated cyber and physical security threats.

Figure 7-5 illustrates the key applications that form an ISO. The functional enhancements which the Smart Grid is expected to yield will require changes in these applications in the following areas:

Communications—The increasing number of external interfaces will necessitate that telemetry data is compatible with IEC 61850. This will require retrofits and upgrades. In addition, the Smart Grid will contain many more transmission line sensors and sensors from wind and solar installations.

Prices-To-Devices—Truly enabling consumers to respond to variations in prices and system constraints will require the ability to broadcast that information to consumers both through the Internet and through secure, automated area networks.

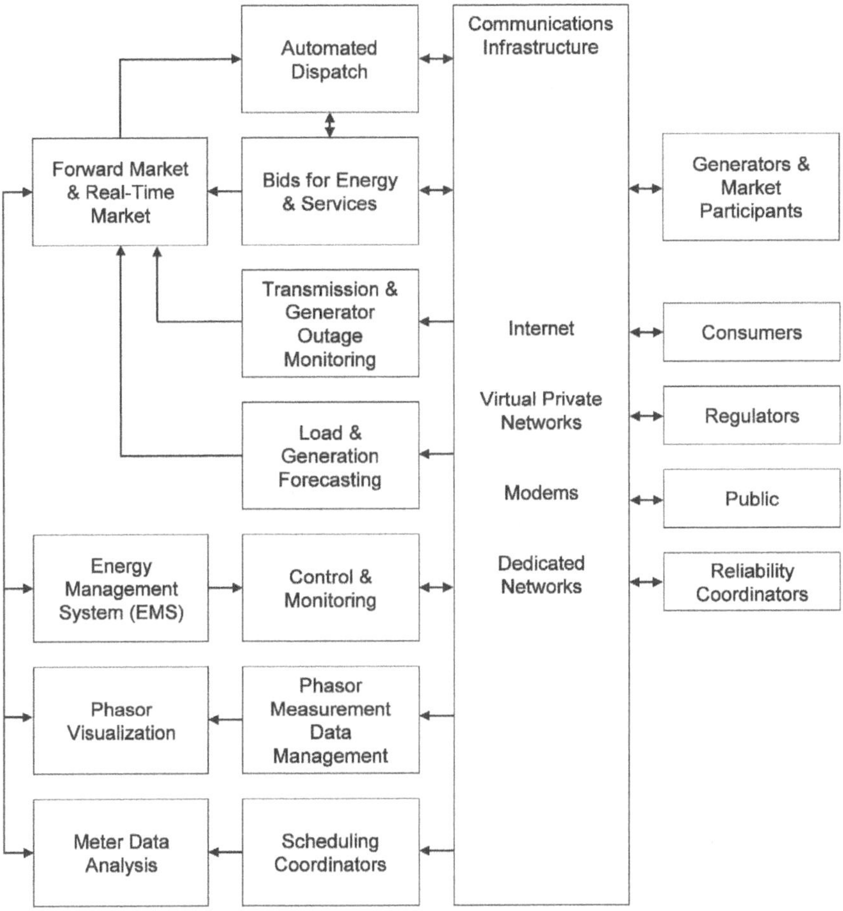

Figure 7-5. Key Components of an ISO Infrastructure

Upgrade to NIST-Proposed Standards—The National Institute for Standards and Technology (NIST) has established a Smart Grid Interoperability Panel (SGIP), and the panel is developing Priority Action Plans (PAP) to encourage adoption of standard business practices and interface specifications. It is likely that these interfaces will require widespread adoption of the Common Information Model (CIM).

Forecasting—Developing far more accurate data to forecast renewable energy production (particularly wind and solar) including very short-term forecasts and ramp rates. In addition, it will be necessary to forecast demand response participation.

Cyber Security—The North American Reliability Council is developing Critical Infrastructure Protection Standards (CIP) to meet Federal Regulatory Commission (FERC) and Congressional requests.

Integrate Synchrophasors—Integrating synchrophasor data sources with visualization and to enable advanced-state estimation and dynamic stability analysis incorporating the data from numerous new distributed generation sources.

Evolve New Markets—The advent of widespread consumer connectivity coupled with increasing participation of third parties allows for new markets to evolve. The ISO's system will need to be sufficiently flexible so as to allow the evolution of new markets.

Other Applications—Individual ISOs may add other functionality like the ability to quickly add new market participants, to improve outage management data and other enhancements.

Each of these and other features will require an ongoing investment in the ISO's applications.

The concept of Grid 3.0 provides a new perspective on how to manage transactions given the nature of the existing and emerging distributed, heterogeneous communications and control network combined with the extensive use of new innovations on the power system. What is needed is an architecture that allows future developers to access this framework as a resource or design pattern for developing distributed software applications, taking into account the core concepts of interoperability and support for multiple operational criteria including business rules.

Grid 3.0 must increase the independence, flexibility and intelligence for optimization of energy use and energy management within local energy networks at the building level, at the local level, at the distribution level, and then integrate local level devices to the Smart Grid. As such, energy sources and a power distribution infrastructure can be integrated at the local level. This could be an industrial facility, a commercial building, a campus of buildings, or a residential neighborhood. Using Grid 3.0, these "local energy networks" are interconnected with different localized systems to take advantage of power generation and storage through the Smart Grid enabling complete integration of the power system across wide areas. This includes control of building energy and thermal storage systems. Localized energy networks can accommodate increasing consumer demands for independence, convenience, appearance, environmentally friendly service, and cost control.

While some of these local energy networks can operate in a stand-along mode, integration into the distribution system using Grid 3.0 allows interconnection and integration with technologies that ultimately enable the next generation of the Smart Grid.

Fundamentally, tomorrow's grid operating system will need to handle millions of intelligent electronic devices located through the power system from the generation switchyard through to end-use energy consuming devices and appliances on the consumer's premises. In addition, tomorrow's grid operating system must have the following functionality:

Geospatial Power System Model Database

Tomorrow's grid operating system must have a hierarchical geospatial data acquisition and maintenance architecture. Geospatial power system model database feeding would allow the power system models to be derived directly from detailed geospatially correct computer-aided design (CAD) models. Increasingly, transmission utilities are analyzing their transmission lines using an optically remote sensing technology known as light detection and ranging (LIDAR) technology in order to measure the dimensions of, and the physical features of, a transmission corridor and its lines. With over 400,000 miles of lines greater than 100 kV in North America, some many decades old, there are discrepancies between design and actual field conditions. Also, many of these lines are in hard-to-reach places. LIDAR permits easy assessment of these lines. This data can be post-processed into CAD models like power-line systems—computer-aided design and drafting (PLS-CADD) tools (see www.powline.com.) These models use the LIDAR data to accurately model lines. These CAD files can then be used as input into line constants computer programs which give accurate transmission line ratings and loading. Geospatial feeding would integrate all of these steps and seamlessly calculate a more precise set of line parameters eliminating many of the assumptions used today. The output would be put into common information model (CIM) formats easily interchanged between systems. These physically precise models could also be used to develop screen interfaces much like those that have been made popular by Google Earth and other geospatial systems.

Integration of Traditional and Non-Traditional Data

Integration of traditional utility operating data along with non-operating data such as smart meters and distributed sensors can be "mashed up" with readily available internet features such as Google Earth. It can provide very useful visual displays of information within geospatial context. Figure 7-6 illustrates an analysis of an actual outage and the response of the Smart Grid meters to loss of power and restore events shown on the multiple screen Google Earth video wall. Figure 7-7 is a Google Earth illustration of 1000,000 meter voltages on a hot summer day.

Advanced Protection and Control Functions

Tomorrow's grid operating system must have advanced protection and control functions which shall include coordination, execution, and confirmation of all necessary actions required to prevent the power grid from entering into an irreversible degradation in performance. The process of advanced protection and control used today is becoming unman-

Figure 7-6. Illustrates an Actual Outage

Figure 7-7. Illustrates 100,000 Meter Voltages, an example of Grid 3.0 functionality

ageable and will not support many of the smart-grid concepts. Grid Operating System 3.0 migrates protection properties further down into the hardware and software on the power system and greatly simplifies the technical skills needed by protection engineers. The existing system of using many, greatly distributed intelligent agents are unmanageable for the long term. Hardware is obsolete by the time normal upgrade cycles of 10 to 15 years are completed on large systems. Protection schemes in Grid 3.0 need to be aligned with the current NERC philosophy regarding special protection schemes.

State Measurement with Look-Ahead Capability

Tomorrow's grid operating system must have a forward-looking state measurement and a decision support tool that can help grid operators manage severe operating condition with future situational awareness. As technology enables synchronized measurements to be collected, power systems shall lead to a linear, non-iterative state measurement—as opposed to the conventional non-linear and iterative—state estimator which will be free of convergence problems. The state measurement is also expected to eliminate or minimize the effects of missing measurements, erroneous measurements, errors in network topology, etc. Furthermore, since synchronized measurements can be obtained much more frequently than conventional measurements, state measurement must have the capability of capturing slow dynamics associated with the states and predict state trajectory which possibly indicates some instability scenarios.

Cyber Security

Cyber security will be an essential element of Grid 3.0. The North American Electric Reliability Corporation (NERC) has created eight Critical Infrastructure (CIP) Standards. These include standards for Critical Cyber Asset Identification (CIP002) and Security Management Controls (CIP003) as well as others. Meeting these standards will be part of Grid 3.0 including emerging security standards like NIST's Smart Grid Interoperability Standards Framework and AMI-SEC System Security Requirements, for end-to-end security of the Smart Grid. Grid 3.0 will likely use a system of systems approach to cyber security by deploying International Organization for Standardization and International Electrotechnical Commission (ISO/IEC), National Security Agency InfoSec Assessment Methodology (NSA IAM), Information Systems Audit and

Control Association (ISACA), and International Information Systems Security Certification Consortium (ISC2). Cyber security will be part of every input/output (I/O) function for interfaces with IEDS and systems as part of Grid 3.0 including:

- — Market Participants including third party Demand Response providers
- — Central station generation resources
- — Advanced metering infrastructures
- — Plug-in electric vehicle (PEV) management systems
- — Distributed generation and storage
- — Distribution operating systems including distribution automation and Local Energy Network Controllers
- — Substations which are automated and fully instrumented
- — Fully instrumented transmission lines and corridors
- — Detection and prevention services (IDS/IPS) as well as security information event management (SIEM).

Enable Active Participation by Consumers

The grid operating system must facilitate interaction with and include customers who are an integral part of the electric power system. Tomorrow's consumer must be informed, modifying the way they use and purchase electricity. They have choices, incentives, and disincentives to modify their purchasing patterns and behavior facilitated by the grid operating system.

Accommodate All Generation and Storage Options

The grid operating system must accommodate all generation and storage options. It must support large, centralized power plants as well as distributed energy resources (DER). DER may include system aggregators with an array of generation or storage systems or individual consumers with a windmill and solar panels.

Enable New Products, Services, and Markets

The grid operating system must enable a market system that provides cost-benefit tradeoffs to consumers by creating opportunities to bid for competing services. As much as possible, regulators, aggregators and operators, and consumers can modify the rules of business to create opportunity against market conditions. A flexible, rugged market infrastructure exists to ensure continuous electric service and reliability, while also providing revenue or cost reduction opportunities for market participants. Innovative products and services provide third-party vendors opportunities to create market penetration opportunities and consumers with choices and clever tools for managing their electricity costs and usage.

Optimize Asset Utilization and Operate Efficiently

The grid operating system optimizes assets and operates efficiently. It dispatches an array of generation, storage and load technologies to ensure the best use of assets. Assets operate and integrate well with other assets to maximize operational efficiency and reduce costs.

Anticipate and Respond to System Disturbances (Self-Heal)

Tomorrow's grid operating system independently identifies and reacts to system disturbances and performs mitigation efforts to correct them. It incorporates software designed to enable problems to be isolated, analyzed, and restored with little or no human interaction. It performs continuous predictive analysis to detect existing and future problems and initiates corrective actions. It will react quickly to disruptions in service and optimizes restoration exercises.

Operate Resiliently Against Attack and Natural Disaster

The grid operating system must resists attacks on the cyber-structure (markets, systems, software, and communications). Constant monitoring and self-testing are conducted against the system to mitigate malware and hackers.

Effectively Integrate Local Energy Networks

The grid operating system must effectively integrate local energy networks with central station power generation and the Smart Grid.

DEVELOPING 3.0

Several organizations have work underway which relates to Grid 3.0 which will be very helpful in the development of this system.

The International Electrotechnical Commission

The International Electrotechnical Commission has several activities and working groups which relate directly to Grid 3.0.

Technical Committee 57

Technical Committee 57 (IEC TC 57) is responsible for international standards regarding Power Systems Management control equipment for power systems control equipment and systems including EMS (Energy Management Systems), SCADA (Supervisory Control And Data Acquisition), distribution automation, teleprotection, and associated information exchange for real-time and non-real-time information, used in the planning, operation and maintenance of power systems. That working group is coincidentally working on standards which build off several IEC standards which will be helpful in designing Grid 3.0.

EC 60870-5 is a general protocol definition developed by the International Electrotechnical Commission (IEC) Technical Committee 57. It is an outline for the structure of a protocol and can only be implemented with a companion standard (profile) to specify options such as one of five link layer formats. Triangle MicroWorks has implemented Source Code Libraries for all of the currently defined profiles: 101, 102, 103, and 104. The 101 profile has been adopted by the Substation Committee of the IEEE Power Engineering Society (1379-2000) as a recommended practice for RTU-IED communication.

The CIM User Group

The CIM Users Group (CIMug) was formed in 2005, as a subgroup of the UCA International Users Group, to provide a forum in which users, consultants, and suppliers could cooperate and leverage the IEC CIM international standards to advance interoperability across the utility enterprise. The primary purpose is to share technology basics, best practices and technical resources while Advancing Interoperability for the Utility Enterprise.

The Common Information Model (CIM) is an abstract information model that provides data understanding through the identification of the relationships and associations of the data within a utility enterprise. This enhanced data understanding supports the exchange of data models and messages and increases the ability to integrate applications both within the enterprise and with trading partners. These trends go beyond exchange or updates of network models to the exchange of specific dynamic data within transactional messages in a real-time environment. The CIM companion standards provide extensions and specifications that, when used in conjunction with the CIM models, provide a framework for the exchange of static models, transactional messages and full enterprise integration.

North American ISO/RTO Council

The ISO/RTO Council (IRC) is advancing the standardization of communications associated with Smart Grid resources in their Enterprise Architect Standardization (EAS) project in wholesale energy markets and transmission system operations. Starting from existing standards, such as the Common Information Model, as well as more recent developments, such as the North American Energy Standards Board's business practice standards for demand response, they have developed the following:

- Common information exchanges to facilitate incorporation of demand-side resources and storage technologies on the wholesale level.
- Information models containing well-defined data elements for use in developing the next generation of Smart Grid software.
- Common definitions, business processes and information exchanges associated with Smart Grid technology.

This effort includes the development of business process diagrams and use cases illustrating the standard interactions between a System Operator and various Market Participants for the administration and deployment of demand response resources in organized wholesale electric markets. The wholesale Demand Response end-to-end business process flow for the communications between the System Operator (SO) and the Service Provider (SP) is comprised of four major functions: Enrollment & Qualification, Scheduling & Award Notification, Deployment & Real-Time Communications, and Measurement & Performance. Two additional processes, Market Participant Registration and Settlements, are shown in, but are considered outside the scope of what is required for the communications flow for demand response.

The IRC has also developed a consolidated data elements list which provides a detailed list of data indentified for the interactions illustrated in the business process diagrams for uses cases. The data elements list is provided in alphabetical order by data element and also by data element category ID and element ID within each data element category. A short description and long description, where applicable, are provided for each data element.

The IRC has also developed an Enterprise Architect® Information Model incorporates all business information in a rigorously consistent tool. NIST and the IEC use this tool to maintain the Common Information Model (CIM). The information model is available in its original EA format (EAP), and in HTML format.

The IRC Smart Grid Project provided recommendations for data exchange standards and APIs supportive of Smart Grid technologies and corresponding ISO/RTO products or services based on Demand Response Information Models. These recommendations include key functional areas, diverse markets and NERC Cyber Security and other relevant reliability standards. For each data flow, we developed a proposed set of nonfunctional requirements organized around the four major requirements categories recommended in IEC/PAS 62559.

CIGRÉ

CIGRÉ has published a set of information technology (IT) requirements which will be useful in outlining some of the communications and IT requirements for Grid 3.0. CIGRÉ has outlined the real-time and near-real-time systems which will be part of the future developments (CIGRÉ 2011). Table 7-1 lists the design tenets which CIGRÉ suggest become part of the system architecture for Grid3.0.

Figure 7-8 Illustrates CIGRÉ's resultant business function decomposition.

The Roadmap for Grid 3.0

The industry has a unique opportunity in developing Grid Operating System 3.0. Rather than each control area operator, independent system operators, and regional transmission operators separately contracting with different EMS vendors, this development could begin with the collaborative engagement of stakeholders to develop the functional and design requirements on an open source basis. In this process, there is a need to engage experts from all domains. Specifically, this entails characterizing what all types of input and output communications need to be and to identify all of the actors in the ultimate operation of the grid operating system. These actors are the suppliers of data and the users

Table 7-1. Design Tenants for Grid 3.0. (CIGRÉ 2011)

Tenet No.	Tenet	Sort Name
1	Services are implemented as modular components with business significant granularity	Modular
2	Services are configurable	Configurable
3	Services are customizable	Customizable
4	Services are open	Open
5	Services interact abstractly with other services	Abstract
6	Services are loosely coupled	Separate
7	Services are technologically neutral	Neutral
8	Services are secure	Secure
9	Services provide encapsulation	Encapsulated
10	Services provide unique non-trivial functionality	Unique
11	Services are instrumented	Instrumented

of data and information. Actors can include devices like individual sensors, smart meters, reclosers and PMUs. This process will identify "use cases" which clearly state the characteristics of data and information and its intended parameters. Once developed, individual vendors can offer bids which respond. By making these uniform requirements available, it is expected that innovative adaption of new systems will result as well as economies in the cost of development. In particular, the ultimate developers will need a combination of traditional EMS and power system vendors coupled with those having expertise in networking, data management and security.

Early Developments

There are early signs of the development of some of the advanced attributes needed in Gird Operating System 3.0 in pilot projects underway in several U.S. states as well as a few European countries. These are

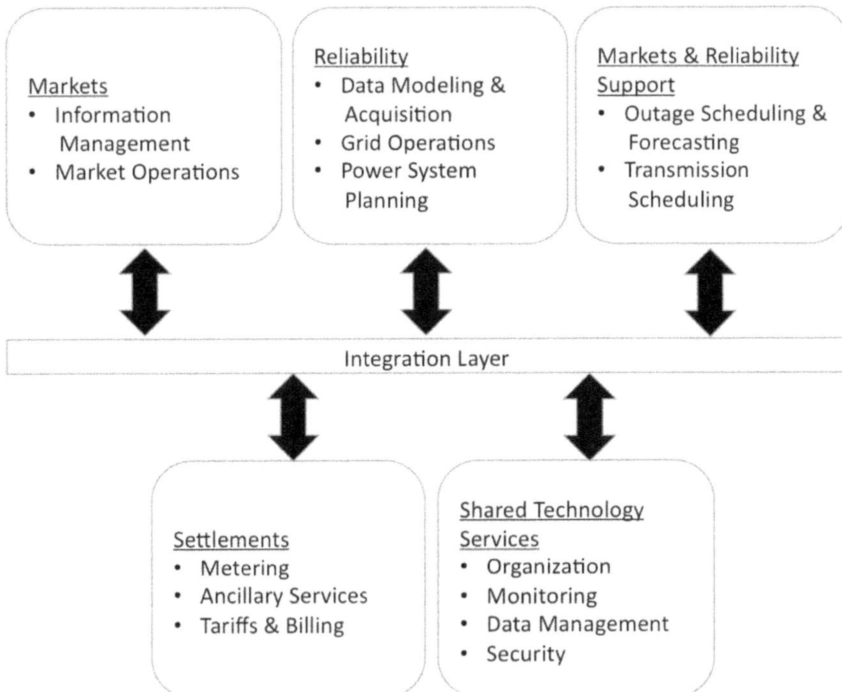

Figure 7-8. Examples of Business Function Decomposition (CIGRE 2011)

early stages of Grid 3.0 in action in pilots and in distribution-level EMS systems. Examples include Hawaii, Illinois, and a series of demonstrations being coordinated by EPRI.

Hawaii

On the Hawaiian island of Maui, the General Electric Company, in cooperation with Hawaiian Electric Company and its subsidiary, Maui Electric Company, Ltd., are testing its smart-grid technology. This application involves development of renewable and distribution system technology integration. It includes an intermittency management system, demand response, wind turbines, and dynamic systems modeling. The distribution operating system under development will include the standard applications of network management, SCADA, outage management, and network analysis, but will also be enhanced with features which enable demand management, Volt/VAR control and transmission-level grid support while enabling two-way communication with

Local Energy Networks and distribution IEDs including buildings, distributed energy resources, capacitor banks, load-tap changers (LTC), voltage regulators, and wind turbine generators.

Illinois

At the campus of Illinois Institute of Technology (IIT) in Chicago, a Local Energy Network is being created which incorporates advanced (smart) metering and sub-metering, an intelligent "perfect power" controller, and an on-site gas-fired combustion turbine power generator, a demand-response controller, and an uninterruptible power supply with electric energy storage. Perfect power is defined by IIT as always on, digital-grade power needed for its various laboratories. The features included in the perfect power controller will contain many of those needed by the master controllers needed to support Local Energy Networks operating as part of Grid 3.0.

EPRI Demonstrations

EPRI has been involved in a massive Smart Grid Demonstration Initiative involving a number of ongoing projects to demonstrate the potential for integrating distributed power generation, storage, and demand response technology into "virtual power plants." Demonstrations include both utility-side and customer-side technologies and are intended to address the challenges of integrating distributed energy resources (DER) in grid and market operations, as well as in system planning. This includes aspects required for Grid 3.0, including:

- Demonstrating effective operational strategies for integrating different forms of distributed resources.
- Demonstrating multiple levels of integration and interoperability among various components.
- Exploring existing and emerging information and communication technologies required for tomorrow's grid operating system.

The demonstrations are taking place at a number of U.S. locations and will include a variety of feeder constructions, climate zones, and technologies. Individual demonstrations are focused on the integration of specific feeder types used in residential neighborhoods, in a mixture of residential and commercial customers, and in areas with mostly commercial customers.

These developments are in their infancy in terms of how such distributed systems will become part of a hierarchical-based system by seamlessly integrate the different domains as highlighted in the Smart Grid Conceptual Model developed by EPRI for the National Institute of Science and Technology (NIST.)

Transitioning to Grid 3.0

Replacing the operating system in today's increasingly complex grid is like replacing an engine on a jet plane while it is flying at 30,000 feet. Much like the plane, the grid needs to be "on" all of the time. There are two issues to look at regarding the power system being on all of the time. The first one relates to the inherent complexity and potential to disrupt the system during transitions from one operating version to a newer version. While many aspects of upgrading the power system have become routine, such as new line extensions and public improvement projects, other projects are more complex, such as upgrading a control system or its many parts that are critical to the minute-to-minute operation. The other issue is the ability to simulate or model the power system at a sufficient scale (e.g., in the eastern interconnection) and behavior (daily, weekly, seasonal modes) so that extensive recursion testing of system can be accomplished prior to actually putting a system in service in a real operating environment. This very issue has been one of the roadblocks to widespread phasor measurement unit deployment.

References

- "EMS for the 21st Century, System Requirements," CIGRÉ—International Council on Large Electric Systems, Working Group D2.24, February 2011, ISBN 978-2-85873-141-1
- "The ElectriNet," C. Gellings and P. Zhang, Electra, CIGRÉ, June 2011.
- "Power to the People: The ElectriNet," Enlargement Europe, Power and Energy, 2004.

Chapter 8

The Grid As a Terrorist Target

PHYSICAL SECURITY

Perhaps no other infrastructure in the world is as visible and at the same time vulnerable as the electric power system. As such, its security is often challenged from a variety of high impact, low-frequency events of all kinds. This chapter addresses one type of such events—that which involves events which threaten the physical security of the smart grid. These threats range from shots fired at insulators by hunters disgruntled over an unsuccessful day, to seemly random high-powered rifle projectiles fired at a substation. The industry has also witnessed a variety of physical threats to the power systems such as unbolting legs of a corner tower transmission structure in order to disable part of a line in order to halt the lines construction.

The most serious concern regarding physical security is that from an organized attack, most likely from terrorists. By addressing concerns regarding terrorism, the power system planner will coincidentally make the system more resilient. In addition to securing the power system against terrorists, increased resiliency offers a measure of security to all manner of high-impact low-frequency events, especially including severe western events.

Terrorists could damage components of the electric power systems, causing local or even regional blackouts. It is unclear how serious a threat terrorists pose to infrastructures such as the electric power system (NRC 2002).

Several elements of the Smart Grid are potential choke points which could become terrorist targets. Substations are often located in suburban and remote locations with no staff on site. They are typically fenced in so as to prevent humans and animals from easily entering. Many of the facilities are several decades old, and security considerations were not part of their design when they were built. Complete destruction of substation will render all the lines connected to that substation inoperative. The vulnerability substations results from the reality that substation transformer and circuit breakers being easily identifiable; that space components are seldom on hand; and that these are extensive lead times to replace transformers. Within substations, transformers and circuit breakers can be attacked with any device that will penetrate the steel tank and allow coolants to drain or gases to escape, attacks often require complete replacement of the equipment. Access to substations can be easily gained and projectiles fired from outside the fence, or explosives used near it, can also cause significant damage.

NERC SECURITY GUIDELINE FOR THE ELECTRICITY SECTOR: PHYSICAL SECURITY

The North American Electricity Reliability Council has developed security guidelines for the electricity sector regarding physical security. (WWW.NERC.com) NERC believes that it is in the public interest for them to develop guidelines that are useful for improving the reliability of the bulk electric system. NERC guidelines "provide suggested guidance on a particular topic for use by bulk electric system entities according to each entity's facts and circumstances and not to provide binding norms, establish mandatory reliability standards, or be used to monitor or enforce compliance."

Physical security practices and measures are intended to help ensure the safety and security of a company's personnel, assets, and information. By providing the basic concepts of physical security and common practices and measures available to electric sector organizations, guidelines enable companies to develop a physical security plan that matches the level of accepted risk for each of their facilities. NERC guidelines apply primarily to facilities and functions that support the reliability of the electricity infrastructure, as well as the business processes and the overall operation of the individual organization.

In addition, attended, essential facilities typically require many more support assets such as uninterruptable power supply, water, redundant external power supply, environmental controls, access controls, and communication infrastructure that the typical unattended facility would not require. Since these support assets are fundamental to the reliable operation of the essential facility, they are themselves essential assets and appropriate physical protection should be considered. NERC suggests that the following facilities should be considered:

- Substations, including Switching Stations
- Generating Plants
- Control centers and relevant administrative sites
- The Transmission Infrastructure

NERC defines an effective physical security program as one which incorporates each of the following objectives:

- **Avoidance**: Ensure electric power system integrity and availability by promoting security policies, standards, and procedures.
- **Assurance**: Ensure electric power system integrity and availability by promoting the regular evaluation of security measures.
- **Detection**: Protect electric power systems through monitoring, identification, reporting and analysis of threats and incidents.
- **Recovery**: Promote methods for timely investigation of security incidents and rapid recovery and restoration of services supporting the delivery of electric power services. Lessons Learned from this layer are incorporated into the other tiers.

The concept of physical security for most businesses generally includes systems for detecting intruders, monitoring control systems, establishing security guards, mitigating access by vehicles and protection against use of projectiles from outside. (CPNI 2010) For electric power systems these concepts require an added layer or two of complexity. The first line of physical security for electric power systems is to deter against physical attacks by the use of visible barriers and security measures which deter intrusion. Secondly, electric power systems must employ detection (sensors and communications) which enable prompt action to respond so as to mitigate damage or at least to delay intrusion. Finally, electric power systems must recover to normal operation and apply knowledge which helps prevent similar events from occurring in the future.

NERC recommends combing these concepts in order to apply a consistent "systems approach" to designing and implementing physical security measures that will mitigate the impact on assets should a physical attack occur. A popular approach to a systems approach is a Defense in Depth (DID) approach. Figure 8-1 illustrates the DID approach. DID is a strategy that seeks to delay rather than prevent the advance of an attacker, buying time by yielding space. Rather than defeating an attacker with a single, strong defensive line, defense in depth relies on the tendency of an attack to lose momentum over a period of time allowing time to respond. Figure 8-1 illustrates the DID approach.

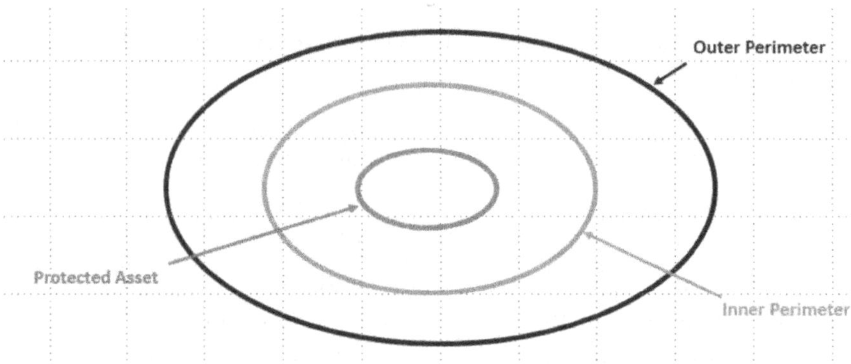

Figure 8-1. The Defense In Depth Approach (Based on NERC Protection in Depth Guideline)

In the figure the outer perimeter, inner perimeter and protected assets may apply to power plants, sub stations or control centers as outlined in Table 8-1.

Substations are "relative soft targets"; they are vulnerable to attacks from nearby as well as penetration attacks from within by those compromising the perimeter fence within the Smart Grid, substations are the most probable terrorist targets. Substations are particularly difficult to protect. Their replacement parts are expensive, not typically stored nearby and difficult to obtain. In addition, when stored on-site the components are equally susceptible to damage.

PHYSICAL BARRIERS

Physical barriers such as fences, walls, and vehicle barricades are the outer perimeter of security. Prevent, or delay, attacks, and define the perimeter of the facility. From a terrorist's view they give the impres-

Table 8-1. Elements of Critical Assets

Critical Asset	Power Plant	Substation	Control Center
Outer Perimeter	Ouside perimeter fence[26]	Outside perimeter fence or substation wall	Outside Perimeter fence or bulding infrastructure in cases where the control center is embeeded in an administration building
Inner Perimeter	Inside perimeter fence routinely guarded and protected with sensors and other intrusion devices	Inside perimeter fence often guarded and protected with sensors and other intrusion devices	Secondary building area controlled and limited to employees and visitors with approved access
Protected Asset	Power plant wall or secure grounds immediately surrounding critical generation equipment, fuel supply and cooling water facilities	Substation control room and relay house	Control room proper and all IT and communications equipment

*In Major power plant installations there are often two outside perimeter fences.

sion of and making intrusions more difficult. Popular physical barriers include fences with razor wire surrounding the perimeter of a property adorned with warning signs. Where fences are impractical such as in urban office buildings, the outer security perimeter are the walls and doors of the structure itself.

In the case of protecting buildings such as control centers, architects can incorporate natural surveillance into the building's design by making the structure open and visible to security personnel and authorized users, so that attackers are unable to perform unauthorized activity without being seen.

Security Lighting

Security lighting is an effective form of deterrence for substations and power plants. Intruders are less likely to enter well-lit areas. Driveways, doors, gates, and the perimeter of the property of critical assets should be well lit to allow close observation of people entering and exiting. When lighting the grounds of a facility, widely-distributed low-intensity lighting is generally superior to small patches of high-intensity lighting, because the latter can have a tendency to create blind spots for security personnel and CCTV cameras. Security lighting should be placed so that it is difficult to tamper with.

Access Systems

Access control systems such as locks must be placed at gates and doors. In large facilities with rapid turnover, managing keys can be a problem, mandating the adoption of electronic access control. Electronic access control manages large user populations, controlling for user lifecycles times, dates, and individual access points. Another form of access control mandated for control centers an applicable elsewhere is the use of policies, processes and procedures to manage the ingress into the restricted area.

Security Personnel

Security personnel play key roles in all layers of security. The technological systems employed to enhance physical security require a security force that is trained in their use and maintenance. Security personnel also know how to properly respond to breaches in security. Security personnel may also patrol at checkpoints, administer access control, respond to alarms, sensors and video feeds.

SUBSTATIONS

Substations include switching stations and both transmission and distribution substations as well as generation switch yards. These are facilities which contain medium or high voltage switches, circuit breakers, circuit buses, and transformers for switching power circuits and often intended to transform power from one voltage to another or from one system to another. Switching stations are substations where transmission lines are connected without power transformers. Substations are common elements in the power system necessary for the transmission and distribution of electric power and are often unattended. Most substations have a control room or building within the perimeter fence containing batteries, relays, and control and communication equipment utilized for operation and protection of the power system.

The control room or building often contains low-voltage or medium-voltage switchgear for station service. Typical designs provide a basic level of physical security. Physical security of substations are typically designed to standards outlines in the Institute of Electrical and Electronics Engineers (IEEE) Standard 1402, Guide for Electric Power Substation Physical and Electronic Security.

Not only are substations a potential target to terrorists and vandals, but with copper prices at an all-time high, it's no surprise that substations have become an increasingly attractive target for copper theft causing expensive losses while compromising the safety of the substation itself. (Sightlogix 2015)

TRANSMISSION LINES

High-voltage transmission lines, and the associated support towers, are often located in isolated areas. These lines and towers are referred to as "Transmission Line Corridors" (Corridors.) They are particularly vulnerable to physical attack and can present significant risk and physical security challenges. Defense in depth principles can be applied to transmission lines and at areas surrounding the base of the towers. Redundant transmission lines and secure replacement are typically considered as common practice. Some transmission corridors have "zones" which are especially sensitive. These include areas: where line come to a 90 degree angle; intersection with other critical infrastructure such as rails, rivers, and pipelines.

Transmission lines are historically a terrorist target in countries experiencing insurrection or civil unrest. Transmission lines can be easily disabled. There are, however, often quickly repaired. Popular tactics to disable transmission lines include: taking out a tower where two lines cross so as to disable both circuits at once. Long transmission lines are often transverse rural areas and make easy unprotected targets.

GENERATING PLANTS

Fossil fuel and nuclear plants typically involve extensive operations consisting of numerous structures spread over several acres with large numbers of entry points onto the site. Hydro plants tend to be contained within or adjacent to dams and have a lesser amount of unauthorized access. Combustion turbine and combined cycle plants consist of numerous separate structures spread over a more open space.

Generating plants are almost always continually staffed and cover a substantial amount of area and utilizing a widely dispersed work force. Occasionally, stand-alone switch-houses or control buildings, which are part of the generating plant configuration, are not routinely attended or patrolled. Generally the entire property perimeter of power plants are fenced. As security issues have risen in concern, asset owners and operators are paying special attention to water-based approaches, as many plants reside on or very near large bodies of water or rivers.

All generating plants generally have generation switchyards associated with them. These are controlled by a stand-alone switch-house or control building, or their operations may be incorporated into the plant's generation control center. Physical security concepts should also be applied to areas in power plants that contain hazardous chemicals and unstaffed areas such electronic or electrical control compartments.

Nuclear power plants enjoy an even greater degree of physical security which follow IEEE Standard 692, Standard Criteria for Security Systems for Nuclear Power Generating Stations.

CONTROL CENTERS AND CONTROL SYSTEMS

Accessing control centers can offer attackers opportunities to destroy rotating equipment or SCADA systems; cause harm to operating personnel; access the grid to introduce malware, or they can hold the facility hostage. In addition to control centers, administrative offices should be protected as any office structure within your company would be by giving due consideration to the inherent risks and the assets at its location.

Electrical systems cannot function without system controls, sensors, communications and computational ability are continuously used to monitor major system elements. The system is designed to take remedial measures against routine disturbances automatically and to report these measures to control center. Disturbances often require quick decisions and reactions on the part of control center operators to prevent widespread outages.

System-wide control centers contain control and communications equipment supported by system operators. Attacks by terrorists that would damage such control center, would also impair the operation or restoration of a system. Many control centers have been extensively hardened, some have complete redundant twin centers and have excellent access control and security. The greatest threat to control centers is from sabotage by insiders.

Control centers could be desirable terrorist target.

SENSING, COMMUNICATION, AND CONTROL

Electric power system operation depends on systems of sensors and automated and manual controls, and computational ability (computers), all of which are tied together through communication systems. Physical destruction of substations, or power lines may be the most obvious strategy for disabling the Smart Grid and causing blackouts, activities that compromise communications and control through cyber incursion could also disrupt the system, cause blackouts, physical damage.

As the Smart Grid provides the interconnections with electronic office and business systems through other layers of communications have created vulnerabilities. This problem is now well understood and corrective actions are being taken. A number of cyber security intrusions, as well as penetration red-team attacks, have been conducted by EPRI, electric utilities, commercial security consultants, and others. These "attacks" have uncovered a variety of cyber vulnerabilities including unauthorized access, penetration, and hijacking of control.

EPRI has conducted a survey of electric utilities to identify their concerns about grid security, cyber security, and communications security (EPRI, 2000). Figure 8-2 ranks the perceived threats to utility control centers. The most likely threats identified were bypassing controls, integrity violations, and authorization violations, with 40 percent of respondents rating the seriousness of each as either a 5 or a 4 on a scale of 0 to 5. Concern about the potential threats generally increased as the size of the utility peak load increased.

DISTRIBUTION SYSTEMS

Distribution systems are complex and consist of numerous components. In addition, distribution spare parts are generally in great supply. Distribution system components can cause troublesome outages, but the magnitude of the outages are not spectacular enough to warrant terrorist attention.

An expanded definition of "national security reliability" is being used in the planning of the electric power industry. The North American

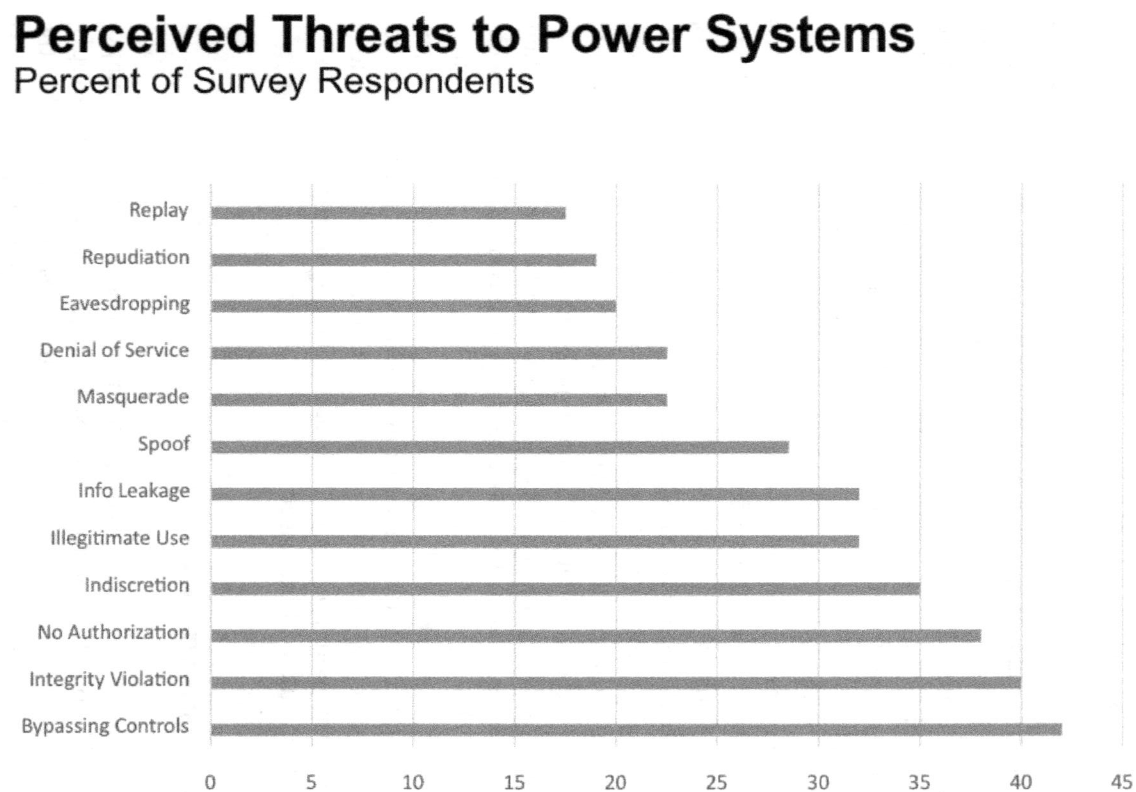

Figure 8-2. Perceived threats to power system control centers as reported in a survey of electric utilities conducted by EPRI in 2000.

Electric Reliability Corporation (NERC), with the Federal Energy Regulatory Commission (FERC) Security protocols and mitigation measures have been deployed through cooperation between government and industry to provide protection against terrorist threats. Many critical elements of today's Smart Grid now involve advanced security protocols and mitigation measures have been deployed through cooperation between government and industry to provide protection against terrorist threats. May critical elements of today's Smart Grid now involve advanced security technologies that include digital CCTV, fiber optics, smart cards and biometric IDs and card keys, as well as fencing design and manufacturing improvements.

Figure 8-3 provide a simplified schematic diagram of the sensing, communication, and control elements of a modern system. The elements of the system depicted in Figure 8-3 are defined below. (EPRI 1002604)

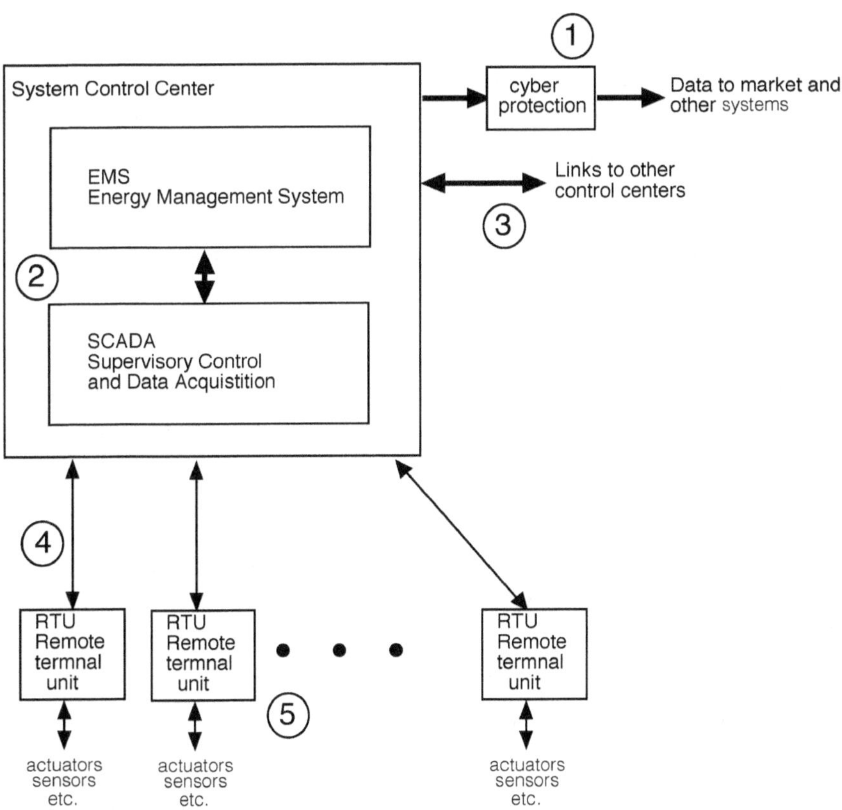

Figure 8-3. Sensing Communications and Control (EPRI 1002604)

EMS: Energy Management System

The emergency management system (EMS) analyzes the real-time data gathered by the Supervisory Control and Data Acquisition (SCADA) system. That data helps operators determine the reliability of the grid to alert the operator to any vulnerabilities and to calculate possible operational changes that could improve the operational condition.

Supervisory Control and Data Acquisition (SCADA)

SCADA systems provide data acquisition, supervisory control, and alarm display.

(RTUs) Remote Terminal Units

RTUs are special-purpose microprocessor-based elements that located at substations to interface with all the substation equipment.

(PLGs) Programmable Logic Controllers

Programmable logic controllers (PLCs) have been used extensively to other industries and are now being used to implement relay and control systems in substations.

Protective Relays

Protective relays are digital controllers designed to respond to system faults and short circuits.

Advanced Metering Infrastructure (AMI)

AMI is designed to upload residential and/or commercial electric meter data. These data can then be automatically downloaded and transmitted.

Control System Vulnerabilities

The North American Reliability Corporation (NERC) in Control Systems Security Working Group addressed 10 control system vulnerabilities (NERC, 2006).

The three most critical vulnerabilities listed by NERC:

1. Inadequately secured wireless communication so as to identify the best areas for wireless use and ensure that leakage is minimized.
2. Use of non-dedicated communications channel for command and control such as use of Interest-based SCADA.

3. Unauthenticated command and control data which do not authenticate who's making the change and authorize that the change is allowed for that user through the controller

One additional control system vulnerability is the potential impact of a High Altitude Electromagnetic Pulses (HEMP). HEMP are a potential nation-state terrorist attack which could severely damage the Smart Grid. In particular the coupling of sensors in the transformer to cables inside of control rooms is most vulnerable to HEMP pulses. Supervisory Control and Data Acquisition (SCAD) systems in nearly all power grid operations rely heavily on digital control systems (DCS) and programmable logic controllers (PLC) networked together over the grid. A large-scale EMP could disrupt operations of entire centers and inter-center communications, either directly or through cascade-type events. In addition, voltages as low as 0.6kV are sufficient to destroy programmable logic controllers used in the flow of fuel and other power generation processes.

TECHNOLOGIES TO ENABLE PHYSICAL SECURITY

Sensors

As detailed in Chapter 2, the transmission system of the future will utilize a synergistic concept for the instrumentation of electric power utility towers with sensor technology designed to increase the physical security of the transmission system. The system concept consists of sensors embedded in structures and towers (EPRI 1016921). This system scope is limited to transmission line applications (i.e., 69 kV and above), not distribution, with the focus on steel lattice and pole structures, not wooden.

In this concept, the addition of wiring to interconnect and/or power distributed sensors is not viable because of electromagnetic susceptibility concerns and labor intensive installation. Consequently, sensor concepts will rely on wireless and/or fiber optic technology.

Automated Security Systems

Several types of automated security systems are now available which detect and send alarms about security incidents in real time. These can include the use of video cameras, fence detection, motion detection and the use of photo beams, fiber sensing and microwave detectors.

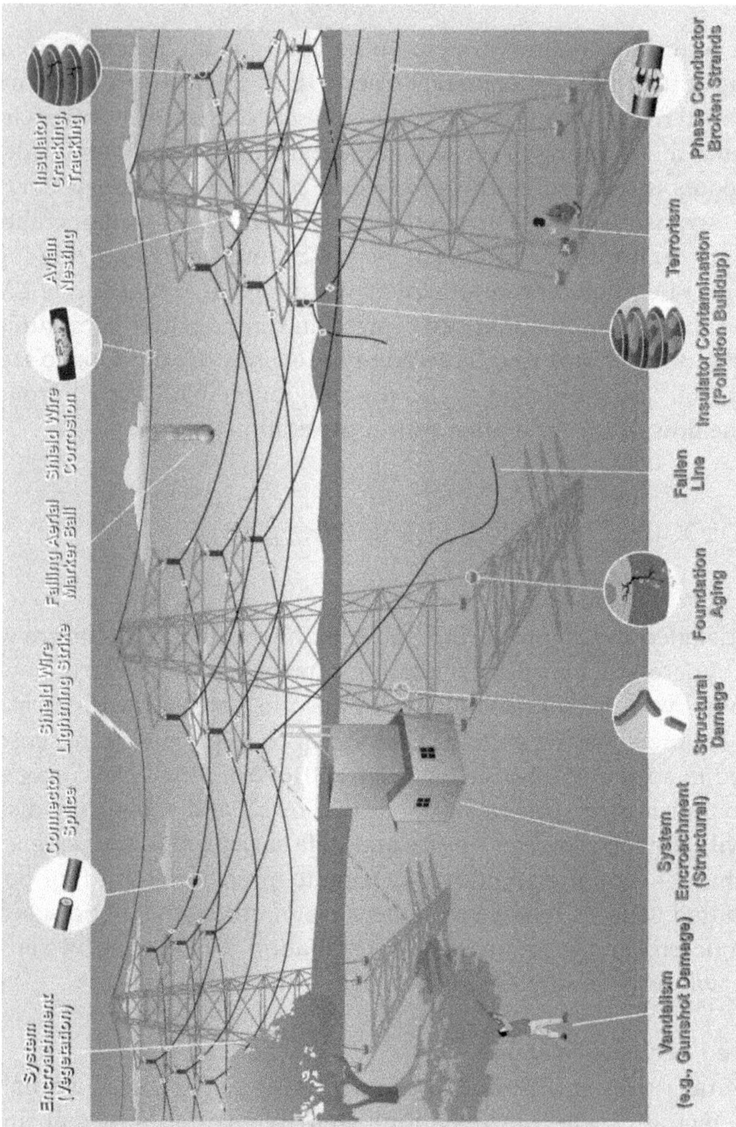

Figure 8-4. Illustration of Sensor Needs for Transmission Lines and Towers (EPRI 1016921)

Video camera systems, including infrared cameras are equipped with video analytics that increase the probability of intruder detection while greatly reducing nuisance alarms. These systems can cover large areas and detect intruders in dark, bright sunlight and poor weather. They can also provide a buffer zone beyond the fence line. In a manned facility these systems come equipped with pan/tilt/zoom cameras for alarm assessment by security personnel. When the systems detects an intruder, an alarm is triggered to alert personnel and the target's location is communicated to systems software which automatically steers cameras to the area of intrusion.

These systems can incorporate the use of substation robots as illustrated in Figure 8-5.

Additional detection technologies can incorporate a series of sensors also focused on the facility's perimeter. Motion detection can be utilized with software that is programmed to filter out signals from animals and birds. Fiber sensing cables can be installed on fencing which will alarm when fences are approached, climbed or cut. Sensors or cables can

Figure 8-5. Substation Robot with Online Infrared and Visual Cameras (EPRI)

be buried or embedded in walls. These systems incorporate fiber optic sensing cables which conduct light which is sensitive to any disturbance. They operate on the principle that any motion, vibration or pressure on the cable will alter the propagation of light through the fiber. (WWW. WESCO.com)

Preventing Projectile Damage

Several alternatives are being researched which could potential mitigate damage to components by projectiles. Figure 8-6 illustrates the use of innovative materials to protect substation equipment.

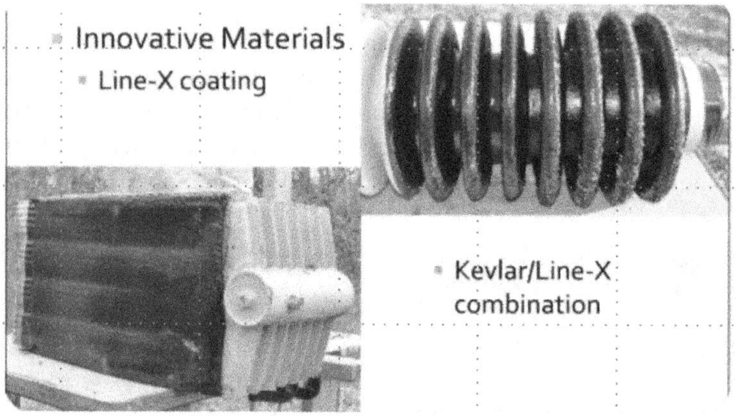

Figure 8-6. Explosive Protection and Coatings (University of Kentucky)

Security Basics

There are some basic actions that all power system owners and operators must take to protect the power system.

Physical Barrier—Physical barriers used for power system facilities include fences and walls. As described earlier, these prevent or delay the process of an attack. At substations and power plants they almost always include fences. A costly alternative to substation and power plant protection utilized by some utilities involves the use of walls. This can range from enclosing the facility completely to constructing walls which inhibit intrusion. These "walled" facilities have the added advantage of mitigating noise and can be viewed as more aesthetic than alternatives. Figure 8-7 and 8-8 illustrate some examples of these techniques.

Figure 8-7. Indoor Substation one completely enclosed on all sides (ConEd)

Figure 8-8. Possible use of a Wall to Protect a Substation (ConEd)

Security Lighting—Security lighting is an effective means of deterring intruders. When carefully designed, these systems can be designed with low intensity, but uniform lighting.

Mechanical Access Control—Mechanical access control ranges from keeping track of where the keys are and who has them to more sophisticated access systems.

Security Guards—Use of guards are an effective means to operationalize security.

POWER GRID HARDENING AND MITIGATION AGAINST HEMP

In its 2008 report, the EMP Commission concluded that protecting the entire power grip system, or even all high value components, from HEMP pulse attacks is impractical because the costs, time, and effort in proportion to the likelihood of a HEMP enhanced nuclear strike make it prohibitively expensive. The report recommendation is to prepare for the widest possible scenarios of GMD and EMPs with a focus on quick response and repairs, as well as conduct further research into cost-effective mitigation and preparation. Work has already begun in these areas, including efforts outlined in the following subsection.

Shielding

The U.S. military, for example, specifically under the NIL-STD-188-125 standard, deploys technology for protecting fixed and mobile installation from EMP effects. Ground-based shielding detailed in MIL-STD-188i-125 refers to "subscriber terminals and data processing centers, transmitting and receiving communications, stations and relay facilities for both new construction and retrofit of existing facilities" (see Figure 8-4). This standard calls for EMP mitigation for all military command, control, communications, and computer systems, including the EM energy point of entry (POE) to these systems. Adhering to this standard would greatly enhance the protection of North American power grid reliance on its current and future (SCADA) operations, including smart grid components and systems, but would be extremely costly.

The major requirement of the standard relies on Faraday cage implementation(s) around specific components, POEs, or entire facilities.

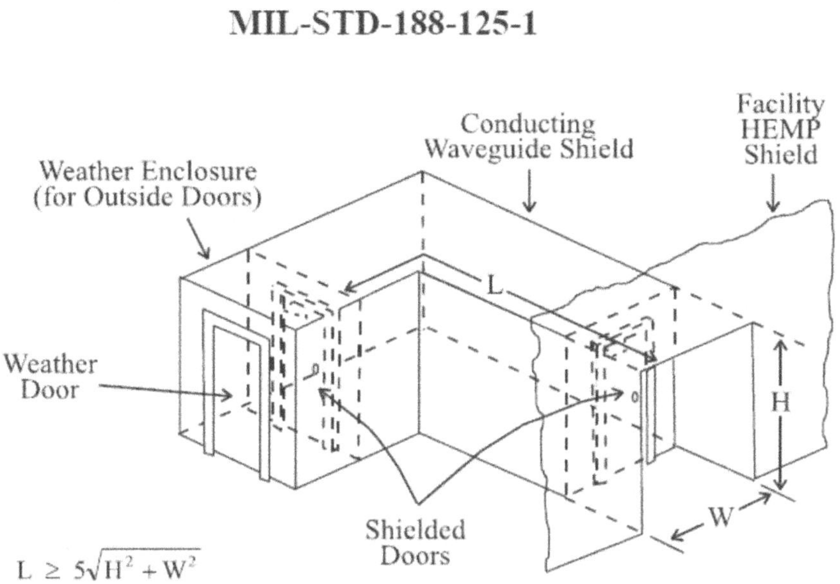

Figure 8-9. HEMP Shielding Around Control Facility

The cage is composed of conductive mesh materials and an insulation barrier around components or entire facilities that absorb pulse energy and channel is directly to ground. The underlying electrical components enclosed in the cage remain intact and unaffected.

Stored and Shared Spare Components

An essential component in the recovery process is the availability of components needed in the event of an emergency. These components could be shared among all stakeholders and geographically dispersed for quick access. The industry can assess the vulnerability to EMP exposure of various components and systems and the inventory of available equipment to [plan sufficient quantities of backup components where needed. Transformers, generation equipment, SCADA, aboveground relays and controls, and other embedded processors are examples of potential spare part candidates for stockpiling.

The goal of a NERC Spare Equipment Database (SED program is to provide a means to securely connect entities that need replacement transformers with entities that have such spares available in the event of simultaneous loss of multiple transformers. The Edison Electric Institute's (EEI) Spare Transformer Equipment program (STEP Program), launched in 2006 in response to the 9/11 terrorist attacks, addresses the need to manage resources in response to a terrorist attack. About 50 transmission providers that represent about 70% of the U.S. Transmission grid currently participate in this program. Defined stakeholders could take responsibility for the inventory, research, and securing of hardened facilities for these parts depots, including a coordinated and clear command and distribution network for part deliveries and repairs. Realizing this vision would require cost and feasibility studies, as well as research that identifies the most vulnerable components under different EMP scenarios (the fact that many components are now manufactured overseas should also be taken into account as further complication).

Recovery Transformers

The U.S. Department of Homeland Security (DHS) and EPRI are engaged in a project that is an important part of an overall strategy to mitigate the potential impacts of high-impact, low-frequency (HILF) events. Working as a complement to power system operational measures, power system equipment hardening, and other approaches, the Recovery Transformer—when fully deployed—can enhance the ability to respond and recover rapidly after a HILF event. This section describes the Recovery Transformer project, including the recent successful deployment of its prototype in Texas, its role vis-à-vis other mitigation approaches, and next steps for the project.

High voltage transformers are critical to the continued reliable operation of power systems. More than 90% of consumed power passes through high-voltage transformers at some point. These large devices, weighing hundreds of tons each, are potentially vulnerable to the effects of HILFs such as GMD, EMPs/HEMPs, natural disasters, and coordinated attacks. This vulnerability is compounded by the fact that many U.S. high-voltage transformers are approaching or exceeding their design lives. Due to their size and complexity, these transformers are expensive and time consuming to replace in the event of failure due to a HILF. In the event of failure, few spares exist, and those that do are not easily installable at other locations besides those originally intended.

The North American power grid, for example, has built-in redundancy to accommodate simultaneous failure of a small handful of high-voltage transformers. The North American Electric Reliability Corporation (NERC) N-1 reliability standard (contingency) is designed such that failure of one high-voltage transformer may strain the power system but not cause a major outage or cascading failure. The concern is that simultaneous failure of a higher number of high-voltage transformers (e.g., as a result of a HILF) could lead to a major failure.

This adds up to a potentially significant vulnerability, and risk, in the power system. One approach is to attempt to retrofit these transformers with devices that harden them against various HILFs. This approach shows promise; various designs have been proposed that block or reduce GIC flow in transformers and lines to mitigate Geomagnetically Disturbances (GMDs), including series compensation, use of blocking capacitors in the neutral ground, and use of neutral resistors to reduce GIC flow. These same devices have the potential to shield against Electromagnetic Pulse (EMP) attack.

This is a challenging approach due to the sheer number of high-voltage transformers, various different designs and sizes, and need to ensure that any retrofits do not adversely affect normal operation. With regard to the number of transformers, according to a 2004 ABB study for the Federal Energy Regulatory Commission (FERC) updated in 2010, there are 1547 high-voltage transformers in the U.S. alone (i.e., high-side voltage of 345 kV and above). With regard to design variation, impedance of most units ranges from 9-15%, MVA ratings vary from 150-750 MVA, and other design variations include single-phase versus three-phase; shell form versus core form; three-, five-, or seven-leg models; and others.

Various operational measures, such as reducing load on some high-voltage transformers in advance of the HILF, will certainly help to mitigate transformer damage. However, depending on the severity and character of the HILF (e.g., HILFs with little or no warning), such measures may not completely protect all high-voltage transformers from overload, damage, or failure.

Another approach is to make available to other utilities in the event of a HILF the limited number of spare transformers that do exist. This approach makes sense and is a key recommended "proposal for action" in NERC's 2010 assessment of HILF risk to the North American bulk power system [NERC 2010]. The goal of the NERC GMD Spare Equipment Database (SED) program is to provide a means to securely connect entities that need replacement transformers with entities that have such spares available. In the event of a HILF-type event, such as a significant GMD, access to information about available spares to match particular needs at a specific substation location would help speed power restoration. NERC's GMD Spare Equipment Database Task Force (SEDTF) is spearheading this effort. This program is not intended to replace or supersede any existing transformer sharing programs, such as the Edison Electric Institute (EEI) Spare Transformer Equipment Program (STEP) or other regional or neighboring utility sharing arrangements [FAMA 2010]. EEI's STEP Program, launched in 2006 in response to the 9/11 terrorist attacks, addresses the need to pool resources in response to a terrorist attack. About 50 transmission providers that represent about 70% of the U.S. transmission grid currently participate in this program.

The spare transformer sharing approach may be limited by the differences in high-voltage transformer designs (listed above), as well as design needs of specific installations (e.g., tap ratios, cooling system variations, use of gas-filled bushings, and others). These and other complexities would almost certainly make transportation, installation, and energization of spare transformers (in locations where they were not originally designed to be installed) a time consuming process in the event of a HILF.

Rapid Response Capability

If a HILF severely damages more than a handful of high-voltage transformers simultaneously, a rapid response capability would be needed. This need would be especially acute if the damaged transformers are in the same geographical area, which would be a likely objective of an intentional, coordinated attack HILF. At the very least, such a rapid response capability would complement other measures in place.

The U.S. Department of Homeland Security (DHS), EPRI, and the U.S. Department of Energy worked together to demonstrate such an approach that, when fully deployed, would provide just such a rapid response capability for high-voltage transformers. The primary objective of this Recovery Transformer project was to enhance resilience to power outages due to natural disasters or deliberate attacks. The project team successfully designed, tested, and installed a prototype Recovery Transformer that can be stored in a centralized location and rapidly transported and installed at a remote utility location.

The particular Recovery Transformer designed in this prototype project—constructed by contractor ABB—is applicable to more than 90% of all U.S. 345 kV/138 kV transformers. And this high-side/low-side voltage combination is the most common one in the U.S., representing NERC regions use transformers with this voltage combination, and the

345 kV/138 kV transmission grid covers a significant portion of the U.S. geographically.

(without Generator Step-Up Transformers)

High Side Voltage kV	Low Side Voltage - kV										TOTAL
	500	345	287	230	161	138	120	115	69	34	
765	4	25	0	3	0	11	0	0	0	0	40
500		26	1	225	41	33	0	29	4	0	359
345			0	136	99	689	30	173	17	4	1148

TOTAL 1547

Highlights 79% of Transformers
Highlights 12% of Transformers
Total Highlighted is 91%

Figure 8-10. The most common combination of high-side and low-side voltage for high-voltage transformers in the U.S. is 345 kV/138 kV—the combination selected for the prototype Recovery Transformer project.

The three single-phase transformer prototypes were successfully transported to, installed, and energized at a CenterPoint Energy (CNP) substation outside of the Houston area over a five and one-half day period in March 2012 (see Figure 8-11). The host utility for this project, CNP serves over 5 million customers in the Houston area. The completion of this deployment exercise represents a quantum leap for the electric power industry to rapidly replace high-voltage transformers in emergency situations.

The Recovery Transformer Program

The crux of the recovery transformer is rapid deployment and broad applicability at a range of utility substations. Under normal circumstances, replacement transformers are typically custom ordered and built, requiring about one year for scheduling, design, and construction, and can take four weeks or more to deliver and install. Typical installations require rail transportation, and a heavy duty crane, along with the associated scheduling. Conversely, Recovery Transformers are designed to be emergency replacements for the transformers found at utility substations across the U.S. Recovery Transformers will be deployed rapidly during the recovery time period after transformer major outages (when

Figure 8-11. Installation of the three single-phase prototype high-voltage Recovery Transformers at CenterPoint Energy in March 2012.

the transformer cannot be repaired in a few days). At present, high-voltage transmission systems in different regions operate at various voltage levels, making standardized transformer design difficult. The Recovery Transformer is designed to replace as many high-voltage transformers as practical. The Recovery Transformer Program advances conventional transformer technology to build a prototype that is small enough and light enough to be easily transported and installed at substations across the country in a few days.

Recovery Transformer Functional Requirements

The recovery transformer functional requirements that follow are consistent with requirements to mitigate the effects of HILF events (see Figure 8-12).

- Applicability: Because the recovery transformer will need to replace as many transformers as possible, the applicability rate is set at 90% or higher for the 345/138 kV voltage class.
- Manufacturing: Because the recovery transformer needs to support recovery in acts of war, natural disaster, or other HILFs, independent manufacture in the U.S. is a requirement.
- Cost to Manufacture: The Recovery Transformer cost should be 20% less than a conventional equivalent.
- Storage Functionality: The storage cost should be low, and the life of the transformers and ancillary equipment should be long.
- Overall Time to Energize from Call to Action: The estimated elapsed time from the call to action until the transformer is energized is four weeks for a conventional high-voltage transformer. The recovery transformer requirement is less than one week. The components of the total time to energize are described as follows:
  - Transportation Time: A conventional transformer is estimated to take one week to transport to the substation. The goal of the Recovery Transformer program is to reduce transportation time to 24 hours from the storage site to the substation.
    - Shipping Size: A conventional transformer is often too tall and in some cases too wide to pass standard bridges and underpasses. The Recovery Transformer requirement is for the transformer to be transportable by road and the specific goal is to reduce the size by 25%.
    - Shipping Weight: A conventional single phase unit weighs 225,000 to 250,000 lbs (102,058 to 113,398 kg). The Recovery Transformer requirement for shipping weight (including trailer) is to be low enough for road transportation on a standard lowboy trailer, which is about 120,000 lbs (54,431 kg).
  - Installation Time Once on Site (includes startup time): Once on site, a conventional transformer takes about one week to install and energize, including about four days to fill with insulating oil and commission. The Recovery Transformer requirement is 60 hours or less, including 12 hours or less to fill and commission.
- Site Preparation Requirements: For a conventional transformer, there are typically considerable site preparations to build a pad to

support the transformer and to build firewalls and oil containment pits to accommodate the oil cooling system. For Recovery Transformers, these preparations are required to be minimal.

- Installation Requirement for Special Equipment: A conventional transformer may require special crane or lifting equipment to install as well as special equipment to load gas filled bushings and to pull a vacuum for oil insulation. The Recovery Transformer requirement is to minimize or eliminate special installation requirements.
- In-Service Factors
  - Operational Reliability: Conventional transformers have an excellent operational record and are proven reliable in service. The Recovery Transformer requirement is to equal this reliability record.
  - Overload Capability: The Recovery Transformer requirement is to provide overload capability consistent with a conventional transformer, which is two times nameplate capacity. Overload capability can only be sustained for short time periods without reducing the transformer life and risking premature failure.
- Operational Efficiency: Conventional transformers are highly efficient (99.8%). The initial Recovery Transformer goal was to achieve 99.00% efficiency in service. The prototype Recovery Transformer has an efficiency of 99.78%.
- Safety: Conventional transformers exhibit safety risks associated with potential damage from highly pressurized oil escaping as well as the fire risk if the oil escapes the transformer tank. The Recovery Transformer requirement is to not increase safety risks.

Protection from GMD

One means of protection against GMDs is to install 2.5-7.5 ohm neutral resistors in the ground connection of high-voltage transformers to make the equipment more resilient to the impact of severe GMDs (see Figure 8-13). This has been called a neutral grounding resistor or a neutral resistor/blocking device (NRBD), even though, strictly speaking, it does not fully block current flow. The reduction in GIC levels depends on the resistances—ground resistance, transformer winding resistances, and transmission line phase resistance—for the specific situation where

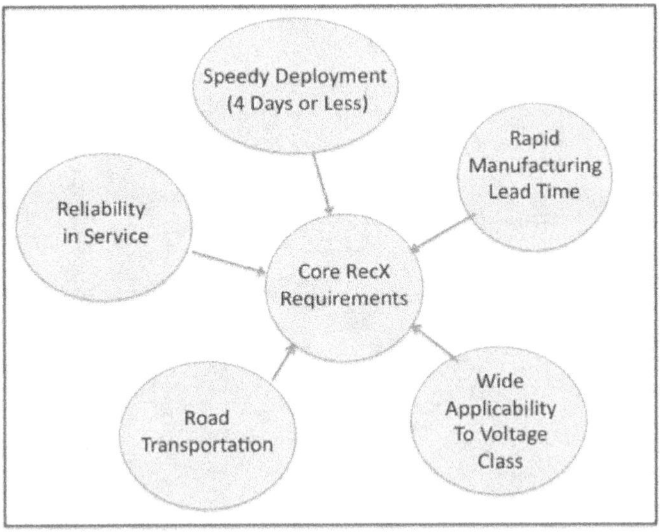

Figure 8-12. Core Recovery Transformer requirements are consistent with those of technologies needed for HILF event mitigation. (EPRI 2012)

it is applied, but typical levels that can be achieved are 60-70% reduction. In this approach, the utility must select a ground resistance value based on fault current and relaying requirements.

The factors that influence the decision include: the cost to install the neutral resistors; the cost to repair the transformer and restore service if the resistors are not installed and a severe GMD causes damage; the number and type of customers affected if an outage occurs; the cost of customer outages; and so forth.

The sources of risk include the probability of a severe GMD occurring, the extent of the damage to the transformer caused by a severe GMD, the duration of interruption as a result of the damage to the transformer, and to what extent (if any) installation of the neutral resistors could adversely impact normal operation.

In performing sensitivity analyses, the factors that have the most influence on the results are identified by quantifying an influence diagram (see an example for a different analysis in Figure 8-14. For the neutral resistor project analysis, these factors include the cost of installing the resistors, the cost of repairing the damaged system if the resistors are not installed, the cost to customers of a lengthy outage due to massive system repairs, and so forth.

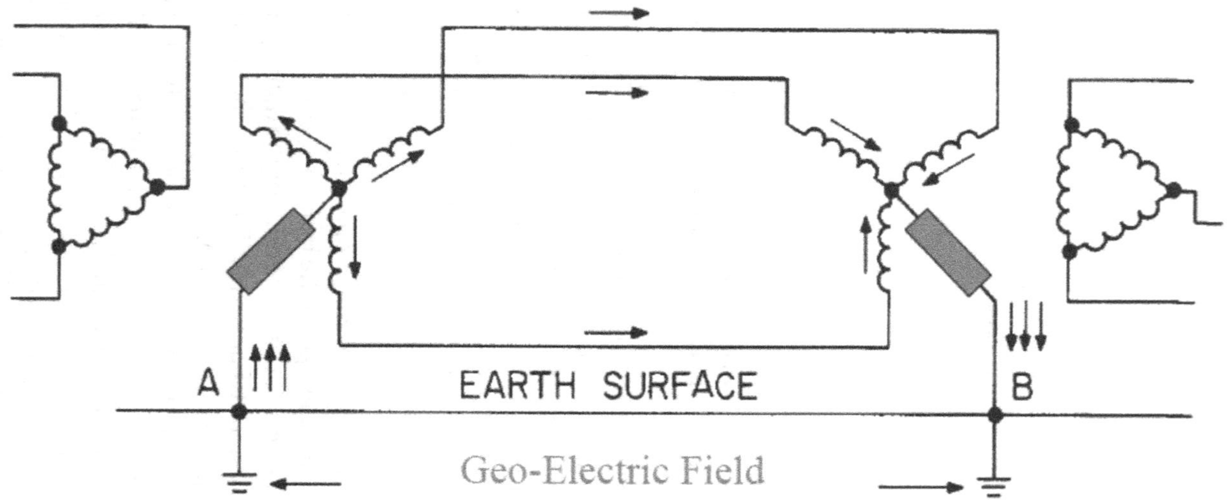

Figure 8-13. One approach to reduce Geomagnetically Induced Currents (GICs) in transformers: ohmic resistance in the ground connection.

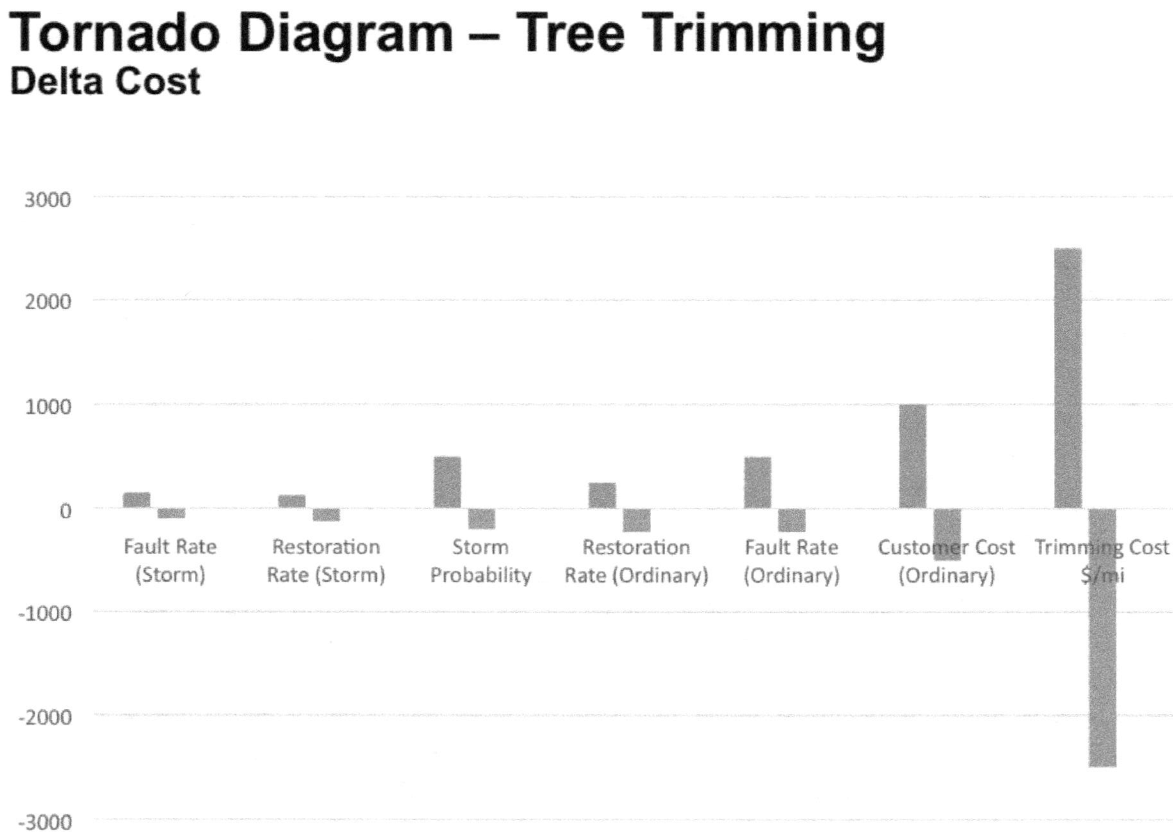

Figure 8-14. Sensitivity analyses can be used to identify the factor(s) that most affect the results of an analysis. In this case, the tree-trimming cost has the greatest impact on the decision whether or not to trim trees to maintain high reliability.

Because of the rarity of severe GMDs, limited historical data is available for these events. In these cases, judgment must be used in the absence of historical data. As new information becomes available when the industry conducts research in various GMD-related areas, the analysis can be refined. For example, the potential adverse impact of neutral resistor installation on normal operation of transformers is not currently known with certainty. Until this uncertainty is resolved, engineering judgment can be used to speculate that if any adverse impacts arose, they could be addressed by modifying the design of the neutral resistance system. Then, after field demonstrations provide more information, the analysis can be refined to incorporate this data. For example, the neutral resistors may be found to pose no ill effects for certain designs of transformers, and hence these transformers can be analyzed with certainty, with respect to this particular aspect of the analysis.

A risk analytical model can be applied to the risks associated with a severe GMD. A diagram could be created that would show the results of a probabilistic analysis. For example, a plot of probability versus total cost would show results with and without the neutral resistors installed. An example of such a diagram for another analysis is shown in Figure 8-15. The range of costs in both scenarios can be compared, and the portion of the probability where total costs are higher can be compared for both scenarios. In this case, total cost could include the sum of all costs quantified in the analysis.

The risk of whether to install the resistors could be evaluated for a utility's entire portfolio of high-voltage transformers. Whether or not to install the resistors on various types of high-voltage transformers, depending on their relative vulnerability to severe GMDs, could be considered. The utility could also consider the relative critical nature of each of the high-voltage transformers when prioritizing. In this way, an entire portfolio of projects can be evaluated.

Due to the rarity of severe GMDs and the novelty of the blocking devices, engineering judgment is required to estimate GMD risk and NRBD effectiveness. This uncertainty suggests another dimension to the risk management strategy: identification of the appropriate investment to reduce uncertainty. Given finite mitigation resources, the project needs to determine how much to spend reducing uncertainty and how much to spend on mitigation. This enables the team to optimize the risk management strategy.

Extending this analysis further, the utility could also consider vari-

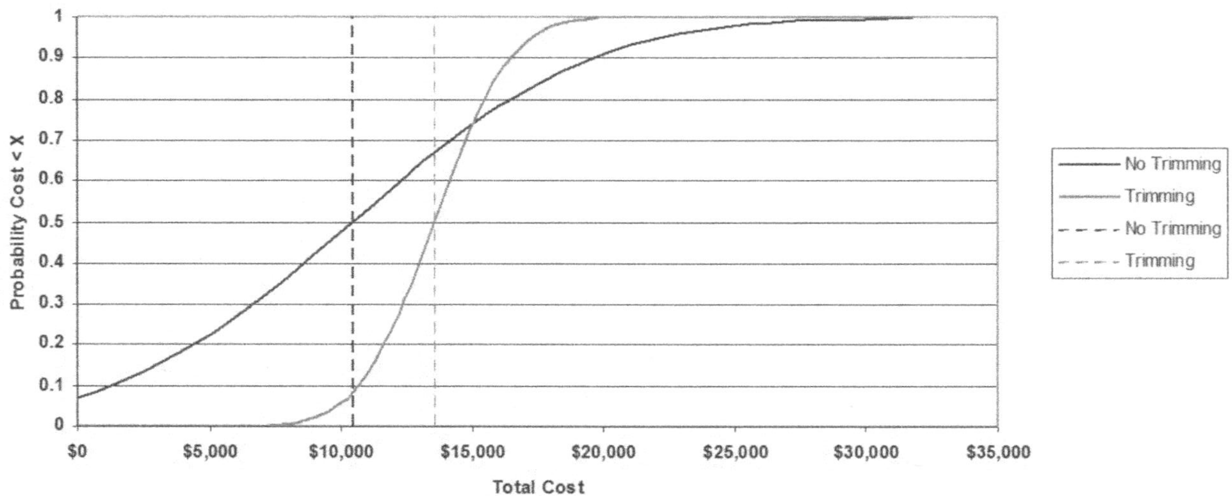

Figure 8-15. An analytical risk model shows the results of a probabilistic analysis, compared with the deterministic expected values (vertical lines).

ous other retrofits to harden its high-voltage transformers against severe GMDs, consider retrofits to protect other equipment from severe GMDs, consider retrofits to equipment for all HILFs, consider operational strategies to mitigate against GMDs or all HILFs, and compare all equipment and operational strategies to identify the optimum combination of strategies.

POWER GRID HARDENING AND MITIGATION

In its 2008 report, the EMP Commission concluded that protecting the entire power grid system, or even all high value components, from HEMP pulse attacks is impractical because the costs, time, and effort in proportion to the likelihood of a HEMP-enhanced nuclear strike make it prohibitively expensive. The report recommendation is to prepare for the widest possible scenarios of GMDs and EMPs with a focus on quick response and repairs, as well as conduct further research into cost-effective mitigation and preparation. Work has already begun in these areas, including efforts outlined in the following subsection.

Stored and Shared Spare Components

Another essential component in the recovery process is the availability of components needed in the event of an emergency. These components could be shared among all stakeholders and geographically dispersed for quick access. The industry can assess the vulnerability to EMP exposure of various components and systems and the inventory of available equipment to plan sufficient quantities of backup components where needed. Transformers, generation equipment, SCADA, aboveground relays and controls, and other embedded processors are examples of potential spare part candidates for stockpiling.

The goal of a NERC Spare Equipment Database (SED) program is to provide a means to securely connect entities that need replacement transformers with entities that have such spares available in the event of simultaneous loss of multiple transformers. The Edison Electric Institute's (EEI) Spare Transformer Equipment Program (STEP Program), launched in 2006 in response to the 9/11 terrorist attacks, addresses the need to manage resources in response to a terrorist attack. About 50 transmission providers that represent about 70% of the U.S. transmission grid currently participate in this program. Defined stakeholders could take responsibility for the inventory, research, and securing of hardened facilities for these parts depots, including a coordinated and clear command and distribution network for part deliveries and repairs. Realizing this vision would require cost and feasibility studies, as well as research that identifies the most vulnerable components under different EMP scenarios (the fact that many components are now manufactured overseas should also be taken into account as a further complication).

RESEARCH NEEDED

EPRI recommends the framework shown in Figure 8-16 to guide next research steps in the area of EMPs. The first step is to define EMP characteristics (e.g., gamma yield and altitude of detonation) for further study. The next step is to perform modeling and simulation of the impact of these scenarios on wide area and local power systems. The vulnerability of these power systems should then be assessed, with particular emphasis on the vulnerability of new and emerging smart grid systems. Mitigation is a major area of research, including development of response plans, equipment spares strategies, equipment hardening, and others. Inter-agency cooperation is a key part of this and other recommended research efforts. Risk management involves integrating the assessment and management of risks from EMPs, other HILF threats, and non-HILF threats that utilities face. Methods developed in this process can be refined and improved as more information on scenarios, EMP-generating devices, grid modernization equipment vulnerabilities, and mitigation methods is obtained.

Enhanced communication and education of all stakeholders on specific EMP threats and their impacts is needed for an effective defensive strategy to be set in place. Education efforts could, for example, recommend ways of implementing practical and phased steps in mitigating and recovering from the most likely EMP events.

The framework shown in Figure 8-16 to guide next research steps in the area of EMPs. The first step is to define EMP characteristics (e.g., gamma yield and altitude of detonation) for further study. The next step is to perform modeling and simulation of the impact of these scenarios on wide area and local power systems. The vulnerability of these power systems should then be assessed, with particular emphasis on the vulnerability of new and emerging smart grid systems. Mitigation is a

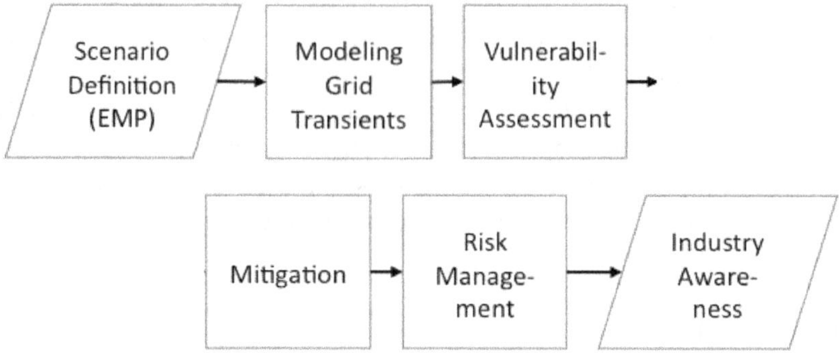

Figure 8-16. Overview of recommended EMP research activities

major area of research, including development of response plans, equipment spares strategies, equipment hardening, and others. Inter-agency cooperation is a key part of this and other recommended research efforts. Risk management involves integrating the assessment and management of risks from EMPs, other HILF threats, and non-HILF threats that utilities face. Methods developed in this process can be refined and improved as more information on scenarios, EMP-generating devices, grid modernization equipment vulnerabilities, and mitigation methods is obtained.

Enhanced communication and education of all stakeholders on specific EMP threats and their impacts is needed for an effective defensive strategy to be set in place. Education efforts could, for example, recommend ways of implementing practical and phased steps in mitigating and recovering from the most likely EMP events.

References

- Defense Science Board Task Force, "Nuclear Weapon Effects Test, Evaluation and Simulation," page 15, Office of the Under Secretary of Defense, Washington, D.C. (2005), http://www.fas.org/irp/agency/dod/dsb/nweffects.pdf.
- Bainbridge, K.T., Trinity (Report LA-6300-H), Los Alamos Scientific Laboratory, May 1976, http://www.fas.org/sgp/othergov/doe/lanl/docs1/00317133.pdf.
- U.S. Department of Energy, "United States Nuclear Tests, July 1945 through September 1992," December 1994, http://www.fas.org/nuke/guide/usa/nuclear/usnuctests.htm.
- Vittitoe, Charles N., Did High-Altitude EMP Cause the Hawaiian Streetlight Incident? Sandia National Laboratories, June 1989, http://www.ece.unm.edu/summa/notes/SDAN/0031.pdf.
- Oak Ridge National Laboratories, "Electric Utility Experience with Geomagnetic Disturbances," September 1991, http://www.ornl.gov/~webworks/cpr/v823/rpt/51089.pdf.

- "Electromagnetic compatibility (EMC)—Part 2: Environment—Section 9: Description of HEMP environment—Radiated disturbance," Basic EMC publication, IEC 61000-2-9.
- "Report of the Commission to Assess the Threat to the United States from Electromagnetic Pulse (EMP) Attack," Volume 1: Executive Report, 2004, http://empcommission.org/docs/empc_exec_rpt.pdf.
- Longmire, Conrad L., "Justification and Verification of High-Altitude EMP Theory," Part 1, Lawrence Livermore National Laboratory, pp 14-16, June 1986. http://www.ece.unm.edu/summa/notes/TheoreticalPDFs/TN368.pdf.
- E. Savage et al., "The Early-Time (E1) High-Altitude Electromagnetic Pulse (HEMP) and Its Impact on the U.S. Power Grid Meta-R-320," Metatech Corporation, Report META-320, 2010.
- U.S. Army, "Nuclear Environment Survivability," April 15, 1994, http://www.dtic.mil/dtic/tr/fulltext/u2/a278230.pdf.
- Cho, Johee, "North Korea Nears Completion of Electromagnetic Pulse Bomb," ABC News, March 9, 2011, http://abcnews.go.com/International/electronic-warfare-north-korea-nears-completion-electromagnetic-pulse/story?id=13081667#.Tzkrg8Xy98E.
- Schneider, Dr. Mark, United States Nuclear Strategy Forum. "The Emerging EMP Threat to the United States," The National Institute Press, November 2007, http://www.nipp.org/National%20Institute%20Press/Current%20Publications/PDF/EMP%20Paper%20Final%20November07.pdf.
- Integrated Risk Management Strategies for HILFs. EPRI, Palo Alto, CA: 2012.
- NERC, "2012 Special Reliability Assessment: Effects of Geomagnetic Disturbances on the Bulk Power System," February 29, 2012, http://www.nerc.com/fileUploads/File/News/MR_GMD29FEB12.pdf.
- Monitoring and Mitigation of Geomagnetically Induced Currents. EPRI, Palo Alto, CA: 2008. 1015938.
- Metatech, John Kappenman, "Low-Frequency Protection Concepts for the Electric Power Grid: Geomagnetically Induced Current (GIC) and E3 HEMP Mitigation," Prepared for Oak Ridge National Laboratory, Meta-R-322, January 2010, http://www.ornl.gov/sci/ees/etsd/pes/pubs/ferc_Meta-R-322.pdf.
- NERC, "High-Impact, Low-Frequency Event Risk to the North American Bulk Power System,,"" A Jointly-Commissioned Summary Report of the North American Electric Reliability Corporation and the U.S. Department of Energy's November 2009 Workshop, June 2010, http://www.nerc.com/files/HILF.pdf.
- NERC, "Task Force Whitepaper Report: Spare Equipment Database: Interim Report to Planning Committee—June 7-8, 2011 Meeting," June 2011, http://www.nerc.com/docs/pc/sedtf/DRAFT_SEDTF_Interim_Whitepaper_June_2011.pdf.
- James P. Fama, Edison Electric Institute, "It's Time to Get in STEP," Transmission and Distribution World, April 1, 2010, http://tdworld.com/substations/power_time_step/.
- Lawrence Livermore National Laboratory. 2006—"The Jericho Option: Al Qaeda and Attacks on Critical Infrastructure." UCRL-;SR-224072, June 2006.
- Scheuer, Michael. 2004. Imperial Hubris. London: Brassey's.
- U.S. Department of State. 2006. "Country Reports on Terrorism 2005." Washington, April 2006.
- Future Inspection of Overhead Lines, EPRI, Palo Alto, CA 2008. 1016921
- Protecting Against Terrorism, Centre for the Protection of National Infrastructure (CPNI), United Kingdom, 2010, www.cpni.gov.uk
- Substation Security, Sightlogix, www.sightlogix.com, 2015

- EPRI. 2004. Supervisory Control and Data Acquisition (SCADA) System Security Guide. EPRI Report 1002604. Palo Alto, CA: EPRI. Available at http://www.epri.com.
- EPRI 2000. Communication Security Assessment for the United States Electric Utility Infrastructure. EPRI Report 1001174. Palo Alto, CA: EPRI.

Chapter 9

Assuring Cyber Security

THE RISK OF A CYBER ATTACK

The United States has embarked on a major transformation of its electric power delivery system into an advanced, digital infrastructure with two-way capabilities for communicating information, controlling equipment, and distributing energy. This evolution is increasing the dependence on information technology and telecommunications infrastructures to ensure the reliability and security of the electric grid. As utilities deploy smart grid technologies, the penetration of this new technology will continue to grow. To realize the full potential of the smart grid, cyber security measures must be designed and implemented to protect it from attacks by terrorists and hackers, as well as to strengthen its resilience against natural disasters and inadvertent threats such as equipment failures and user errors. Security must be included in all phases of the system development life cycle, from the design phase through implementation, maintenance, and disposition/sunset. (EPRI 1024573)

The nation and all of its citizens depend on a stable, safe and resilient power system. As technology has evolved that system is entirely dependent on information technology and a resilient cyberspace.

The Grid is Vulnerable

Terrorists could execute a coordinated attack on multiple assets within the bulk power system to effectively disable the system. This risk could assume the form of physical attacks, cyber-attacks, or a combination of the two. (EPRI 3002001935)

There are numerous examples where cyber-attacks have been successful:

- An unauthorized access in 2000 to the control system of a wastewater plant in 2000 released 1 million gallons of raw sewage. (Sirdharan 2012)

- Intrusion by the "Slammer" worm in 2003 into the Davis-Besse Nuclear Power Plant in Ohio disabled the safety monitoring system. (Sirdharan 2012)
- An infection by a worm into train control and signaling bringing trains in Washington, DC to a standstill. (Sirdharan 2012)
- A 2009 Florida blackout in Florida caused by hackers. (Sirdharan 2012)
- A 2010 worm which damaged Iranian's centrifuges. (Gaines 2012)
- Heartland [Credit] Card Payment Systems—100 million credit card/debit cards exposed (Gaines 2012)
- Sony Entertainment interruptions in service—PlayStation offline 24 days (Gaines 2012)

Cyber attacks could include:

- Distributed denial-of-service attack in which attackers flood network resources to render physical systems unavailable or less responsive
- Rogue devices used to access the system to manipulate it or provide incorrect data to system operators
- Reconnaissance attacks that probe a system to provide attackers with information on system capabilities, vulnerabilities, and operation
- Eavesdropping attacks that violate the confidentiality of communications within a network
- Collateral damage that occurs as an unplanned side effect of cyber attacks
- Unauthorized access attacks in which the attacker obtains some control over the system, and accesses and manipulates assets without authorization
- Unauthorized use of assets, resources, or information that would feed system operators incorrect information presumed to originate from a "trusted" source
- Malicious code or malware through viruses, worms, and Trojan horses

The distributed nature and diversity of the power system presents some unique security challenges. Limited on-site supplies and the challenges of securing replacement system components would hinder a smooth recovery from these types of attacks. Attacks on the bulk power system workforce also present a risk. Cyber vulnerability has increased as the industry relies more on automation, and remote monitoring and control.

Cyber-attacks can lay dormant, be widely distributed, and can be executed at a specific day and time controlled or preset by the attacker. Once executed, the adverse impacts of the attacks may be difficult to detect. While physical attacks result in physical manifestations such as equipment destruction, cyber-attacks lead to unseen damage in operation, information, and control systems. This means that conducting an inventory of damage may take much longer and require more resources than such an inventory after a physical attack. A large amount of infrastructure may need to be evaluated for impacts. This may require highly specialized diagnostic and repair crews that may not be available at all sites. This factor may lead to new ways for utilities to approach the design of systems and workforce training.

The electric industry has adopted significant steps to help mitigate these security issues with the development of The North American Electric Reliability Corporation (NERC's) Critical Infrastructure Protection (CIP) Standards. According to a June 2010 NERC report, "Better technology solutions for the cyber portion of the threat should be developed, with specific focus on forensic tools and network architectures to support graceful system degradation that would allow operators to 'fly with fewer controls.' Component and system design criteria should also be reevaluated with respect to these threats with an eye toward designing for survivability. Prioritization of key assets for protection will be a critical component of a successful mitigation approach (NERC 2010.)

Electric Sector Trends Impacting Cyber Security

As the smart grid drives the evolution of the electric sector, several trends have impacted the role of cyber security:

- Increased connectivity of control systems to other networks, particularly enterprise networks, to obtain productivity improvements and information sharing;
- Widespread deployment of open industry standard protocols that

replace vendor specific proprietary communication protocols;

- Deployment of standardized computer platforms with common operating system to support control system applications;
- Increased reliance on external communications, such as public communication systems and the Internet; and
- Increased capabilities of field equipment, for example, "smart" sensors and controls with enhanced capability and functionality.

These trends have created many new challenges for securing the electric sector.

Changes to the Way the Grid is Managed

In the power industry, the historic focus has been on implementation of equipment that could improve power system reliability and quality. Communications and information technology (IT) equipment were seen typically as supporting power system reliability. The various departments, particularly the IT and grid operations organizations, often worked in silos—with limited coordination and with different cyber security requirements.

As the smart grid evolves, the electric, communications, and IT sectors are converging to support power system reliability and privacy of consumer information. From a practical perspective, this means the integration of corporate IT systems, legacy components and devices (with limited or no cyber security), and grid operations systems into the smart grid. When all these systems are integrated into the smart grid, cyber security mitigation strategies must be developed to address the security weaknesses. Cyber security in the smart grid must include a balance of both power and cyber system technologies and processes in IT and power system operations and governance.

Securing the Grid

The smart grid will have additional vulnerabilities due to its complexity and the increased use of new technologies. In addition, the IT and communication sectors have known vulnerabilities that will need to be assessed in the context of the smart grid.

As indicated by National Association of Regulatory Utility Commissioners (NARUC), cyber security is a concern wherever computer systems and data exist. "With the advent of smart grid technologies, which layer software on top of utility operations and computer systems, threats become increasingly likely and relevant." (NARUC 2013) While a smarter grid is generally more reliable, new vulnerabilities appear that must be managed as grids become two-way exchanges of kilowatt hours, as well as network and customer-usage data that may be valuable and desirable to bad actors. (NERC 2013)

Security must be "built in" to every smart grid system. For Utilities, it is not always clear how security will be integrated into the life cycle of each smart grid project. Cyber security may call for securing entire networks, in addition to devices on those networks. The meters within a smart grid system can be fortified against attack, but in order to ensure the entire network of the smart grid is secure, the components linking those meters, as well, as every other component in between, must be secured as well. If an attack occurs at one meter, the rest of the system linked to that meter is not also at risk because the components linking them have been protected. (NARUC 2013)

Many utilities must comply with cyber security and privacy requirements from multiple sources. For example, the majority of states have privacy breach laws that address personal information. These laws are not consistent and over 45 states have their own privacy laws. Utilities must also comply with federal and state regulations; for example, utilities that are federal agencies must comply with the Federal Information Security Management Act (FISMA) requirements that include cyber security requirements. The National Institute of Standards and Technology (NIST) has developed the majority of these federal cyber security requirements. In addition, state public utility commissions (PUCs) regulate essential utility services and specify public policy for investor owned utilities (IOUs). Finally, the Federal Energy Regulatory Commission (FERC) and the North American Electric Reliability Corporation (NERC) have responsibility for the reliability of the bulk power system. NERC is responsible for the development of the Critical Infrastructure Protection (CIP) mandatory standards. One concern is the focus on compliance to the NERC CIPs, rather than on cyber security. Compliance is an important step, but compliance does not ensure that security has been adequately addressed. Finally, there are rural electric cooperatives, public power districts, and community owned electric utilities that are outside the jurisdiction of FERC, NERC, and the state PUCs.

Because of the identification of new vulnerabilities and attacks almost on a daily basis, the cyber security community is frequently in a "reactive mode" rather than being proactive. This approach needs to be reevaluated for critical infrastructures such as the electric sector. Also, the IT and communications sectors perform firmware/software updates and patch management on a regular schedule, often as frequently as once a week. The systems within the electric sector that support the more typical IT and communications functions may be maintained in the more typical IT update cycle. In contrast, updates and patch deployment to the operational control systems, such as transformers, generators, and substation devices, are done infrequently and require significant testing—that can take from six months to several years to perform.

A Presidential Order

The importance of cyber security was highlighted in a [U.S.] Presidential Order in 2013 referred to as Executive Order 13636 titled Improving Critical Infrastructure Cyber security. The order acknowledged that there have been repeated intrusions into critical infrastructures in the United States and that "the cyber threat to critical continues to grow and represents one of the most serious national security challenges we must confront." In the order, the President directs the National Institute of Standards and Technology (NIST) to lead the development of a framework to reduce cyber risks to critical infrastructure (called a Cyber security Framework.) (Federal Register 2/19/2013)

In response NIST developed a framework in collaboration with industry intended to provide guidance to an organization on managing cyber security risk. NIST believed that by relying on the practices outlined in its framework to develop, and manage cyber security, the framework will evolve with technological requirements and business requirements. The framework intends to encourage organizations to consider cyber security risks as a priority similar to financial, safety and operational risk. (NIST 2013) The framework is a risk-based approach consisting of three parts: The Framework Core; the Framework Profile: and the Framework Implementation Tiers.

The Framework Core—is a set of cyber security activities and references that are common across critical infrastructure sectors organized around particular outcomes. The Core presents standards and practices in a manner that enables communication of cyber security risk across an organization from the executive suite to the implementation and operations levels. The Core consists of five critical functions:

1. Identify—Develop the institutional understanding to manage cyber security risk.
2. Protect—Develop and implement safeguards.
3. Detect—Develop and implement the appropriate activities to identify cyber security events.
4. Respond—Develop and implement the appropriate activities to take action in response to a cyber security event.
5. Recover—Develop and implement activities which assure that capabilities that were impaired are restored.

These five functions can provide a critical overview of an organizations cyber security risk management. The Framework Core comprises four elements: Functions; Categories; Subcategories; and Informative References as depicted in Table 9-1.

These elements can be described as follows:

- Functions organize basic cyber security activities at their highest level as described above.
- Categories are the subdivision of Functions into groups of cyber security outcomes such as "Asset Management," "Access Control," and "Detection Process."
- Subcategories divide a Category into high level outcomes such as "Physical devices are catalogued," and "Alarms are investigated."
- Informative Resources are standards, guidelines and practices which apply to the Subcategories.

Table 9-1. The Framework Core (NIST 2013)

Functions	Categories	Subcategories	Informative Resources
Identify			
Protect			
Detect			
Respond			
Recover			

A Framework Profile—represents the outcomes that a particular organization has achieved as specified in the framework. The profile is the alignment of industry standards and best practices to a particular implementation scenario.

Framework Implementation Tiers—describe how cyber security risk in managed by an organization.

Risk Management and the Cyber security Framework

Risk Management is the process of identifying, assessing, and responding to risk so as to lessen or mitigate the risk entirely. With the electric power systems this infers that an organization should understand the likelihood that a risk event will occur and provide an estimate of the potential resulting impact. As a result, participants in the generation and delivery of power can determine the acceptable level of risk for Information and Communications Technologies (ICT) and systems expressed as their level of risk tolerance.

Coordination of Cyber security Implementation

Figure 9-1 illustrates the typical flow of information and decisions within most organizations at the senior executive level, at the business/process level, and at the implementation/operations level. The senior executive level is focused on organizational risk and acts to make risk—informed decisions and to inform priorities. The senior executives inform the business/process level of mission priorities and their risk appetite and budget. The business/process level is focused on critical infrastructure risk management. Their actions include selecting risk profiles and allocating budgets. They develop the Framework Profile which is shared with the Implementation/operations level. The Implementation/operations level is focused on securing critical infrastructure and implementing the Framework Profile. The Implementation/operations level informs the enterprise of their progress, changes in assets, vulnerabilities and threats.

The Energy Independence and Security Act of 2007

It was the Energy Independence and Security Act of 2007 (EISA) that first prompted NIST to engage formally in cyber security regarding the smart grid. EISA gave NIST "...primary responsibility to coordinate development of a framework that includes protocols and model standards for information management to achieve interoperability of Smart

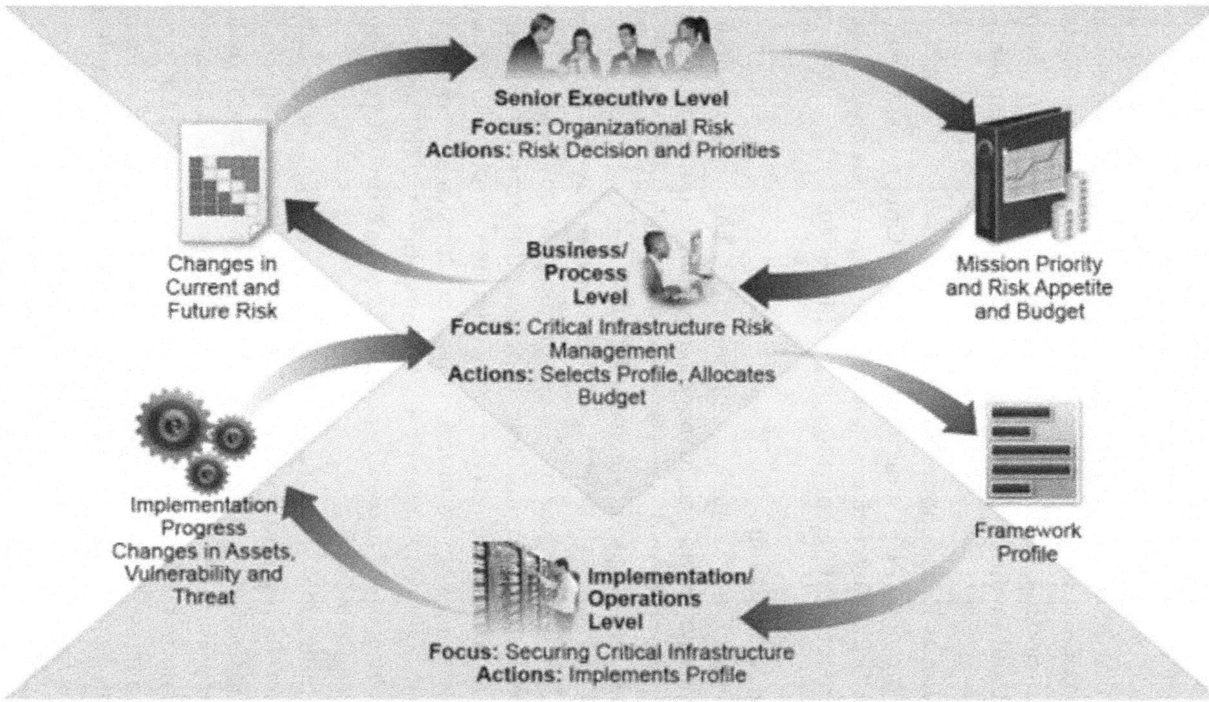

Figure 9-1. Notational Information and Decision Flows within an Organization (NIST 2013)

Grid devices and systems…" (EISA Title XIII, Section 1305) EISA further stated that the framework should accommodate "traditional centralized generation and distribution resources" and facilitate incorporation of new, innovative Smart Grid technologies, such as distributed renewable energy resources and energy storage. However, no funding was provided until, in February of 2009, The American Recovery and Reinvestment Act of 2009 (ARRA) providing funding. Shortly thereafter, NIST retained EPRI to conduct workshops and provide the material which led to the initial Framework and Roadmap for Smart Grid Interoperability Standards.

EPRI's Vision as Recommended to NIST

EPRI provided NIST a definition of Smart Grid, an outline of benefits, Smart Grid characteristics, and challenges. The vision is presented as it appeared in EPRI's report to NIST (EPRI 2009).

Smart Grid Networking

The Smart Grid is a network of networks. That is, many networks with various ownership and management boundaries are interconnected to provide end-to-end services between stakeholders, and in-and-among intelligent electronic devices (IEDs).

Figure 9-2 is a high level view of the information network for the Smart Grid. It handles the two-way communication between the network end points residing in their respective domains. By domain, we mean the unique distributed computing environments in which communicating end points can be found. Thus, any domain application could communicate with any other domain application via the information network, subject to the necessary network access restrictions and quality of service requirements.

The applications in each domain are the end points of the network as shown on the top and bottom of Figure 9-2. For example, an application in the Customer domain could be a smart meter at the customer premise; an application in the Transmission domain could be a phasor measurement unit (PMU) unit on a transmission line or in a Distribution domain at a substation; an application in the Operation domain could be a computer or display system at the operation center. Each of these applications has a physical communication link with the network. The smaller clouds within the network represent sub-networks that may be implementing unique functionality. The networking function in the Operations, Market, Service Provider domains may not be easily differenti-

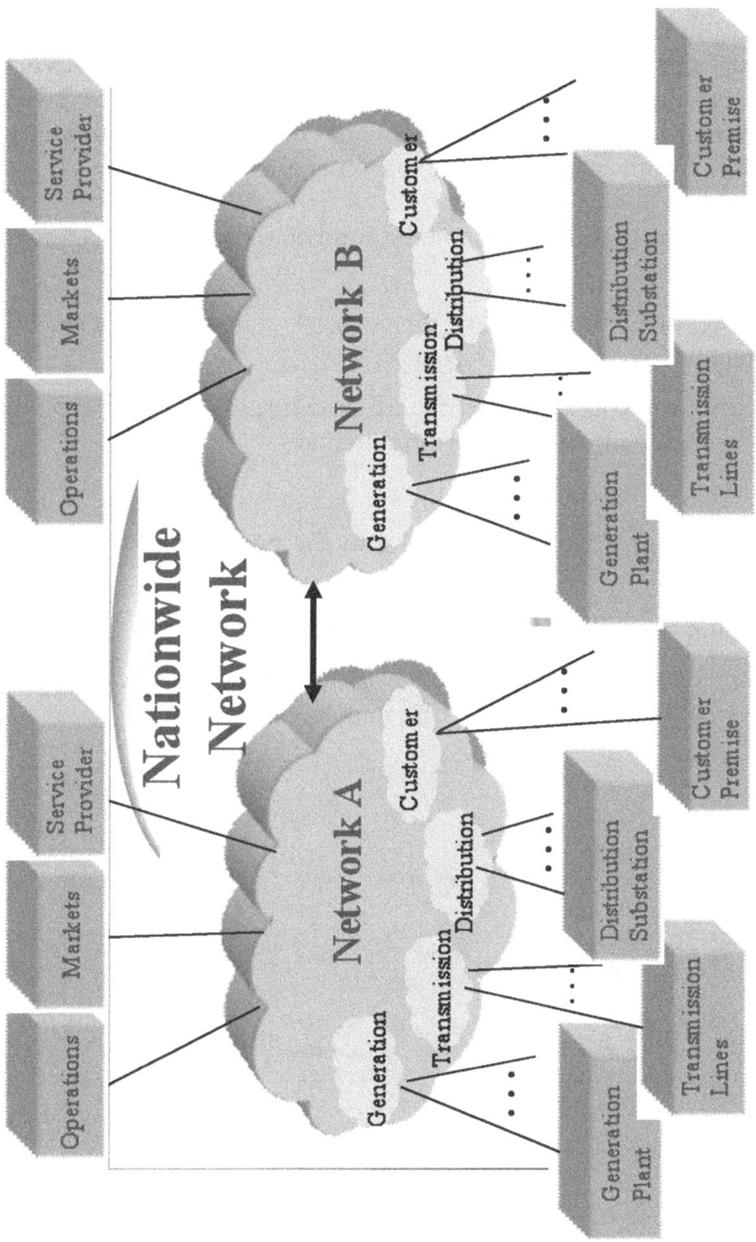

Figure 9-2. Smart Grid Networks for Information Exchange (EPRI Reports to NIST, 2009)

ated from normal information processing networks; therefore no unique clouds are illustrated.

This information network may consist of multiple interconnected networks as shown in Figure 9-2, where two backbone networks, A and B, are illustrated. The physical links within these two networks and between the network and the network end points could utilize any appropriate communication technology currently available or yet to be developed.

Additional requirements for the information network are as follows:

- Management functionality for network status monitoring, fault detection, isolation, and recovery;
- Secure protocols to protect Smart Grid information in transit and authenticate infrastructure components;
- Cyber security countermeasures;
- Addressing capability to entities in the network and devices attached to it;
- Routing capability to all network end points;
- Quality of service support for a wide range of applications with different latency and loss requirements.

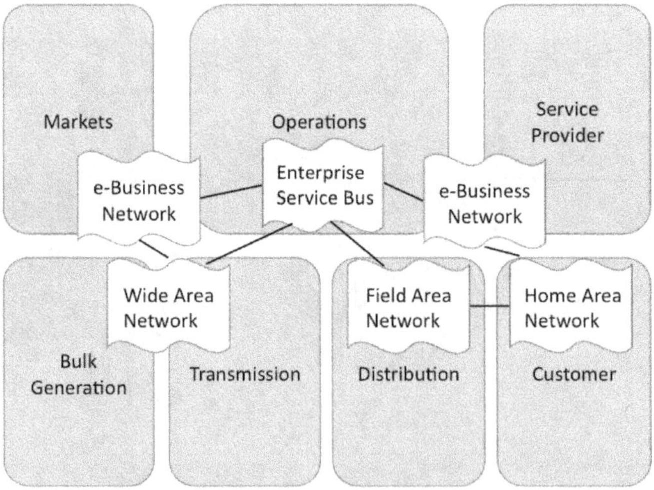

Figure 9-3. Smart Grid Information Exchange

The Smart Grid Interoperability Panel (SGIP)

As part of the development of the Framework, NIST undertook, with considerable industry support, the development of an organization and consensus process. This launched the creation of the Smart Grid Interoperability Panel. The primary mission of the SGIP was to support NIST responsibilities under EISA by; undertaking coordination of interoperability standards; identifying the necessary testing and certification requirements; overseeing the performance of these activities and maintaining momentum; educating stakeholders on interoperability; and establishing global interoperability alignment.

As of 2012, nearly 800 companies and organizations are members of the SGIP. Hundreds have standards have been considered for inclusion in the SGIP Catalog of Standards, the primary reference guide for Smart Grid interoperability. The SGIP has standing working groups on Cyber security, Architecture, Testing & Certification, and Implementation experiences.

North American Electrical Reliability Corporation (NERC)

As implied above, NERC has been active on behalf of the North American Power Industry in addressing cyber security in the energy infrastructure. The original guidelines addressing cyber security, known as NERC 1200 UAS were passed in 2003 in response to the Homeland Security Act of 2002. It was designed to reduce the vulnerability of the bulk electric power system. However, NERC 1200 was only a temporary fix to the problem and was soon followed by NERC 1300 since achieving a consensus on a final set of standards. A year later NERC Critical Infrastructure Protection (CIP) cyber security standards were passed. (NERC.com)

The Electric Reliability Organization Enterprise

In February 2014 The Electric Reliability Organization (ERO) Enterprise was formed. The North American Electric Reliability Corporation (NERC) and eight associated Regional Reliability Entities comprise the ERO Enterprise. The ERO's key programs impact more than 1,900 Bulk-Power System owners and operators and are based on four pillars of success:

- Reliability—to address events and identifiable risks, thereby improving the reliability of the Bulk-Power System.

- Assurance—to provide assurance to the public, industry and government for the reliable performance of the Bulk-Power System.
- Learning—to promote learning and continuous improvement of operations and adapt to lessons learned for improvement of Bulk-Power System reliability.
- Risk-Based Approach—to focus attention, resources and actions on issues most important to Bulk-Power System reliability.

The ERO enterprise is responsible for implementing the NERC standards

NERC CIP Standards

NERC CIP spells out an auditable guide covering a variety of areas related to cyber security. There are eleven separate standards that NERC has passed. Each standard sets out details concerning who the responsible party is, what the requirements are, and what constitutes different levels of non-compliance. The standards are as follows:

- CIP-001 Sabotage Reporting
- CIP-002 Critical Cyber Assets
- CIP-003 Security Management Controls
- CIP-004 Personnel & Training
- CIP-005 Personnel & Training
- CIP-006 Electronic Security
- CIP-007 Physical Security of Critical Cyber Assets
- CIP-008 Incident Reporting and Response Planning
- CIP-009 Recovery Plans for Critical Cyber Assets
- CIP-010 Configuration Change Management and Vulnerability Assessments
- CIP-011 Information Protection

Alignment of NERC CIP with NIST Framework

NERC's CIP Standards align very well with the NIST Framework's Core Disciplines of Identify, Protect, and Detect, Respond and Recover as shown in Table 9-2.

Table 9-2. Alignment of NIST Framework Core Discipline with NERC CIP Cyber Security Standards

NIST Framework Core Discipline	NERC CIP Cyber Security Standard
Identify	NERC CIP-002
Protect	NERC CIP-003, CIP-004, CIP-005, CIP-006, CIP-007, CIP-010 and CIP-011
Detect	NERC CIP-008
Respond	NERC CIP-008
Recover	NERC CIP-009

Figure 9-4 illustrates the NERC reliability enforcement actions of all time as of May 31, 2012. Many of these enforcement actions have occurred in response to violations of NERC CIP standards.

THE SMART GRID NEEDS COMMUNICATIONS

Nearly all of the technologies to be installed on the smart grid will generate new data that will be communicated and available for use. Hence, new infrastructure capable of supporting the higher level of information monitoring, analysis, and control required for Smart Grid operations, as well as the communication infrastructure to support full integration of upstream and downstream operations, will be needed. At the same time, it is necessary to ensure the privacy and security of individual customer information.

For example, the substation of the future will require a wide-area network interface to receive and respond to data from an extensive array of transmission line sensors, dynamic-thermal circuit ratings, and strategically placed phasor measurement units. The smart substation must be able to integrate variable power flows from renewable energy systems in real time, and maintain a historical record or have access to a historical record of equipment performance. Combined with real-time monitoring of equipment, the smart substation will facilitate reliability-centered and predictive maintenance.

The core and distributed IT infrastructure will be able to coordinate the flow of intelligence from critical equipment, such as self-diagnosing

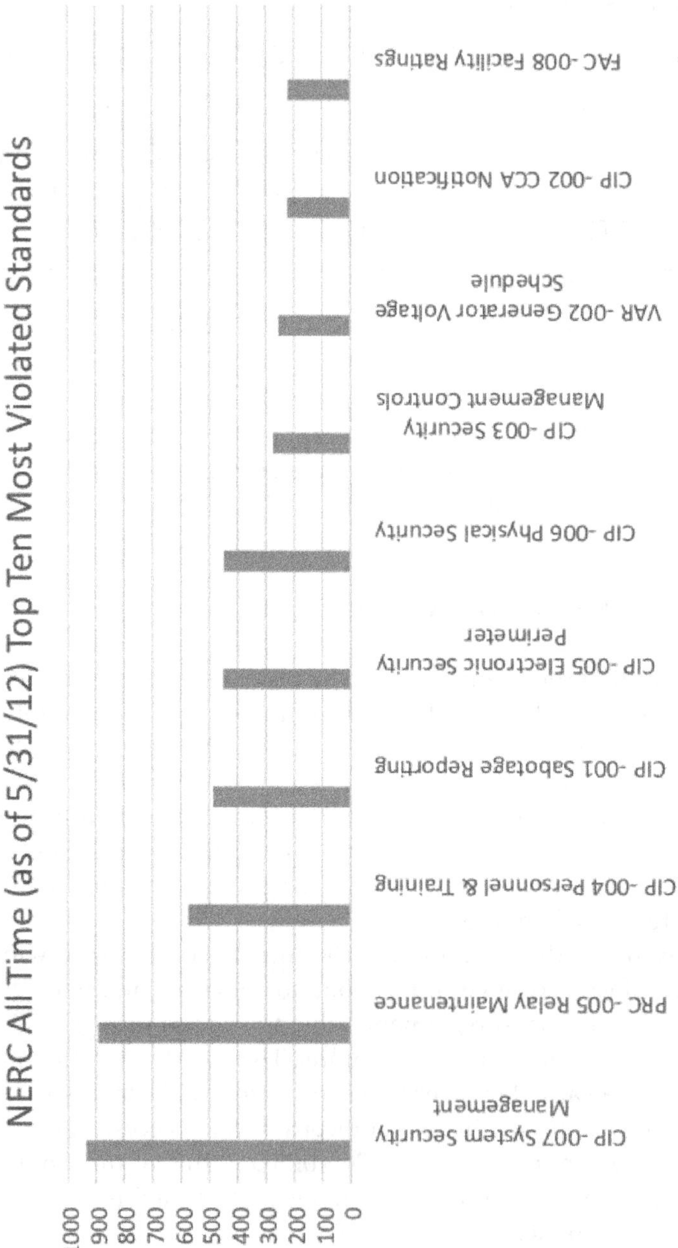

Figure 9-4. NERC Reliability Enforcement Actions (NERC)

transformers, with downstream operations, and be able to differentiate normal faults from security breaches. It will be able to distill and convey critical performance data and maintenance issues to back office systems.

The smart substation will build upon existing platforms. There is already a significant installed base of sensors at substations, but there is still limited bandwidth connecting the substation to the enterprise. Historically, the communications channel to the substation was justified as part of the installation of the energy management system (EMS) and supervisory control and data acquisition (SCADA) systems. A key consideration for the future is that these legacy systems have limited bandwidth.

Communications constitute the critical backbone for integrating customer demand with utility operations. Detailed, real-time information is key to effectively managing a system as large and dynamic as the distribution power grid. Each smart meter in the advanced metering infrastructure (AMI) must be able to communicate with a wide range of user control systems, as well as reliably and securely communicating performance data, price signals, and customer information to and from an electric utility's back-haul system.

No single technology is optimal for all applications. Among the communications media now being used for AMI applications are cellular networks, licensed and unlicensed radio, and power line communications. In addition to the media, the type of network is also an important part of communications design. Networks used for Smart Grid applications include fixed wireless, mesh networks, and a combination of the two. Several other network configurations, including Wi-Fi and Internet networks are also under investigation.

Table 9-3 details the known public/private open and propriety communications which may be used in the smart grid. Ultimately a mixture of solutions will be used to serve an individual utility's needs which are application dependent and based on business drivers. Each approach can require a different cyber security strategy which will vary by scale, supportability, investment drivers and other factors.

Interoperability remains one of the most critical success factors for Smart Grid communications. There is growing interest in the potential use of the Internet Protocol Suite (TCP/IP) as a networking protocol that could run over many different communication technologies. In June, 2009, EPRI submitted a Report to NIST on the Smart Grid Interoperability Standards Roadmap that became the starting point for NIST's own

Table 9-3. Public/Private and Open/Proprietary Communications in the Smart Grid (Assante 2012)

Generation	Transmission	Distribution
Automated Generation Control	System Control Center – SCADA and EMS	Advanced Metering
Plant Systems (PCS/Safety/Historian)	Substation (Control & Data Acquisition and Protection)	System Control Center – SCADA and DMS
Remote Access (3rd parties & Internal)	Phasors, Data Concentrators, & Gateways	Substation (Control & Data Acquisition and Protection)
Market Systems	Remote Access (3rd Parties & Engineering)	Distribution Automation (Remote Disconnect)
Network Connections	CIP Standards Compliance	Intelligence Metering
CIP Standards Compliance		Phasors, Data Concentrators, Gateways
		Remote Access (3rd Parties & Engineering)
		ARRA Security Plan

roadmap, released in September, 2009 (www.nist.gov/smartgrid). In 2014, NIST published Release 3.0 of the NIST Framework and Roadmap for Smart Grid Interoperability Standards, reflecting its collaboration with the Smart Grid Interoperability Panel and the work and input of hundreds of technical professionals.

THE SMART GRID INCREASES CYBER SECURITY RISKS

In September of 2010, The Cyber Security Working Group (CSWG) of the Smart Grid Interoperability Panel published a series of volumes offering "Guidelines for Smart Grid Cyber Security." In that document, the Working Group articulated that the convergence of the information and communication infrastructure with the electric power grid introduces new security and privacy-related challenges. The introduction of these technologies to the electric sector also presents opportunities to increase the reliability of the power system, to make it more capable and more resilient to withstand attacks, equipment failures, human errors, natural disasters, and other threats. The Working Group also highlighted that improved monitoring and control capabilities must include cyber security solutions in the development process rather than as a retrofit. (NIST CSWG 2010)

The Guidelines articulated a few examples of potential risks associated with the evolution of the Smart Grid including:

- Greater complexity increases exposure to potential attackers and unintentional errors;
- Networks that link more frequently to other networks introduce common vulnerabilities that may now span multiple Smart Grid domains and increase the potential for cascading failures;
- More interconnections present increased opportunities for "denial of service" attacks, introduction of malicious code or compromised hardware, and related types of attacks and intrusions;
- As the number of network nodes increases, the number of entry points and paths that potential adversaries might exploit also increases; and
- Extensive data gathering and two-way information flows may broaden the potential for compromises of data confidentiality and breaches of customer privacy, and compromises of personal data and intrusions of customer privacy.

INCORPORATING CYBER SECURITY INTO THE SMART GRID

Electric utilities have been incorporating cyber security features into their operations since the early 2000s. In recent years, cyber security concerns have increased significantly. While such concerns are significant without regard for whether a utility has implemented Smart Grid technologies, Smart Grid has helped focus attention on and support the development of cyber security capabilities. Cyber security is an essential element of the Smart Grid. It is the protection needed to ensure the confidentiality and integrity of both legacy utility systems and the digital overlay that is part of the Smart Grid.

Development of an effective cyber security strategy which includes the smart grid requires a holistic approach to analyzing risk. Risk is the potential for an unwanted outcome resulting from internal or external factors as identified from the likelihood of occurrence and the resulting consequences. Risks is the product of interactions among threats, vulnerabilities and consequences as represented by the following formula: (NIST CSWG 2010)

Threat x Vulnerability x Consequence = Risk

Where:

- Threat is an event, actor, or action with the potential to harm.
- Vulnerability is weakness.
- Consequence is impact.
- Risk is operational, economic, safety or environmental risks.

Technical controls are required to meet the additional cyber security challenges which the smart grid brings to the party. The smart grid can be thought of as having 6 cyber sensitive levels as illustrated in Figure 9-5 (Gaines 2012)

- Level 1 is the Corporate Network
- Level 2 is the Energy Management System (EMS) Network
- Level 3 is Substation Access Control
- Level 4 is the Substation Network
- Level 5 are data collectors from the AMI system
- Level 6 is the Advanced Metering Infrastructure (AMI) System

The technical controls ensure protection as follows:

- Level 2 is protected from level 1 by firewalls and an Intrusion Protection System.
- The security control of level 1 and 2 ensures monitoring and protection of these networks.
- The Energy Management System (EMS)—level 2 and Substation Access Control—level 3 and the Substation Network—level 4 host digital access.
- Access to devices within level 4 is strictly controlled via device in level 3.

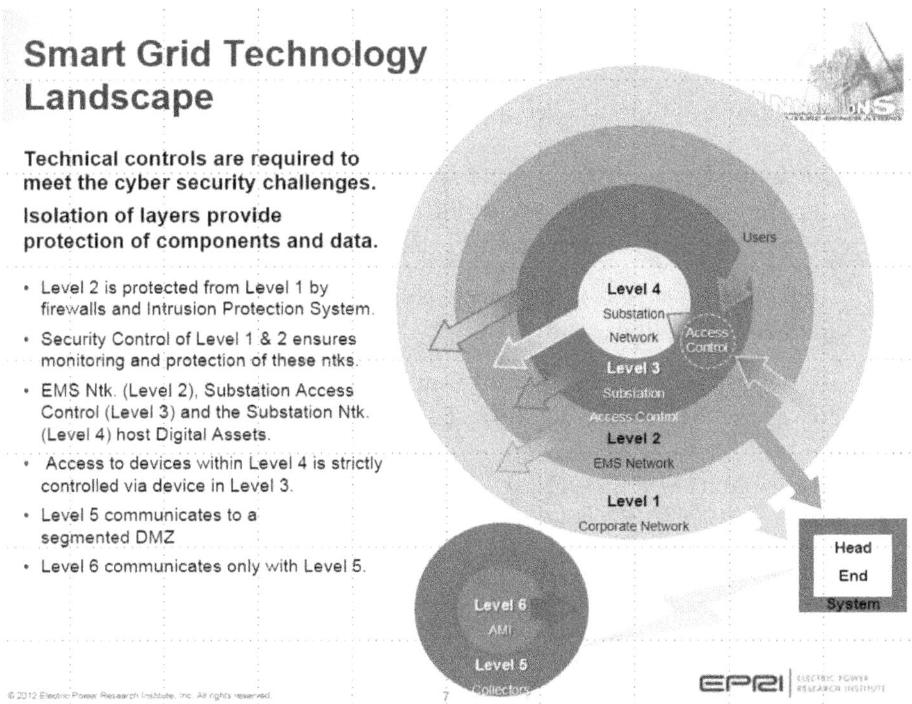

Figure 9-5. Illustrates the Cyber Sensitive Levels of the Smart Grid (Gaines 2012)

- Level 5 communicates to a segmented DMZ.
- Level 6 communicates only with level 5.

PRIORITY CYBERSECURITY PROJECTS

At present, utilities are considering cyber security as part of information technology (IT) projects for:

- Advanced metering infrastructure
- Plug-in electric vehicle (PEV) management systems
- Distribution automation
- Substation automation

- Transmission upgrades
- Demand Response Programs
- Communication with DET devices

One of the key industry needs to prepare for the future is to address cyber security requirements, particularly in the integration framework, reference architecture, and information exchange model. Specifically:

- To develop secure information exchange policy and systems management guidelines, and
- To identify security issues and solutions implemented for communications and information infrastructure to support the integration.

As the Smart Grid drives the evolution of the electric sector, several trends have impacted the role of cyber security:

- Increased connectivity of control systems to other networks, particularly enterprise networks, to obtain productivity improvements and information sharing.
- Widespread deployment of open industry standard protocols that replace vendor-specific proprietary communication protocols.
- Deployment of standardized computer platforms with common operating system to support control system applications.
- Increased reliance on external communications, such as public communication systems and the Internet.
- Increased capabilities of field equipment, for example, "smart" sensors and controls with enhanced capability and functionality.

In the power industry, the historic focus has been on implementation of equipment that could improve power system reliability and quality. Communications and information technology (IT) equipment were seen typically as supporting power system reliability. The various departments, particularly the IT and grid operations organizations, often worked in silos—with limited coordination and with different cyber security requirements.

As the Smart Grid evolves, the electric, communications, and IT sectors are converging to support power system reliability and privacy of consumer information. From a practical perspective, this means the integration of corporate IT systems, legacy components and devices (with limited or no cyber security), and grid operations systems into the Smart Grid. When all these systems are integrated into the Smart Grid, cyber security mitigation strategies must be developed to address the security weaknesses. Cyber security in the Smart Grid must include a balance of both power and cyber system technologies and processes in IT and power system operations and governance. A great deal of research is needed to enhance IT technologies to enable them to withstand attacks and intrusions.

ENTERPRISE RISK MANAGEMENT

An enterprise risk management strategy identifies how an organization frames, assesses, responds to, and monitors risk on an ongoing/continual basis. This overall strategy may be further refined and tailored for specific departments/agencies within an organization and for specific classes or families of systems. Enterprises have developed processes to evaluate risks associated with their business and to address those risks based on organizational priorities and both internal and external constraints. In addition, utilities have developed a variety of risk management methodologies, models, and systems for addressing risks related to safety. This management of all these types of risk is an ongoing process that is part of normal operations.

Although this report focuses on risk management and risk assessment for systems, it is important to consider the overall enterprise risk culture. For example, as specified in the Department of Energy (DOE) Electricity Subsector Cyber security Capability Maturity Model (ES-C2M2):

Institutionalization describes the extent to which a practice or activity is ingrained into an organization's operations. The more deeply ingrained an activity, the more likely it is that the organization will continue to perform the activity over time. (EPRI 3002001181)

The U.S. Department of Energy (DOE) and the Department of Homeland Security (DHS) have sponsored the development of ES-C2M2. The Model is intended to develop capabilities to manage dynamic threats and understand the cyber security posture of the grid. The Model is designed: to enable consistent evaluation and benchmarking of cyber security capabilities; to share best practices; and to enable prioritized actions and cyber security investments. (ES-C2M2@doe.gov)

CAPABILITY MATURITY MODEL

A capability maturity model defines a set of characteristics and attributes that may be used to assess current capabilities and define capabilities for the future. This allows an organization to determine its current state, define a target state, and identify the capabilities that will be needed to meet the target state. The model may reference standards, policies, and best practices. The ES-C2M2 is organized into 10 domains. Each domain is further divided into objectives and each objective contains multiple practices. Included below is a summary of the ten domains extracted from the ES-C2M2.

RISK MANAGEMENT

Establish, operate, and maintain an enterprise cyber security risk management program to identify, analyze, and mitigate cyber security risk to the organization, including its business units, subsidiaries, related interconnected infrastructure, and stakeholders.

Asset, Change, and Configuration Management

Manage the organization's IT and OT assets, including both hardware and software, commensurate with the risk to critical infrastructure and organizational objectives.

Identity and Access Management

Create and manage identities for entities that may be granted logical or physical access to the organization's assets. Control access to the organization's assets, commensurate with the risk to critical infrastructure and organizational objectives.

Threat and Vulnerability Management

Establish and maintain plans, procedures, and technologies to detect, identify, analyze, manage, and respond to cyber security threats and vulnerabilities, commensurate with the risk to the organization's infrastructure (e.g., critical, IT, operational) and organizational objectives.

Situational Awareness

Establish and maintain activities and technologies to collect, analyze, alarm, present, and use operational and cyber security information, including status and summary information from the other model domains, to form a common operating picture (COP).

Information Sharing and Communications

Establish and maintain relationships with internal and external entities to collect and provide cyber security information, including threats and vulnerabilities, to reduce risks and to increase operational resilience, commensurate with the risk to critical infrastructure and organizational objectives.

Event and Incident Response, Continuity of Operations

Establish and maintain plans, procedures, and technologies to detect, analyze, and respond to cyber security events and to sustain operations throughout a cyber security event, commensurate with the risk to critical infrastructure and organizational objectives.

Supply Chain and External Dependencies Management

Establish and maintain controls to manage the cyber security risks associated with services and assets that are dependent on external entities, commensurate with the risk to critical infrastructure and organizational objectives.

Workforce Management

Establish and maintain plans, procedures, technologies, and controls to create a culture of cyber security and to ensure the ongoing suitability and competence of personnel, commensurate with the risk to critical infrastructure and organizational objectives.

Cyber security Program Management

Establish and maintain an enterprise cyber security program that provides governance, strategic planning, and sponsorship for the organization's cyber security activities in a manner that aligns cyber security objectives with the organization's strategic objectives and the risk to critical infrastructure.

Maturity Indicator Levels

Finally, as indicted above, the practices are organized into three maturity indicator levels (MIL1 through MIL3).

Figure 9-5 illustrates the domains in ES-C2M2. For example, referring to the Asset, Change, and Configuration Management Domain:

- Access control activities may include:
  - Unique accounts vs. shared accounts
  - Access limited to specific assets, during a specified maintenance interval
  - Use single or multifactor authentication
- Example MIL Progression for "Control Access"
  - MIL1—Access requirements, including those for remote access, are determined (which types of entities are allowed to access the asset, the limits of allowed access, and authentication parameters)
  - MIL2—Root privileges, administrative access, emergency access, and shared accounts receive additional security and monitoring
  - MIL3—Anomalous access attempts are monitored as indicators of cyber security events

Using templates for each domain the range of activity of any organization can be assessed and compared to industry norms and targets identified for improvement.

Risk assessment approaches used by an organization reflect its culture (the values and norms of its leaders, management, and staff that influence their actions) in dealing with and communicating risks. Recognizing, and addressing these influences help achieve effective risk management. The ES-C2M2 evaluations measure the institutionalization of various cyber security and risk management practices and thus can serve as a useful tool to understand the organizational culture with regard to risk management.

Risk Management Process

The following discussion on risk management is drawn from the Department of Energy (DOE) risk management document, Electricity Subsector Cyber security Risk Management Process (RMP) published in May 2012 and tailors the content to systems, rather than organizations. (DOE RMP 2012)

The RMP provides a scalable risk management process that is specific to the risks inherent in operating information technology (IT) and

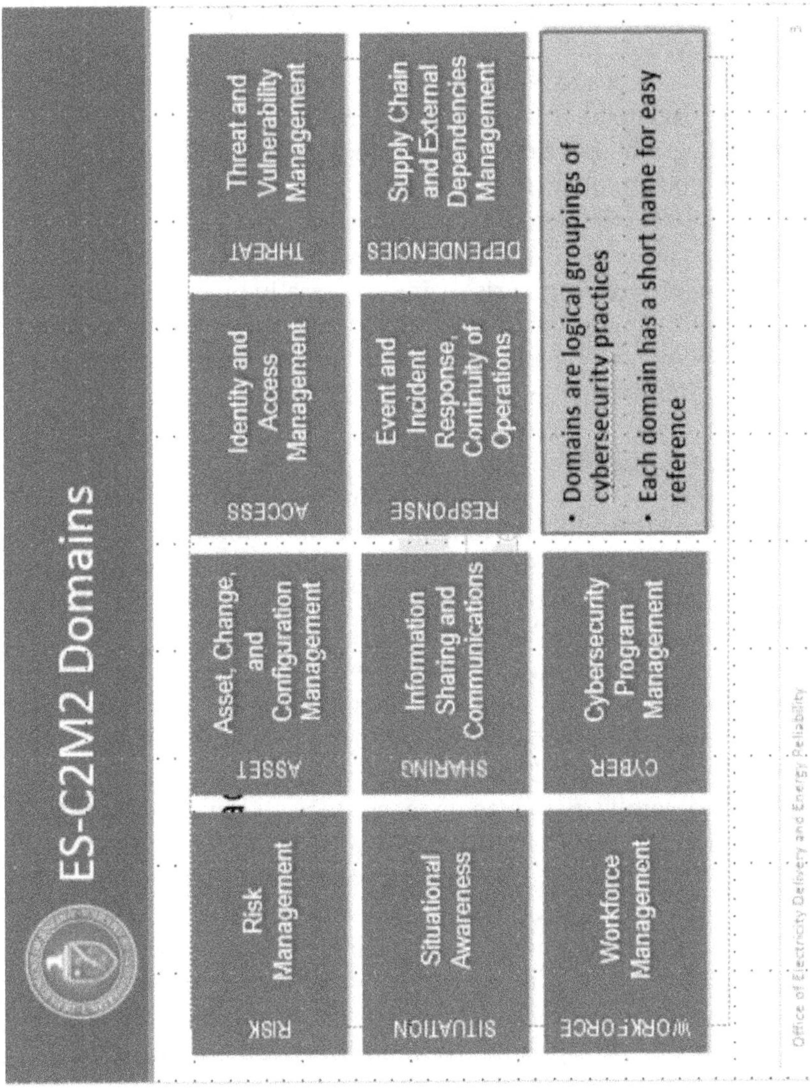

Figure 9-6. ES-C2M2 Domains (DOE)

industrial control systems (ICS). The term risk management refers to the program and supporting processes used to manage cyber security risk to an organization's operations, its assets, and individuals. In implementing the RMP, organizations have the flexibility to determine how best to conduct the activities, including the sequence, degree of rigor, formality, and how the results or outputs of each activity are captured and shared across the organization and between organizations. The RMP is meant to supplement an organization's existing risk management framework and provide flexible guidelines that may be leveraged as needed.

The risk management cycle includes four phases. These phases require utilities to (1) frame risk (i.e., establish the context for risk-based decisions), (2) assess risk, (3) respond to risk once determined, and (4) monitor risk on an ongoing basis, using an iterative feedback loop for continuous improvement in the risk-related activities of organizations. The risk management cycle and the four phases are illustrated in Figure 9-6 and further defined below. (EPRI 3002001181)

Security Posture Using ES-C2M2

Initially, the ES-C2M2 was intended for organizations to determine their maturity levels in the ten domains. The next step is for organizations to apply the ES-C2M2 to information technology (IT) and Operational technology (OT) systems and determine the maturity levels in the domains. The results of the system assessments define the current security posture. An organization may also develop target security postures for the different systems, identify gaps between the current and target postures, and determine the mitigation strategies that need to be implemented to address the gaps. The selection of mitigation strategies should be based on the overall risk management strategy, the level of acceptable residual risk, and the cyber security risk strategy. (EPRI 3002003332)

The cyber security risk strategy is divided into three methodologies: maturity model, control-based, and compliance. This discussion focuses on the maturity model approach to system assessment as highlighted in Figure 9-5. The maturity model methodology uses the ES-C2M2 document and the ES-C2M2 toolkit in the assessment. The ES-C2M2 is most frequently applied to organizations rather than systems.

The Ultimate Cyber Attack—An Electromagnetic Pulse

An electromagnetic pulse (EMP) refers to a very intense pulse of electromagnetic energy, typically caused by detonation of a nuclear

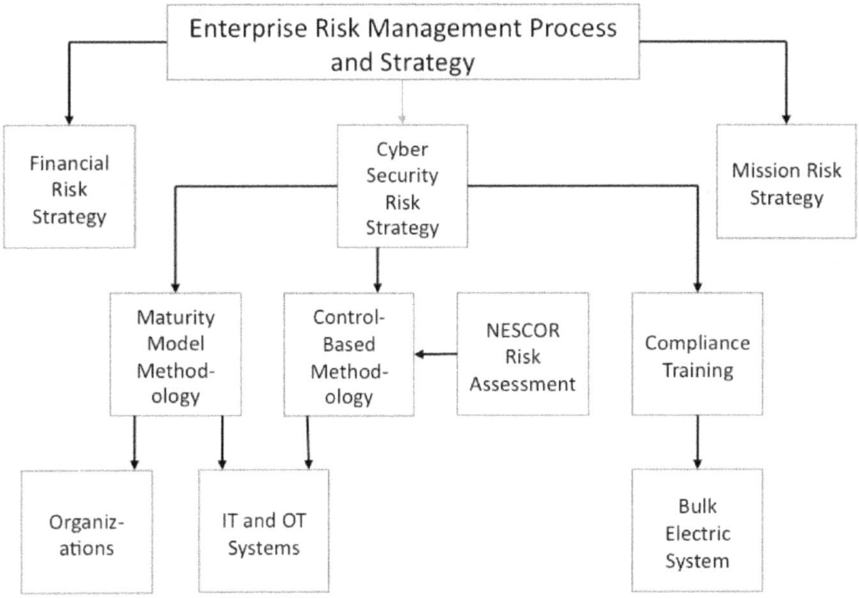

Figure 9-7. Enterprise Risk Management Process and Strategy—Maturity Model Methodology (EPRI 3002003332)

or other high-energy explosive device. HEMP is an EMP detonated at high altitude to produce more widespread effects (an EMP, as opposed to HEMP, can cause more localized but still significant impacts). The HEMP from a high-yield gamma ray weapon can in principle impact the functionality of power grids, communication infrastructures, computing and electronic processing systems, and ground transportation systems dependent on microprocessors or embedded electrical systems that are susceptible to the disruptive effects of large EM perturbations.

Multi-Pulse EMP Components during Nuclear HEMP Events

Unlike the electromagnetic perturbations created by solar storms interacting with the Earth's magnetosphere (geomagnetic disturbances), the HEMP generated by the detonation of nuclear devices creates a three-component pulse, each with distinct characteristics as defined by the International Electrotechnical Commission (IEC). The commission labeled the sequential components E1, E2, and E3.

E1 Pulse—The first pulse component, E1, generates a brief, yet highly energetic, electromagnetic field that may induce extremely high voltages as it couples with electrically conductive structures. The E1 component is responsible for the most damage in any HEMP event, as the high voltages may cause electrical breakdown in devices and components. The damage or disruption occurs practically simultaneously over a very large area.

E2 Pulse—The E2 pulse produced by HEMP explosions is generated by scattered gammas (secondary gammas) and by "neutron scatter gammas" from gammas created when the neutrons produced by the nuclear weapon are scattered by air. The E2 pulse is similar to a conventional lightning strike, beginning approximately a few microseconds to one full second after the onset of the E1 pulse

E3 Pulse—The E3 component is similar to the geomagnetic disturbances (GMDs) that occur when particles from solar flares strike the Earth's magnetosphere.

Research and development related to HEMP devices has been focused on producing a HEMP-enhanced nuclear explosion that can cause potential damage over a large land mass. The damaging effects, besides possibly affecting major power grids in large geographical regions, could also include:

- Disruption of cellular wireless communications
- Failure of computer systems
- Disruption of business and financial computing systems
- Depending on the blast intensity and proximity to the blast point, damage or disruption to a significant portion of the transportation infrastructure
- Disruption of non-military grade (unhardened) video, radio, and internet communications
- Interference with air transportation, as unhardened or unprotected critical aircraft instrumentation could lose functionality
- Damage to petroleum and natural gas delivery systems
- Disruption of essential services dependent on the power grid, such as healthcare facilities, waste treatment, and food distribution systems

Military studies have documented the effects of HEMP attacks as observed in the field or through experimental tests. All available studies confirm the likelihood of the above mentioned HEMP effects on unprotected infrastructure systems. In testimony before the U.S. Congress (Commission to Assess the Threat to the United States from Electromagnetic Pulse (EMP) Attack), for example, military officials have stated that the military facilities are largely prepared and hardened against HEMP disruptions and that equipment and facilities are regularly tested for HEMP vulnerabilities. All critical U.S. military equipment must now conform to the MIL-STD-188-125 standard for mitigating HEMP effects. There are other sectors, for example aerospace, automotive, financial institutions, and government agencies, where the MIL-STD-188-125 standards have been adopted for hardening systems against EMP attacks.

References

- NIST Special Publications 1108R2 Release 2.0, National Institute of Standards and Technology, February 2012
- EPRI Smart Grid Demonstration Projects: Designs, Challenges and Trends. EPRI, Palo Alto, CA: 2011. 1024573
- Federal Register Executive Order 13636—Improving Critical Infrastructure Cyber security, Vol. 78, No. 33 Tuesday, February 19, 2013
- Cyber Security in Power Systems, V. Sridharan, Georgia Institute of Technology, May 2012
- Improving Critical Infrastructure Cyber Security Executive Order 13636: Preliminary Cyber Security Framework, National Institute of Standards and Technology (NIST), Nov. 2013
- Cyber Security Update, Presentation by Bennett Gaines, First Energy at EPRI 2012 Summer Seminar, August 2012
- Assessing the Risks and Seizing the Opportunities, Presentation by Asante, National Board of Information Security Examiners (NBISE), EPRI 2012 Summer Seminar, August 2012
- Introduction to NISTIR 7628 Guidelines for Smart Grid Cyber Security, The Smart Grid Interoperability Panel (SGIP) Cyber Security Working Group (CSWG), National Institute of Science and Technology, September 2010
- Integrating Electricity Subsector Failure Scenarios into a Risk Assessment Methodology. EPRI, Palo Alto, CA: 2013. 3002001181
- Guidelines for Smart Grid Cyber Security: Vol. 2, Privacy and The Smart Grid (NISTIR 7628), National Institute of Standards and Technology, August 2010
- Cyber security Procurement Language for Energy Delivery Systems, Energy Sector Control Systems Working Group (ESCSWG), U:S. Department of Energy
- Cyber security for State Regulators 2.0 (With Sample Questions for Regulators to Ask Utilities) N. Keogh & C. Cody, The National Association of Regulatory Utility Commissioners (NARUC), February 2013.
- Cyber Security and Power System Communication—Essential Parts of a Smart Grid Infrastructure, IEEE Transactions on Power Delivery, Vol. 25, No. 3, July 2010.

- Department of Homeland Security (DHS), http://www.dhs.gov/cybersecurity-overview
- High-Impact, Low-Frequency Events in the Electric Power Industry: Potential Impacts, Mitigation and Risk Management, EPRI 3002001935, August 2013.
- NERC Report "High-Impact, Low-Frequency Event Risk to the North American Bulk Power System," June 2010. www.nerc.com/HILF.pdf
- Grid Strategy 2011: Security in Demonstrations, EPRI 1024573, October 2011
- Steering Committee for Innovation in Smart Grid Measurement Science and Standards, Nation Institute for Standards and Technology (NIST), December 2012
- Electricity Subsector Cyber security Risk Management Process (RMP), U.S. Department of Energy (DOE), May 2012

Chapter 10

The Benefits and Cost of the Smart Grid

ESTIMATING THE BENEFITS AND COST OF THE SMART GRID

The purpose of this chapter is to describe a preliminary estimate of the benefits and cost of the Smart Grid.

WHAT CONSTITUTES THE POWER DELIVERY SYSTEM?

To estimate the benefits and cost of the Smart Grid the author divided the power delivery system into distinct functional areas and then applied a number of assumptions applied about technology development, deployment, and cost over a twenty year period (2010-2030).

In this chapter, the power delivery system includes the busbar located at the generating plant (where the power delivery system begins) and extends to the energy-consuming device or appliance at the end. The power delivery system encompasses generation step-up transformers; the generation switchyard; transmission substations, lines, and equipment; distribution substations, lines, and equipment; distributed energy resources: power quality mitigation devices and uninterruptible power supplies; sensors, and other equipment.

Inadequacies in the power delivery system are manifested in the form of poor reliability, degraded power quality, vulnerability to terrorist attack, the inability to integrate renewables, and the inability to provide basic or enhanced services to consumers.

WHAT DIFFERENTIATES SMART GRID ENHANCEMENT?

Meeting the energy requirements of society will require the application across the entire power delivery system of a combination of current and advanced technologies, including but not limited to the following:

- Automation: the heart of a "smart power delivery system."
- Communication architecture: the foundation of the power delivery system of the future and the enabler of Smart Grid integration.
- Distributed energy resources and storage development and integration.
- Power electronics-based controllers and widely dispersed sensors throughout the delivery system.
- An advanced metering infrastructure.
- A consumer portal that connects consumers and their equipment with energy services and communications entities.
- Power market tools—information systems which enable fluid wholesale power markets.
- Technology innovation in electricity use.
- Appliances and devices which are demand-response ready.

Developing an optimal combination of these technologies will require a significant, sustained investment.

To conduct a preliminary quantitative estimate of the level of investment needed over the next 20 years, the core technologies were separated into four broad areas: transmission and substations, distribution systems and the customer. Estimates were constructed into two segments:

- **Meeting load growth and correcting deficiencies** via equipment installation, upgrading, and replacement to accommodate new customers (new connects), to meet the increasing energy needs of existing customers as their load grows, and to correct deficiencies often due to lack of investment in the power delivery system in the past.
- **The Smart Grid**: The investment needed to develop and deploy advanced technologies needed to enhance the functionality of the power delivery system to achieve the level of a Smart Grid.

The first segment represents investments required to maintain adequate capacity and functioning of the existing power delivery system, while the second segment is the additional cost to elevate this system to that of a Smart Grid.

Key Assumptions

Cost estimates in this chapter are built upon a number of key assumptions:

- The smart grid will incorporate technologies that not only make the electricity delivery system smarter, but also stronger, more resilient, adaptive, and self-healing.
- The smart grid will eventually include every reasonable and cost-effective enhancement to accommodate regulatory mandates:
  - Consistent with the functionality requirements of the Energy Independence and Security Act (EISA) of 2007.
  - Meets reasonable cost-benefit assessment.
  - Meets North American Electric Reliability Corporation (NERC) reliability standards, maintaining or enhancing today's reliability levels (1 day in 10 years loss of load probability or LOLP).
  - Meets System Average Duration Interruption and System Average Interruption Frequency (SADI/SAFI) state guidelines normally suggest 100 minutes SADI/SAFI and power quality (PQ) events to remain at today's levels or to improve.
  - Meets performance rate-making targets.
  - Meets requirements of renewable portfolio standards (RPS).
- The smart grid will incorporate technology and policies that enhance Smart Grid functionality while meeting load growth, expanding and modernizing the power delivery system to:
  - Enable a fully functional power delivery system
  - Enable consumer connectivity and service enhancement; and
  - Enable integration of distributed energy resources
- The smart grid will accommodate expansion of renewable energy resources.

- — Adding 135GW of renewables by 2030.
- — Achieving 20% of wind by 2020.

- The Energy Information Agency's Annual Energy Outlook 2010 projects that the annual growth rate in electricity for the period 2008 to 2035 is projected to be 1.0%. This is as a result of "structural changes in economy—higher prices—standards—improved efficiency" (EIA, 2009). EPRI estimates that the programs and activities which are part of the Smart Grid as incorporated in this chapter have the potential to reduce this growth rate to 0.68% per year. In addition, EPRI estimates that peak demand's growth rate will be 0.53% per year (EPRI 1016987). These growth rates were used in assumptions about the increasing needs for assets to serve consumers.

- This chapter assumes simultaneous deployment of Smart Grid functionality. While deployments will realistically be made along parallel paths and in discrete steps, this study assumes they will occur simultaneously and continuously.

- This chapter assumes steady rate of deployment. Deployment of most technologies will be made at a rate of 1/20th of the maximum assumed penetration each year over the 20-year period. Enhancement and modernization will continue thereafter.
  - — The Smart Grid will never be finished. It will continue to evolve organically, not as a step function and not as a "revolution," but as new technology becomes available.
  - — Investments in Smart Grid will not be made linearly over 20 years, or necessarily even within 20, and will not be uniformly distributed around the country.

- Total power delivery investment costs will exceed Smart Grid investments. They will include investments to meet load growth and to maintain reliability.

- Technology costs are likely to decrease while performance levels increase in unforeseen, and possibly dramatic, ways over the next 20 years. Reasonable estimates have been made, but they are likely to prove conservative given the rapid pace of technological advances. Historically, massive technology advances such as implied by the Smart Grid are invariably driven by a single breakthrough innovation. Smart Grids don't have just one; instead they have a wide range of different ideas—some at the pilot or even experimental stage. In some sense, communications, control, and computational ability is one such set of breakthroughs. However, other advances in storage, power electronics, and sensors are still needed to complete the mosaic this report paints.

- Appropriately considering operating and maintenance (O&M) costs associated with running utilities who deploy Smart Grid technologies is critical. O&M expenses are a substantial part of total costs and are built into rates at estimated levels. The IT and technology O&M aspects of the Smart Grid need to be included in cost estimates.

Smart Grid Costs are Particularly Hard to Estimate

Smart Grids are by their nature difficult to estimate for several reasons:

- They frequently involve the integration of digital technology—Sometimes virtually embedded transmission and distribution assets have different failure rates and life expectancy than the majority of today's grid technologies. These failures and resultant replacement rates must be estimated. Utilizing a reliable component, like a substation transformer, with a 40-year design life and incorporating an information technology with 10, 15 or 20 life forces careful consideration of the costs to upgrade the embedded components.
- The obsolescence of digital technology is rapid. Increasingly complex and expanding communications and computational ability makes it possible to render Smart Grid components obsolete or inoperable with respect to the rest of the information and communications technology (ICT) system well before their end of life. Therefore, reasonable replacement costs must be estimated.
- The improvement in Smart Grid technologies and projected decreases in their costs will occur at a greater rate than "conventional" technology.
- Uncertainty in performance—Many Smart Grid technologies are relatively new and unproven. If their performance is marginal or degrades unexpectedly over time, the entire business plan for the technology could be undermined.

- Smart Grid component costs are declining rapidly. As these technologies mature and as production volumes increase, the marginal costs of Smart Grid technologies have the potential to decline rapidly.

Figure 10-1. Grid Component Costs (Illustrative)

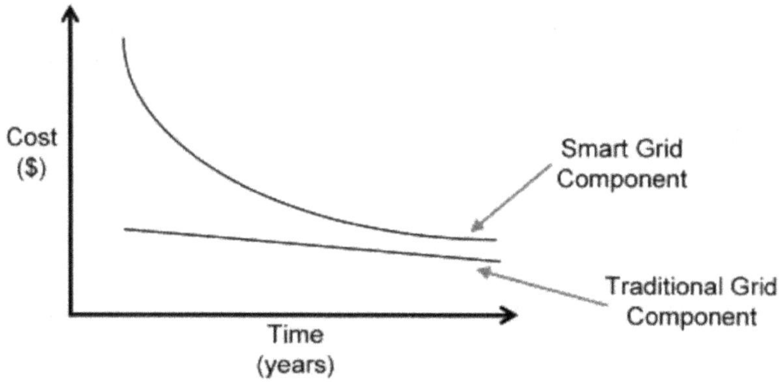

Technology Assessment: What's In and What's Not In?

Table 10-1 summarizes what has been included in the cost analysis describe in this chapter and what has not been included. For the most part, T&D line expansion to meet load growth has been excluded.

Figure 10-2 illustrates the scope of the cost estimates included in this report. Investments traditionally made by customers, such as appliances and hybrid vehicles, have been excluded. However, infrastructure integration costs are included.

To conduct a preliminary quantitative estimate of the level of investment needed over the next 20 years to enable the envisioned power delivery system, transmission, distribution, and customer-related costs are treated separately. This is due to fundamental differences in the nature of the transmission and distribution portions of the power delivery system; and uncertainty if the costs categorized as consumer costs would be borne by utilities, consumers, or third-party service providers. It should be noted that there are, however, substantial areas in which distribution technology enhancements will greatly affect the operation and potentially the configuration of the transmission system. These interactions were not considered in this evaluation. The author also decided to further subdivide the estimating process for transmission and distribution into the following two segments:

Table 10-1. Grid Technologies Included and Excluded in the Cost Estimates

Technologies	Costs Included in This Analysis	Costs Excluded
1. Increased use of digital information and controls technology to improve reliability, security, and efficiency of the electric grid.	Sensors, communications and computational ability	None
2. Dynamic optimization of grid operations and resources, with full cyber-security.	All grid-related	None
3. Deployment and integration of distributed resources: storage and generation, including renewable resources.	All integration costs	New transmission lines including those to integrate renewables; and costs of renewable power generation technology*
4. Development and incorporation of demand response, demand-side resources, and energy-efficiency resources	All integration costs	The cost of energy-efficient devices
6. Integration of "smart" appliances and consumer devices.	All integration costs (see above)	Consumer appliances and devices
7. Deployment and integration of advanced electricity storage and peak-shaving technologies, including plug-in and hybrid-electric vehicles, and thermal-storage air conditioning.	Bulk power storage devices and high-value distributed storage – such as bulk storage for wind penetration; distributed storage for grid support; customer-side-of-the-meter storage for end-use energy management	Low-value distributed storage
Technologies	**Costs Included in This Analysis**	**Costs Excluded**
8. Provision to consumers of timely information and control options.	All enabling costs including cost of consumer display devices	None
10. Identification and lowering of unreasonable or unnecessary barriers to adoption of Smart Grid technologies, practices, and services.	All	None

*It should be noted that the cost of new transmission needed to integrate renewables should ultimately be included in estimating the cost of tomorrow's power delivery system.

- Load Growth. Via equipment installation, upgrading, and replacement, transmission and distribution system owners invest in the power delivery system to accommodate new customers (so-called "new connects") and to meet the increasing energy needs of existing customers as their load grows.
- Power Delivery System of the Future ("Future PDS"). The author estimated the investment needed to develop and deploy advanced

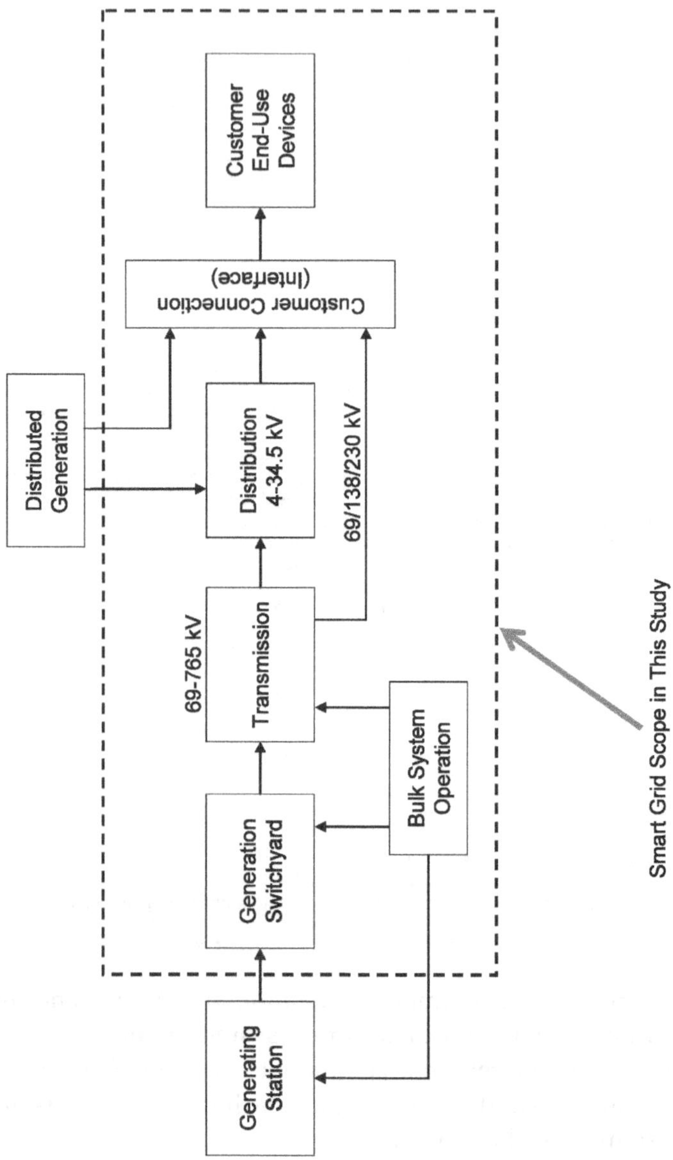

Figure 10-2. Smart Grid Bulk Component Diagram

technologies needed to realize the vision of the power delivery system (both transmission and distribution) described above.

There is a third cost element not considered here—that of maintaining reliability. Transmission and distribution utilities often spend an amount equal to 1 to 2% of their depreciated plant in service on refurbishment so as to maintain reliability. In conducting business as usual, necessary expenditures to accommodate load growth and to maintain reliability will naturally be made with equipment that is compatible with the power delivery system of the future.

Figure 10-3 illustrates how these three cost elements might combine to build the power delivery system of the future.

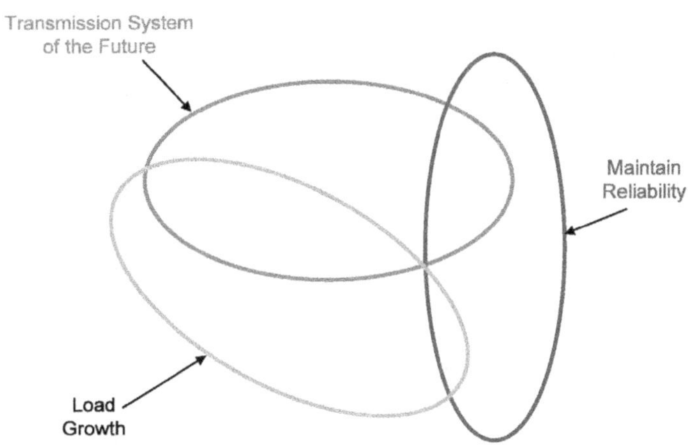

Figure 10-3. Illustration of Synergies in the Three Categories of Needed Transmission and Distribution Investment

In this study, the author separated the expenditures necessary to accommodate load growth from the expenditures directly related to the power delivery system of the future, in order to elucidate the true cost of the Future PDS.

THE BENEFITS OF THE SMART GRID

There have been a number of studies which have estimated some of the benefits of a Smart Grid. Each of them varies somewhat in their approach and the attributes of the Smart Grid they include. None of them provides a comprehensive and rigorous analysis of the possible benefits of a fully functional Smart Grid.

In 2004, EPRI undertook a study to estimate the cost and value (benefits) of the power delivery system of the future. The fundamental approach that was used in the 2004 study involved the identification of attributes of the power system (e.g., cost of energy, capacity, security, quality, reliability, environment, safety, quality of life, and productivity). EPRI then developed the framework to quantitatively estimate the dollar value of improving each of these attributes by a defined amount (i.e., percentage improvement).

Existing, documented data sources were used for this estimation process for each attribute. These sources included the U.S. Energy Information Administration, the U.S. Department of Energy's Policy Office Electricity Modeling System, the Federal Energy Regulatory Commission's transmission constraint study, the U.S. Labor Department's Bureau of Labor Statistics, and many more.

Table 10-2 shows the various types of improvements that correspond to each of the attribute types used in the root study. A key aspect of the value estimation process in general is its consideration of improvements to the power delivery system (see the left column of Table 10-2), as well as improvements that consumers directly realize (see the right column of Table 10-2). This was done to ensure that emerging and foreseen benefits to consumers in the form of a broad range of value-added services addressed in the estimation of value.

The "cost of energy" attribute is the total cost to deliver electricity to customers, including capital costs, O&M costs, and the cost of line losses on the system. Therefore the value of this attribute derives from any system improvement that lowers the direct cost of supplying this electricity. "SQRA" is the sum of the power security, quality, and reliability attributes, because the availability part of SQRA is embedded in the power quality and reliability attributes. The quality of life attribute refers to the integration of access to multiple services, including electricity, the Internet, telephone, cable, and natural gas. This involves integration of the power delivery and knowledge networks into a single intelligent electric power/ communications system, which sets the stage for a growing variety of products and services designed around energy and communications.

To quantify the benefit of these improvements for various attributes, EPRI staff developed various "benefit calculator tools." Figure 10-4 shows the relationship of these benefit calculator tools, the attri-

Table 10-2. Attributes and Types of Improvements Assumed in the Value Estimation of the Future Power Delivery System (Left: Power Delivery System Improvements; Right: Improvements that Consumers Realize)

Power Delivery (Improvements/ Benefits) 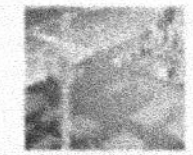	Attributes	Consumer (Improvements/ Benefits) 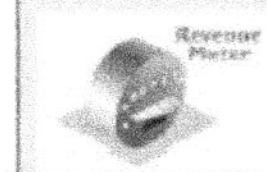
O&M Cost Capital Cost of Asset T&D Losses	*Cost of Energy (Net delivered life-cycle cost of energy service)*	End Use Energy Efficiency Capital cost, end user infrastructure O&M, End User Infrastructure Control/Manage Use
Increased Power Flow New Infrastructure Demand Responsive Load	*Capacity*	Improved power factor. Lower End User Infrastructure cost through economies of scale and system streamlining, expand opportunity for growth
Enhanced Security Self Healing Grid for Quick Recovery	*Security*	Enhanced Security and ability to continue conducting business and every day functions
Improve Power Quality and enhance equipment operating window	*Quality*	Improve Power Quality and enhance equipment operating window
Reduce frequency and duration of outages	*Reliability & Availability*	Enhanced Security Self Healing Grid for Quick Recovery Availability Included
EMF Management Reduction in SF6 (sulfur hexafluoride) emissions Reduction in cleanup costs Reduction in power plant emissions	*Environment*	Improved Esthetic Value Reduced EMF Industrial Ecology
Safer work environment for utility employees	*Safety*	Safer work environment for end-use electrical facilities
Value added electric related services	*Quality of Life*	Comfort Convenience Accessibility
Increase productivity due to efficient operation of the power delivery infrastructure Real GDP	*Productivity*	Improved consumer productivity Real GDP

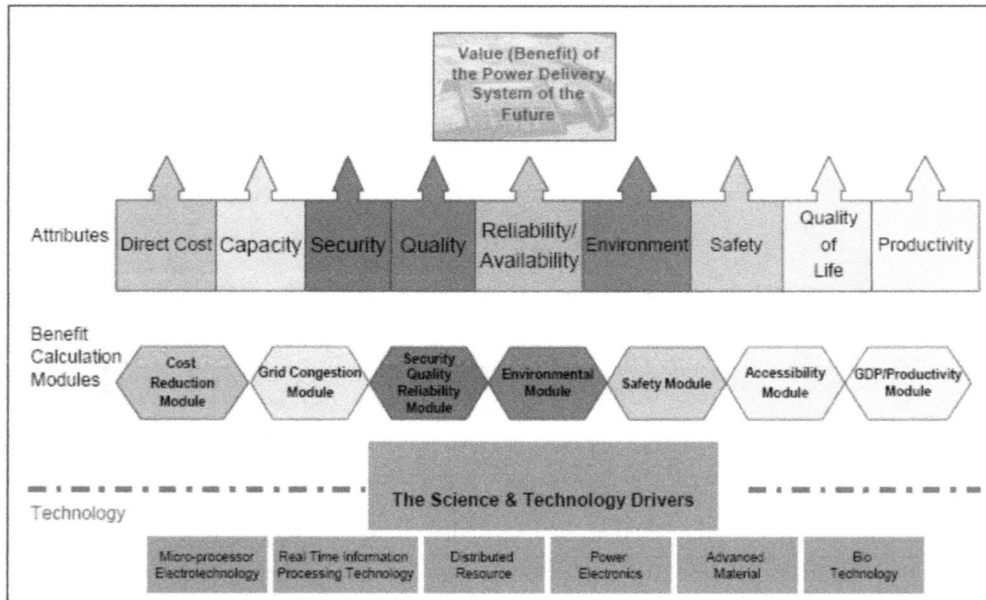

Figure 10-4. Relationship between Value, Attributes, and Benefit Calculator Tools in the Value Estimation Process

butes, and the overall value. Note that science and technology drivers feed generally into the process.

Table 10-2 and Figure 10-4 illustrate the attributes considered in the original estimates of value published in 2004.

The electric power industry is the last industry in the western world to modernize itself through the use of sensors, communications, and computational ability. The combination of these functions allow for a truly interactive power system which can integrate consumer demand with supply interactively.

In addition to the obvious benefits achieved by this enhanced functionality—namely, improved asset utilization, reduced electricity cost, and improved reliability, it is now obvious that other substantive benefits will accrue once the Smart Grid is implemented. Chief among these is the reduction of peak demand, the adoption of electric vehicles, the use of storage, and the increased use of renewable power production.

Table 10-4 lists the major attributes and benefits not included in the original EPRI study.

EPRI developed estimates of the major benefits of a fully functional power delivery system, These are depicted in Table 10-5 using a framework developed by the U.S. Department of Energy (USDOE) and EPRI (EPRI 1020342). The table includes the attributes and benefits explicitly

Table 10-3. Summary of Benefit Calculations Included in Original EPRI Study

Benefit Calculations	Attributes
T&D Cost Reduction	Energy efficiency and T&D losses impact on: • Capital cost • O&M Cost • Administrative and general cost
Congestion	• Transmission congestion cost
Security	• Self-healing infrastructure • Mitigating major outages
Power Quality & Reliability/ Availability	• Reliability • Power Quality
Environmental Impact	• SO_2 • CO_2 • NOx
Safety	• Accidental electrocutions • Building fires caused by electrical infrastructure
Quality of Life	• Access to competing suppliers
GDP/Productivity	• Increase GDP from reduced electricity cost

Table 10-4. Major Attributes and Benefits Not Included in Original EPRI Study

Attributes	Benefits
Demand Response	• Reduced need for generation capacity • Reduced demand for electricity
Facilitating Renewables	• Reduced environmental impact of electricity generation
PEVs	• Reduced environmental impact from displaced
Attributes	**Benefits**
	fossil fuels • Increased system flexibility/ancillary services
Work Force	• Improved utilization of work force
Energy Efficiency	• Generation Capacity deferrals
Enhanced Energy Efficiency (additional energy efficiency)	• Reduced Environmental Impacts
AMI	• AMI-related cost reductions
Distributed Generation	• Facilitating distributed generation
Value-Added Electricity Services	• Comfort and convenience
Synergistic Effects Between Elements	• Compounding between multiple attributes
Storage (various benefits)	• Capacity • O&M • Congestion
Safety	• Personal safety
Transmission O&M	• Ancillary services

included in the DOE/EPRI framework as well as other attributes not included.

As summarized in the table, the total benefit of all attributes for the Smart Grid is estimated to be between $1,294 billion and $2,028 billion for 20-year period. Once all of the attributes and benefits of a Smart Grid are identified and analyzed, estimates of the total benefit will increase even more.

TRANSMISSION SYSTEMS AND SUBSTATIONS COSTS

To estimate the investment needed in the transmission system, a top-down approach was used for the load-growth and correct-deficiencies segments of investment, while a bottom-up approach was used for estimating the elements needed to create a Smart Grid.

Table 10-5. List of Smart Grid Benefits: Based on EPRI/DOE Framework (EPRI 1020342)

Benefit Category	Benefit Sub-Category	Benefit	Included in Original Estimate?	Estimated Value 2010-2030 $Billion		Reference
				Low	High	
Economic	Improved Asset Utilization	Optimized Generator Operation		---	---	not included
		Deferred Generation Capacity Investments	→	→	→	Appendix A
		Reduced Ancillary Service Cost	X →	→	→	included below
		Distributed Generation		---	---	not included
		Storage		48	89	Appendix A
		PEVs as Storage & Load Control		11	11	Appendix A
		Energy Efficiency	X →	→	→	included below
		Demand Response		---	---	not included
		Enhanced Energy Efficiency*		---	---	not included
		Reduced Ancillary Service Cost		---	---	not included
		Reduced Congestion Cost	X →	→	→	included below
	T&D Capital Savings	Distributed Generation		27	27	Appendix A
		Storage		23	65	Appendix A
		Demand Response		192	242	not included
		Energy Efficiency	X →	→	→	included below
		Enhanced Energy Efficiency*		1	3	Appendix A
		Deferred Transmission Capacity Investment	X →	→	→	included below
		Deferred Distribution Capacity Investment	X →	→	→	included below

Table 10-5 (*Cont'd*). List of Smart Grid Benefits: Based on EPRI/DOE Framework (EPRI 1020342)

Benefit Category	Benefit Sub-Category	Benefit	Included in Original Estimate?	Estimated Value 2010-2030 $Billion Low	Estimated Value 2010-2030 $Billion High	Reference
		Reduced Equipment Failures	X →	→	→	included below
	T&D O&M Savings	More Effective Use of Personnel		---	---	not included
		Economic Benefit of Added Personnel		---	---	not included
		Operations Savings from AMI		4	4	Appendix A
		T&D Efficiency	X →	→	→	included below
		Reduced Distribution Equipment Maintenance Cost	X →	→	→	included below
		Reduced Distribution Operations Cost	X →	→	→	included below
	Theft Reduction	Reduced Electricity Theft		---	---	Not included
	Energy Efficiency	Enhanced Energy Efficiency*		0	2	Appendix A
		Electrification (Net Reduced Energy Use)		---	---	Appendix A
		Reduced Electricity Losses	X →	→	→	included below
		Productivity Increase	X →	→	→	included below
	Electricity Cost Savings	Reduced Electricity Cost	X →	→	→	Included below
		Automatic Meter Reading		91	91	Appendix A
		Customer Service Costs (Call Center)		2	2	Appendix A
		Storage		115	199	Appendix A
		Enhanced National Productivity	X →	→	→	included below
---	---	Reduced Restoration Cost	X →	→	→	included below

Table 10-5 (*Cont'd*). List of Smart Grid Benefits: Based on EPRI/DOE Framework (EPRI 1020342)

Benefit Category	Benefit Sub-Category	Benefit	Included in Original Estimate?	Estimated Value 2010-2030 $Billion Low	High	Reference
		Speed of Restoration		---	---	not included
		Storage		2	20	Appendix A
		Reduced Sustained Outages	X →			included below
		Reduced Major Outages	X →			included below
		Accessibility	X →			included below
	Power Quality	Reduced Momentary Outages	X →			included below
		Reduced Sags and Swells	X →			included below
		Storage		1	21	Appendix A
Environmental	Air Emissions	Electrification		21	21	Appendix A
		PEVs		5	123	Appendix A
		Enhanced Energy Efficiency*		1	4	Appendix A
		Storage		10	15	Appendix A
		Facilitate Renewables		10	172	Appendix A
		Reduced CO_2 Emissions	X →			included below
		Reduced SOx, NOx and PM-10 Emissions	X →			included below
Security	Energy Security	Reduced Imported Oil Usage		---	---	not included
		Personal Security		---	---	not included
		National Security		---	---	not included
		Reduced Wide-Scale Blackouts	X →			included below

Table 10-5 (*Cont'd*). List of Smart Grid Benefits: Based on EPRI/DOE Framework (EPRI 1020342)

Benefit Category	Benefit Sub-Category	Benefit	Included in Original Estimate?	Estimated Value 2010-2030 $Billion Low	High	Reference
		Safety	X →	→	→	included below
Previous EPRI Estimates – All included in original estimate			X	564	1,111	
Not included in original estimate				730	917	
Total				1,294	2,028	

*Enhanced Energy Efficiency includes:

- Continuous Commission of Large Commercial Buildings
- Direct Feedback on Energy Usage
- Energy Savings Corresponding to Peak Load Management
- Energy Savings Corresponding to Enhanced M&V Capability

In the U.S., according to the Edison Electric Institute (EEI), there are more than 200,000 miles of high-voltage transmission lines greater than 230 kV. An earlier study by DOE, entitled the *National Transmission Grid Study, 2002*, showed a total of 187,000 miles.

The total cost for enhancing transmission system and substation performance to the level of a Smart Grid is estimated between $56 and $64 billion, as summarized in this section. The cost includes several categories of technology whose functionality overlaps significantly between both the transmission system and substations as well as some elements of the distribution system described later, as well as enterprise level functions, such as cyber security and back office systems.

The number of substations is one of the basic metrics upon which investment costs were determined. This study assumed that there are an estimated 70,000 substations in the U.S. that reduce voltage between the bulk transmission system and the distribution feeder system, and serve as critical hubs in the control and protection of the electricity grid. This figure was derived from FERC data that shows investor-owned utilities (IOUs) operate a total of 40,619 substations at voltage levels ranging from just above 1 kV to 765 kV. Since IOUs represent roughly 70% of all U.S. customers, the number of existing substations was thus calculated to be 58,027 (40,619/.7 = 58,027).

Load was estimated in the study to grow at a rate of 0.68%/year. Compounded over the 20-year period of the study (2010-2030), this would imply an additional 8,423 substations will be required by 2030. Accordingly, the base figure used throughout this report for substations is 58,027 which could potentially be upgraded and 8,423 which will be new. In addition to this base, another 700 substations will be required by 2030 to handle renewable generation.

Other key benchmarks used in the analysis include an estimate of 8 feeders/substation serving lower-voltage customers downstream. Thus, there are 464,216 feeders that are eligible to be upgraded with intelligent electronic devices for a fully functioning Smart Grid. In addition, there are 67,384 new feeders to be added to accommodate load growth. Segments of the feeders that can be isolated electrically in case of faults and/or reconfiguration are called "pods," and for purposes of the Smart Grid, the analysis team used roughly 4 pods/feeder, yielding more than 2,260,000 isolatable pods for purposes of monitoring and control.

Table 10-6 lists these assumptions:

Table 10-6. Number of Substations and Feeders

Substations	Number
Existing substations	58,027
New substations to accommodate load growth (2030)	8,423
New substations to accommodate renewables (2030)	700
Distribution	
Number of existing feeders	464,216
New feeders to accommodate load growth (2030)	67,384

An underlying assumption in the report is that the digital devices to be deployed in the Smart Grid will comply with the International Electrotechnology Commission (IEC) Standard 61850 (IEC 61850). That standard applies to substation automation and protection, distribution automation, distributed energy resources, hydro generation, Supervisory Control and Data Requisition (SCADA) to field devices, and applies to protective relays, SCADA Master, Distributed Energy Resources (DER), PQ meters, fault recorders and other applications.

Cost Components for the Smart Grid: Transmission Systems and Substations

The core components of cost for the transmission and substation portion of the Smart Grid are as follow:

- Transmission line sensors including dynamic thermal circuit rating
- Storage for bulk transmission wholesale services
- Flexible AC Transmission Systems (FACTS) devices and High Voltage Direct Current (HVDC) terminals
- Short circuit current limiters (SSCL)
- Communications infrastructure to support transmission lines and substations
- Core substation infrastructure for Information Technology (IT).
- Cyber-security
- Intelligent electronic devices (IEDs)
- Phasor measurement technology (PMUs) for wide area monitoring
- Enterprise back-office system, including Global Information Systems (GIS), outage management and distribution management
- Other system improvements assumed to evolve naturally include:

— Faster than real-time simulation
— Improved load modeling and forecasting tools
— Probabilistic vulnerability assessment
— Enhanced visualization

Substation upgrades will enable a number of new functions including, but not limited to:

- Improved emergency operations
- Substation automation
- Reliability-centered and predictive maintenance

Dynamic Thermal Circuit Rating (DTCR)

The study team assumed that AC transmission lines rated 115 kV to 230 kV in the U.S. are most susceptible to being thermally limited. They considered lines rated 345 kV and above are more likely voltage-limited. Albeit, there may be a few lines rated at these higher voltages that are thermally limited—they would be an exception and not the rule. Furthermore, of the lines which are potentially thermally limited, only 50% actually are. There are 85,048 miles of lines at 230 kV. To dynamically rate all 85,048 miles would require one unit per 7.5 miles or 11,340 units. The cost of DTCR deployed in quantity is estimated to be $20,000 initially, declining to $10,000.

It should be noted that the above calculation does not include any transmission lines at voltages lower than 230 kV, e.g., 115, 138 and 161 kV. It is expected that these transmission lines will benefit from DTCR as well.

Table 10-7. Cost of Dynamic Thermal Circuit Rating

Technology	Total Units	Units	% Sat	Cost/ Unit Low $	Cost/ Unit High $	Total Cost Low–High $M
Dynamic-Thermal Circuit Rating	11,340	Units/7.5 miles DTCR line	100	10,000	20,000	113.4-226.8

Sensors

By 2030 one-half of all substations will have installed an advanced sensors package on the transmission system, costing roughly $50,000-100,000/substation. The total Smart Grid investment for sensors approaches $1.5 to 2.9 billion. In addition, 100% of new substations built to accommodate load growth would incorporate a suite of sensors at a cost ranging between $421 to 842 million.

Table 10-8. Cost of Sensors

Technology	Total Units	Units	% Sat	Cost/ Unit Low $	Cost/ Unit High $	Total Cost Low–High $M
Transmission Line Sensors	58,027	Number of existing substations	50*	50,000	100,000	1,451-2,901
Transmission Line Sensors	8,423	Number of new substations	100	50,000	100,000	421-842
*Assumes 50% of substations will have sensors listed in Table 5-4 partially deployed.						

Short-Circuit Current Limiters (SCCL)

The installation of the SCCL will begin in ten years and to be phased in slowly. From 2020, penetration will rise to 2% of transmission substations by 2030. The installation cost will be about $500,000 for SCCL at transmission substations, and about $50,000 at distribution substations. Total installed cost for transmission circuit current limiters through 2030 is estimated at $2.03 billion.

Table 10-9. Cost of Transmission Short Circuit Current Limiters

Technology	Total Units	Units	% Sat	Cost/ Unit Low $	Cost/ Unit High $	Total Cost Low–High $M
Transmission Short-Circuit Current Limiters	58,027	Number of substations	2	500,000	500,000	580.3-580.3

Table 10-10. Flexible AC Transmission (FACTS)

FACTS Devices	Number of Units	Cost/Unit		Total Costs	
		Low $	High $	Low $	High $
STATCON & UPFC	20	35,000,000	45,000,000	700,000,000	900,000,000
TCSC	100	10,000,000	12,000,000	1,000,000,000	1,200,000,000
HVDC Terminals	60	22,500,000	27,500,000	1,350,000,000	1,650,000,000
Power Electronic Transformers	25	40,000,000	40,000,000	1,000,000,000	1,000,000,000
Geomagnetic Controllers	25	5,000,000	7,000,000	125,000,000	175,000,000
Total FACTS Cost				**4,175,000,000**	**4,925,000,000**

Storage

Table 10-11—shows the cost for Compressed Air Energy Storage (CAES) running from $810-1045/kW, whereas lead, acid batteries typically exceeds $2000/kW (EPRI 1017813).

Table 10-11. Highest Value Electrical Storage Technologies

Highest Value Storage Market	kW	Best Technology Fit	Cost/kW
Wholesale Services Without Regulation	5,800,000	CAES	$810-1045
Wholesale Services With Regulation	2,800,000	CAES	$810-1045
Home Backup	2.8	Lead-Acid	$2200
Industrial Power Quality & Reliability	1.8	Advanced Lead-Acid	$2300-2400
Transportable Storage Systems (for Distribution Deferral)	1.7	Advanced Lead-Acid	$2180-2900

Lithium Ion (Li-ion) systems may, in the long run, be a potentially low-cost option for grid support. These may be the most compelling for energy durations under four hours with one to three hours being the sweet spot.

The author estimated the installation of 5,800,000 kW of CAES storage capacity for wholesale services in areas without regulation, and 2,800,000 kW of capacity for wholesale services with regulation. The total investment cost over the next 20 years is estimated at $4.7 billion to $6.1 billion, as shown in Table 10-12.

Table 10-12. Cost of Storage Technology

Technology	Total Units	Units	% Sat	Cost/ Unit Low $	Cost/ Unit High $	Total Cost Low–High $M
Storage for Bulk Transmission Wholesale Services Without Regulation	5,800,000	kW	100	810	1045	4,698-6,061
Storage for Bulk Transmission Wholesale Services With Regulation	2,800,000	kW	100	810	1045	2,268-2,926

Recently, it was announced that several Li-ion systems are going in to provide fast regulation services. It is speculated that their costs may be as low as $1,200 per kW. These technologies could provide frequency regulation at lower costs and less lumpy investments than CAES. CAES is likely needed to avoid wind curtailment under high penetration wind scenarios.

Communications and IT Infrastructure for Transmission and Substations

The study team estimates a cost of $50,000-$75,000 per substation is required to achieve the optimal performance level required for the Smart Grid. Substation upgrades will phase in slowly over the next 20 years, reaching an 80% penetration level of all existing substations by 2030. This suggests a cumulative investment between $2.9 billion and $4.2 billion by 2030. All new substations will incorporate communications and IT infrastructure at the time of construction.

Table 10-13. Cost of Communications and IT Infrastructure for Transmission and Substations

Technology	Total Units	Units	% Sat	Cost/ Unit Low $	Cost/ Unit High $	Total Cost Low–High $M
Core Substation Infrastructure for IT: Smart Substations	58,027	Number of existing substations	80	50,000	75,000	2,321-3,481
Communications Infrastructure to Support Transmission Lines & Substations	58,027	Number of existing substations	80	14,400	14,400	668.5-668.5
Total IT & Communications Infrastructure for Existing Substations						2,989.5-4,149.5
Core Substation Infrastructure for IT: Smart Substations	8,423	Number of new substations	100	50,000	75,000	421-632
Communications Infrastructure to Support Transmission Lines & Substations	8,423	Number of new substations	100	14,400	14,400	121-121
Total IT & Communications Infrastructure for New Substations						542-753

Intelligent Electronic Devices (IEDs)

The cost of incorporating IEDs to monitor and control critical functions at substations will cost an average of $110,000/substation. Approximately 80% of the substations would be brought up to Smart Grid levels by 2030 and 100% of new substations would incorporate them.

Table 10-14. Cost of Intelligent Electronic Devices

Technology	Total Units	Units	% Sat	Cost/ Unit Low $	Cost/ Unit High $	Total Cost Low–High $M
Intelligent Electronics & Sensors	58,027	Number of existing substations	80	110,000	110,000	5,106-5,106
Intelligent Electronics Devices (IED) – Relays &Sensors	8,423	Number of new substations	100	110,000	110,000	927-927

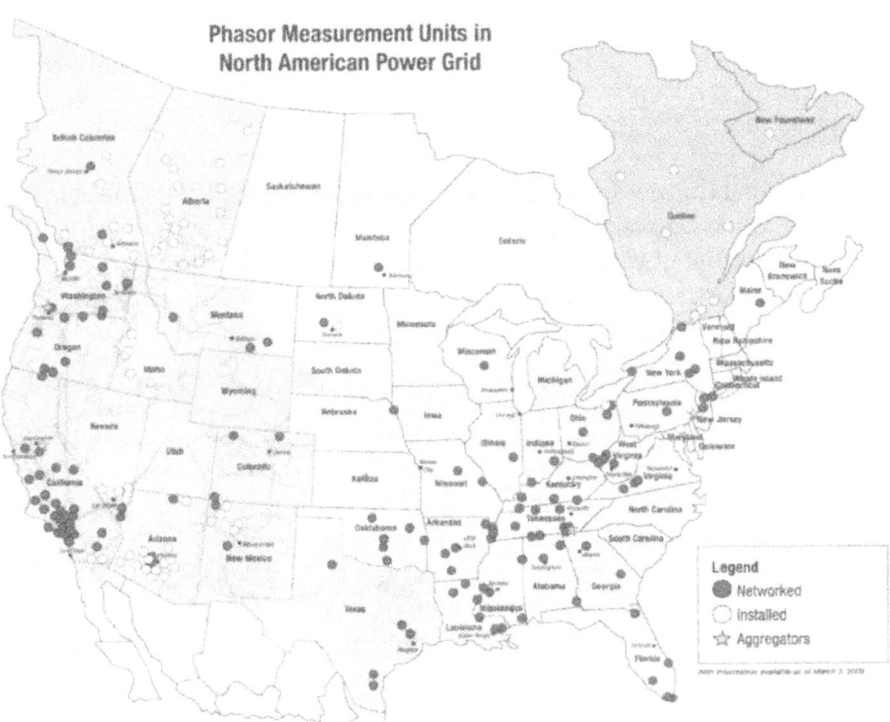

Figure 10-5. Phasor Measurement Units (PMUs) Installed in North America as of September 2009

Phasor Measurement Technology

Utilities are expected to install approximately 1,250 PMUs throughout the grid over the next 20 years at a total cost of $26-39 million.

Cyber Security

There is very little information available as to what actual cyber security costs are new or what they may become. To make estimates, IT suppliers who specialize in cyber security were interviewed. As a result, the estimates in Table 10-16 were developed.

To categorize utilities into small, medium, and large, the study first identified investor-owned utilities (IOU), rural electric cooperatives (Co-ops), municipal utilities (Munis), and Power agencies. To estimate the number of utilities which would need to make provisions to engage in investments to secure their Smart Grid-related cyber information technology activities, Platt's 2010 Directory of Electric Power Producers and Distributors (Platts, 2009) was used.

Platts estimates that there are a total of 342 investor-owned utilities (IOUs). This includes 60 holding companies, 29 transmission companies, and a number of other service and wholesale generation companies. Platts estimated 893 rural electric cooperatives (Co-ops), including both distribution and generation and transmission (G&T) entities. Platts' data for municipal utilities (munis) is divided into two categories: One is municipal and local government utilities; the second is the array of federal, state and district government utilities in the U.S. Table 10-17 summarizes the Platts' estimates. Estimates of the breakdown of size between each of these utility types were made.

Enterprise Back-Office Systems

Enterprise systems will be needed to be upgraded by virtually all utilities. Medium and large utilities will need complete systems of their own. Small utilities may aggregate their needs or use service providers. Table 10-19 summarizes estimates for enterprise back-office systems.

Incremental Ongoing System Maintenance

PMUs and sensors installed on transmission lines and substations would cause additional incremental maintenance of $50,000 per substation.

Table 10-15. Cost of Phasor Measurement Units

Technology	Total Units	Units	% Sat	Cost/ Unit Low $	Cost/ Unit High $	Total Cost Low–High $M
Phasor Measurement Units (PMU)	1,250	Numbers of	100	125,000	125,000	156-156

Table 10-16. Estimates of Cyber Security by Utility Size

	Initial Cyber Costs $K/Utility*	Ongoing Cyber Costs $K/Year
Small	100	10
Medium	400	40
Large	2,200	200
*To be renewed every 10 years		

Table 10-17. Breakdown of Utility Types

Type	Total #	Small		Medium		Large	
		%	#	%	#	%	#
IOU	342	0	--	25	86	75	256
Co-op	893	50	446	40	357	10	90
Muni	2,118	50	1,059	45	953	5	106
Total	**3,353**		**1,505**		**1,396**		**452**

Table 10-18. Cyber Cost Estimates (Note: The smallest utilities are not included)

Size	Number	Cyber Investment $M 2010-2030		Ongoing Cyber Cost $M 2010-2030	
		Per Utility Each 10 Years	Total	Per Utility Per Year	Total
Small	1,505	.10	301.5	.01	30.1
Medium	1,396	.40	1,116.8	.14	111.7
Large	452	2.20	1,988.8	.20	108.8
Total	**3,353**		**3407.1**		**322.6**

Table 10-19. Cost of Enterprise Back Office Systems

Utility Size	Number	Back-Office Investment $M 2010-2030 Each 10 Years	
		Per Utility	Total
Small	1,505	1,000	3,010
Medium	1,396	4,000	11,168
Large	452	20,000	18,080
Total	**3,353**		**32,258**

Enhancements to ISO Systems

The costs of the enhancements to an ISO's system to respond to changes in the applications are very difficult to estimate. Each ISO is in a different phase of development. Each varies as to the projected nature and penetration of distributed resources and as to the extent of other market participants.

Executives from several ISOs including PJM, CAISO, NEISO, and NYISO were interviewed in order to establish estimates for the costs which an ISO will occur in order to accommodate the functionality required to achieve Smart Grid goals. Their informal estimates varied but generally included five or six full-time staff ($2.4 million per year) and at least several million dollars annually in software. For one ISO, this totaled 10% of their annual budget or $12 million per year.

The following are the ten regional power markets in the U.S.

- California (CAISO)
- Midwest (MISO)
- New England (ISO-NE)

Table 10-20. Smart Grid Incremental Maintenance

Technology	Total Units	Units	% Sat		Cost/ Unit Low $	Cost/ Unit High $	Total Cost Low–High $M
Incremental Ongoing Maintenance	58,027	Number of existing substations	50		50,000/yr	50,000/yr	15,232-15,232
Incremental Ongoing Maintenance	8,423	Number of new substations to meet load growth	100		50,000/yr	50,000/yr	4,422-4,422
Incremental Ongoing Maintenance	700	Number of new substations to accommodate renewables	100		50,000/yr	50,000/yr	368-368

- New York (NYISO)
- Northwest
- PJM
- Southeast
- Southwest
- SPP
- Texas (ERCOT)

Table 10-21 summarizes the estimated cost of alignment of ISOs with Smart Grid requirements.

Table 10-21. Cost to Align ISOs with Smart Grids

Technology	Total Units	Units	% Sat	Cost/ Unit Low $	Cost/ Unit High $	Total Cost Low–High $M
ISO Smart Grid	10	Regional markets	100	12/yr	12/yr	2,400-2,400

SUMMARY OF TRANSMISSION AND SUBSTATIONS COSTS

The cumulative cost for bringing the nation's transmission and substations system up to the performance levels required for Smart Grid operation is estimated to cost between $6,312 and $7,280 million by 2030, as shown in the summary Table 10-22. Smart Grid related investment in the transmission system will continue well beyond 2030. This does not include related investments to meet load growth which are estimated to cost between $56,350 and $63,702 million by 2030.

DISTRIBUTION SYSTEMS

Estimates of the cost of individual distribution system components in this analysis were based largely upon utility experience in deploying the first wave of Smart Grid investment. Estimates of Advanced Metering Infrastructure (AMI) and distribution automation investment were drawn, for example, from FirstEnergy, SCE, SDG&E, Dayton Power & Light, FortisAlberta, Inc, and Idaho Power, among others. Representative costs are shown in Tables 10-26 and 10-27 (FirstEnergy, 2009).

Table 10-22. Smart Grid Transmission and Substation Costs

Technology	Total Units	Units	% Sat	Cost/ Unit Low $	Cost/ Unit High $	Total Cost Low–High $M
Dynamic-Thermal Circuit Rating	11,340	Number of substations with one DTCR unit/7.5 miles of line	100	10,000	20,000	113.4-226.8
Substation & Transmission Line Sensors	58,027	Number of existing substations	50	50,000	100,000	1,451-2,901
Transmission Short-Circuit Current Limiters	58,017	Number of substations	2	500,000	500,000	580.3-580.3
Storage for Bulk Transmission Wholesale Services w/o Regulation	5,800,000	kW	100	810	1045	4,698-6,061
Storage for Bulk Transmission Wholesale Services with Regulation	2,800,000	kW	100	810	1045	2,268-2,926
FACTS Devices	330	Numbers of	100	Various	Various	4,175-4,925
Communications: Core Infrastructure for Smart Substations	58,027	Number of substations	80	50,000	75,000	2,321-3,481
Communications to Substations	58,027	Number of substations	80	$1,200/mo	$1,200/mo	668.5-668.5
Phasor Measurement Units (PMU)	1,250	Numbers of	100	125,000	125,000	156-156
Intelligent Electronics Devices (IED) – Relays & Sensors	58,027	Number of substations	80	110,000	150,000	5,106-6,963
Cyber Security – Enterprise-Wide	1,454	Number of utilities	100	100,000*	2,200,000*	3,729.2-3,729.2**
Enterprise BackOffice System–GIS, Outage management, Distribution Management	1,454	Number of utilities	100	1,000,000*	20,000,000*	32,258-32,258
Technology	**Total Units**	**Units**	**% Sat**	**Cost/ Unit Low $**	**Cost/ Unit High $**	**Total Cost Low–High $M**
ISO Smart Grid	10	Regional markets	100	12,000,000/yr	12,000,000/yr	2,400-2,400
Incremental Ongoing System Maintenance	58,027	Numbers of substations	50	50,000/yr	50,000/yr	15,232-15,232
Total						**75,157-82,509**

SCCL will begin in 2020 and reach a 2% penetration rising to 7% by 2030
*Varies by size
**Includes annual cost

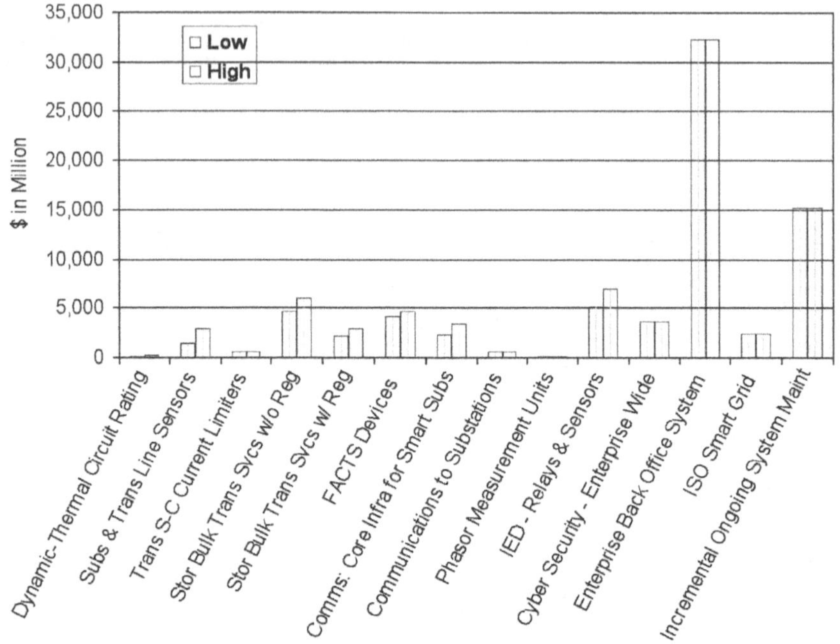

Figure 10-6. Smart Grid Transmission and Substation Costs

Table 10-23.
Smart Grid Transmission and Substation Cost to Meet Load Growth

Technology	Total Units	Units	% Sat	Cost/ Unit Low $	Cost/ Unit High $	Total Cost Low–High $M
Substation & Transmission Line Sensors	8,423	Number of new substations	100	50,000	100,000	421-631.7
Communications: Core Infrastructure for Smart Substations	8,423	Number of substations	100	50,000	75,000	421-632
Transmission Systems & Communications to Substations	8,423	Number of substations	100	14,400	14,400	121-121
Intelligent Electronics Devices (IED) – Relays & Sensors	58,027	Number of substations	80	110,000	150,000	927-1,264
Incremental Ongoing System Maintenance	8,423	Number of substations	100	50,000	50,000	4,422-4422
Total						**6,312-7,280**

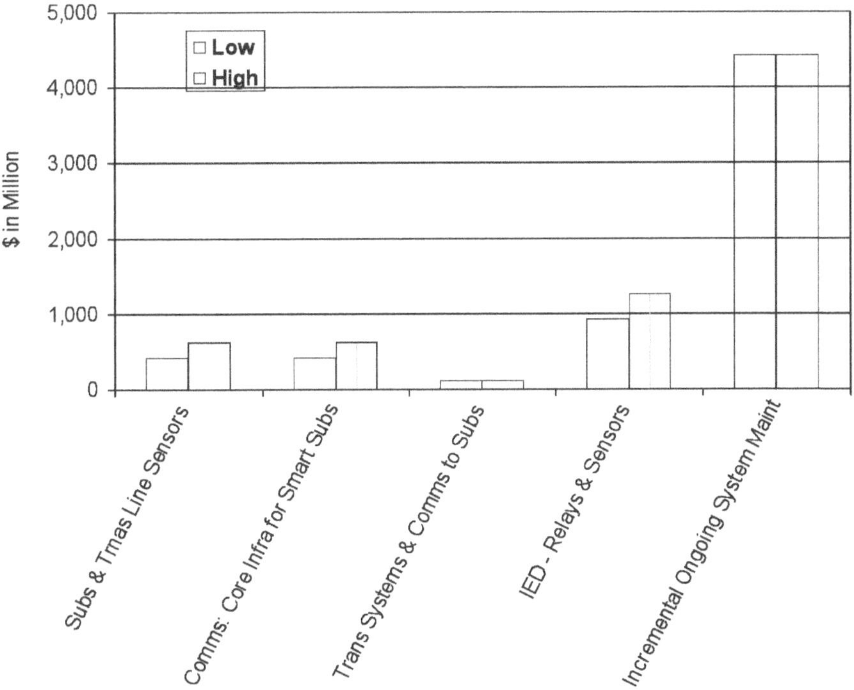

Figure 10-7. Smart Grid Transmission and Substation Cost to Meet Load Growth

Table 10-24. Smart Grid Transmission and Substation Costs to Meet Renewables

Technology	Total Units	Units	% Sat	Cost/ Unit Low $	Cost/ Unit High $	Total Cost Low–High $M
Communications: Core Infrastructure for Smart Substations	700	Number of substations	100	50,000	75,000	35-53
Transmission Systems & Communications to Substations	700	Number of substations	100	14,400	14,400	10-10
Phasor Measurement Units (PMU)	700	Number of substations	100	125,000	125,000	88-88
Intelligent Electronics Devices (IED) – Relays & Sensors	700	Number of substations	100	110,000	150,000	77-105
Incremental Ongoing System Maintenance	700	Number of substations	100	50,000/yr	50,000/yr	368-368
Total						**577-623**

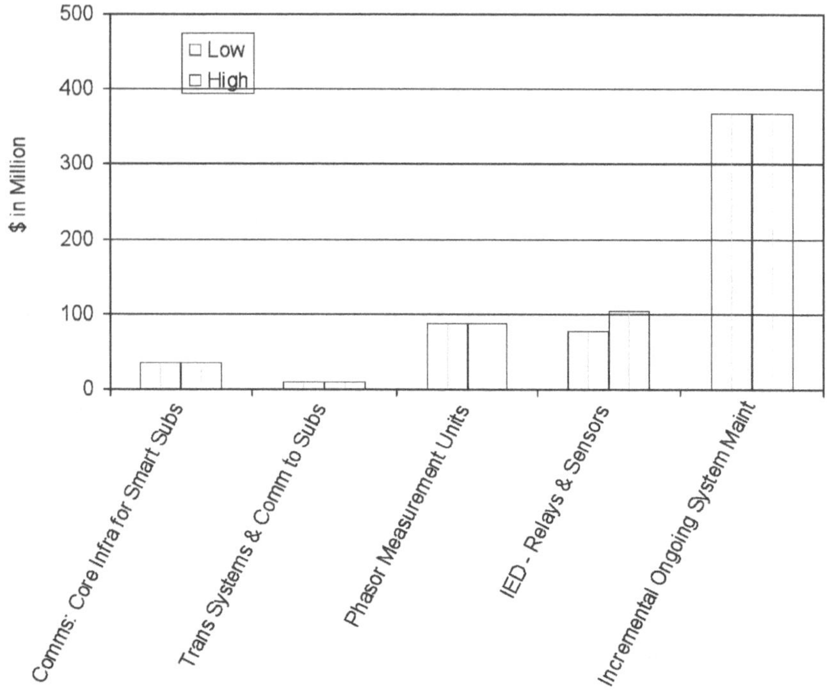

Figure 10-8. Smart Grid Transmission and Substation Costs to Meet Renewables

Table 10-25. Total Smart Grid Transmission and Substation Costs

Costs to Upgrade the Existing System ($M)		
	Low	**High**
Transmission and substations	75,157	82,509
Costs to Embed Smart Grid Functionality While Accommodating Load Growth ($M)		
	Low	**High**
Transmission and substations	6,312	7,281
Costs to Embed Smart Grid Functionality While Accommodating Large-Scale Renewables ($M)		
	Low	**High**
Transmission and substations	577	623
Total	**82,046**	**90,413**

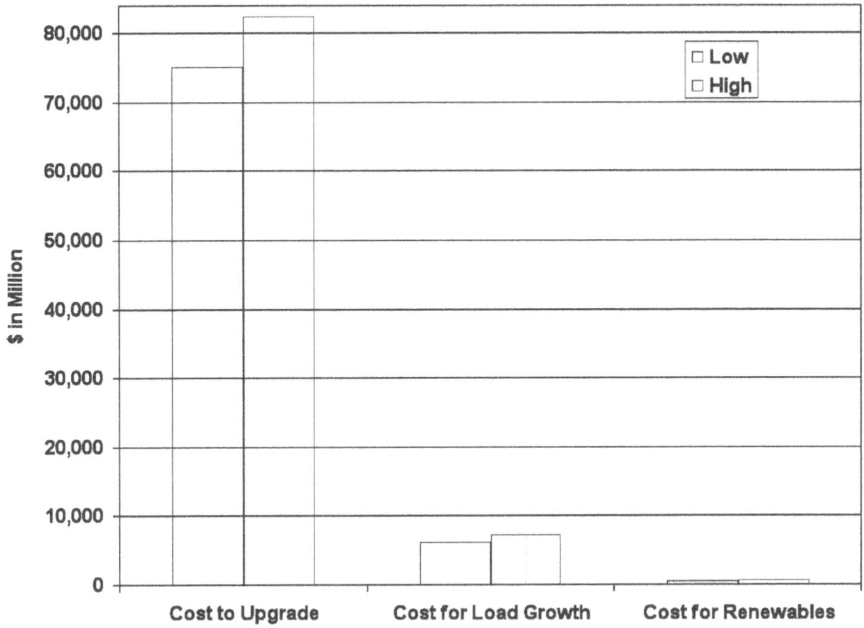

Figure 10-9. Total Smart Grid Transmission and Substation Costs

Table 10-26. Distribution Cost of Estimates Per Feeder (First Energy, 2009)

Function	No. of Feeders	Total Cost	Cost/Feeder
Distribution Automation	59	$18.2 M	$308,000
Volt/VAR Control	33	$8.5 M	$258,000

Table 10-27. Distribution Cost Estimates per Customer (First Energy, 2009)

Function	No. of Customers	Total Cost	Cost/Customer
Direct Load Control	34,000	$24.6 M	$728
AMI with DR	44,000	$41.2 M	$940

Cost Components for Distribution Systems

Smart Grid investments in the distribution system entail high bandwidth communications to all substations, intelligent electronic devices (IED) that provide adaptable control and protection systems, complete distribution system monitoring that is integrated with larger asset management systems, and fully integrated intelligence to mitigate power quality events and improve reliability and system performance. The key cost components for the distribution portion of the Smart Grid are as follows:

- Communications to feeders for AMI and distributed smart circuits
- Distribution automation
- Distribution feeder circuit automation
  - Intelligent reclosers and relays at the head end of feeders
  - Power electronics, including distribution short circuit current limiters
  - Voltage and VAR control on feeders
- Intelligent universal transformers
- Advanced metering infrastructure (AMI)
- Local controllers in buildings, on microgrids, or on distribution systems for local area networks

Communications

Communications to feeders for AMI and distribution smart circuits were estimated to cost about $20,000 per feeder, and to be fully installed on 80% of existing feeders and 100% of new feeders by 2030, for a total cost of nearly $9 billion.

Distribution Automation

By 2030, an estimated 55% of all existing distribution feeders will be integrated with advanced distribution automation systems at a cost of $308,000 per feeder, and 100% of all new feeders will be equipped by 2030. In this analysis, there are additional costs assumed for automation of the feeders themselves. The Smart Grid investment of distribution automation is estimated at nearly $96 billion, as shown in Table 10-29.

As shown in Table 10-30, the total investment required for distribution feeder automation exceeds $92 billion from 2010 to 2030.

Table 10-28. Cost of Communication to Feeders for AMI

Technology	Total Units	Units	% Sat	Cost/Unit Low $	Cost/Unit High $	Total Cost $M
Communication to Existing Feeders for AMI & Distribution Smart Circuits	464,216	Number of feeders	80	20,000	20,000	7,427
Communication to New Feeders for AMI & Distribution Smart Circuits	67,384	Number of feeders	100	20,000	20,000	1,348

Table 10-29. Cost of Distribution Automation

Technology	Total Units	Units	% Sat	Cost/Unit Low $	Cost/Unit High $	Total Cost $mill
Distribution Automation	464,216	Number of existing feeders	55	Varies	Varies	124,134-177,008
Distribution Automation	67,384	Number of new feeders	100	308,000	308,000	20,754-20,754

Intelligent Universal Transformers

The cost Intelligent Universal Transformers (IUT), as shown in Table 10-30, are expected to decline dramatically over the next 20 years, from $1.50 to 2.00/W today to $0.20/W in 2030. As a result, deployment is expected to grow rapidly, from for example, 10,000 25-kW units in 2015 to 1 million in 2030.

Total Smart Grid investment in IUTs through 2030, based upon the expected deployment of three million units, ranging in cost from $7,500 to $100,000, is $76 to $131 billion, as shown in Table 10-33.

Note: It is assumed that 50% of IUTs will be installed on existing feeders and 50% will be installed on new feeders added for load growth.

Advanced Metering Infrastructure (AMI)

The following are Cost Assumptions related to AMI.

- Residential meter costs are based more on volume than other factors
  - Meter + AMI — $40-80/unit
  - Meter + AMI + Disconnect — $70-130/meter
  - Meter + AMI + Disconnect+ HAN — $80-140/meter

Table 10-30. Cost of Distribution Feeder Automation for Existing Systems

Technology	Total Units	Units	% Sat	Cost/ Unit Low $	Cost/ Unit High $	Total Cost Low–High $M
Communications to Feeders for AMI & Distribution Smart Circuits	464,216	Number of feeders	80	20,000	20,000	4,428-4,428
Head End of Feeders – Intelligent Reclosers & Relays	464,216	Number of feeders	70	50,000*	50,000*	16,248-16,248
Power Electronics, Include Distribution Short-Circuit Current Limiters	58,027	Number of substations	5	80,000	80,000	2,321-2,321
Smart Switches, Reclosers, Monitored Capacitor Banks, Regulators & Circuit Improvement	464,216	Number of feeders	55	308,000	308,000	78,638-78,638
Voltage &VAR Control on Feeders	464,216	Number of feeders	55	60,000	258,000	15,319-65,873
Intelligent Reclosers	464,216	Number of feeders	25	100,000	150,000	11,605-17,408
Remotely Controlled Switches	464,216	Number of feeders	5	50,000	75,000	1,161-1,741
Direct Load Control (not integrated with AMI)	123,949,916	Number of customers	5 declining to 0	100	100	1,859-1,859
ElectriNet Controllers	464,216	Number of feeders	10	50,000	100,000	2,321-4,642
Total Distribution Feeder Cost						**124,134-177,008**

*Incremental cost – i.e., excludes the cost of the "switch" itself.

- Commercial and Industrial meter costs are based more on features selected than other factors.
  - Meter + communications $120-150/meter
- GT&D
  - Meter $1500-5000
- Installation costs
  - Residential $7-10/meter
  - Commercial and industrial $20-65/meter
  - AMI network and backhaul equipment $3-11/endpoint
  - Head end software and integration $4-10/endpoint
  - System initiation and management $2-4/endpoint
- Ongoing maintenance $3-11/year/endpoint

Table 10-31. Cost of Distribution Feeder Automation for New Feeders

Technology	Total Units	Units	% Sat	Cost/Unit Low $	Cost/Unit High $	Total Cost Low–High $M
Communications to Feeders for AMI & Distribution Smart Circuits	67,384	Number of feeders	100	20,000	20,000	1,348-1,348
Head End of Feeders – Intelligent Reclosers & Relays	67,384	Number of feeders	100	50,000*	50,000*	3,369-3,369
Power Electronics, Include Distribution Short-Circuit Current Limiters	8,423	Number of substations	100	80,000	80,000	6,738-6,738
Smart Switches, Reclosers, Monitored Capacitor Banks, Regulators & Circuit Improvement	67,384	Number of feeders		308,000	308,000	20,754-20,754
Voltage &VAR Control on Feeders	67,384	Number of feeders	100	258,000	258,000	4,043-17,385
Intelligent Reclosers	67,384	Number of feeders	25	100,000	150,000	1,685-2,527
Remotely Controlled Switches	67,384	Number of feeders	5	50,000	75,000	169-253
ElectriNet Controllers	67,384	Number of feeders	25	50,000	100,000	842-1,684
Total Distribution Feeder Cost						**38,948-54,059**

*Incremental cost – i.e., excludes the cost of the "switch" itself.

Table 10-32. Declining Costs and Growing Deployment of the IUT

IUT	2010	2015	2020	2025	2030
$/Watt With Storage Integration Option	$1.50 - $2.00	$.75	$.50	$.35	$.20
25 kW No. Units	Demo	10×10^3	50×10^3	200×10^3	1×10^6
50 kW No. Units	Demo	5×10^3	25×10^3	100×10^3	500×10^3
$/watt With PV Inverter	$3.00 - $1.00	$.75	$.50	$.35	$.20
25 kW No. Units	Demo	10×10^3	50×10^3	200×10^3	1×10^6
50 kW No. Units	Demo	5×10^3	25×10^3	100×10^3	500×10^3

Table 10-33.
Cost of Intelligent Universal Transformers

Technology	Total Units	Units	% Sat	Cost/Unit Low $	Cost/Unit High $	Total Cost Low–High $M
Intelligent Universal Transformer With Storage	1,500,000	Number of	various	37,500	100,000	12,563-12,688
Intelligent Universal Transformer With PV Inverter	1,500,000	Number of	various	7,500	50,000	12,437-12,937
Total IUT Cost	**3,000,000**	**Number of**	**various**			**25,000-25,625**

AMI Costs for the Smart Grid

Based upon these unit costs and the assumption of an average of 83% saturation, the total costs for the AMI portion of Smart Grid investment from 2010 to 2030 ranges from $15 to $42 billion, as shown in Table 10-34.

One of the key advantages of the new architecture is the efficiency that can be achieved in terms of energy savings and tons of avoided emissions. Estimated energy savings by 2030 are between 56-203 billion kWh, with a corresponding reduction in annual carbon emissions of 60-211 million metric tons of CO2. On this basis, the environmental value to the U.S. is equivalent to converting 14 to 50 million cars into zero-emission vehicles each year (EPRI 1016905).

Controllers for Local Energy Network

As shown in Table 10-36, the estimated cost of local energy network (LEN) controllers by 2030 is roughly $3-6 billion.

Summary of Distribution Systems Costs

The cumulative cost for bringing the electrical distribution system up to the performance levels required for the Smart Grid is estimated at $167 to $249 billion by 2030. Smart distribution investment will continue well beyond 2030 and will be influenced by the increasing functionality and lower costs of future technology.

Table 10-34.
Cost of Advanced Metering Infrastructure (AMI) for Existing Customers

Technology	Total Units	Units	% Sat	Cost/Unit Low $	Cost/Unit High $	Total Cost Low–High $M
Advanced Meter Infrastructure (AMI) Residential Meters	123,949,9166	Number of customers	80	70	140	7,437-13,387
Installation of Residential Meters	123,949,9166	Number of customers	80	7	15	694-1,487
Advanced Meter Infrastructure (AMI) Commercial & Industrial Meters	18,170,986	Number of customers	100	120	500	2,240-9,284
Installation of Commercial & Industrial Meters	18,170,886	Number of customers	100	20	65	364-1,184
Other AMI Costs	142,121,652	Number of customers	83	Various	Various	1,062-2,949
Ongoing System Maintenance	142,121,652	Number of customers	83	3/year	11/year	3,716-13,624
Total AMI Costs	**164,982,450**	**Number of customers**	**83**			**15,513-41,915**

Table 10-35.
Cost of Advanced Metering Infrastructure (AMI) for New Customers

Technology	Total Units	Units	% Sat	Cost/Unit Low $	Cost/Unit High $	Total Cost Low–High $M
Advanced Meter Infrastructure (AMI) Residential Meters	19,978,760	Number of customers	80	70	140	1,498-2,697
Installation of Residential Meters	19,978,760	Number of customers	80	7	15	140-300
Advanced Meter Infrastructure (AMI) Commercial & Industrial Meters	2,800,932	Number of customers	100	120	500	360-1,493
Installation of Commercial & Industrial Meters	2,800,932	Number of customers	100	20	65	59-190
Other AMI Costs	22,907,634	Number of customers	100	Various	Various	586-1,523
Ongoing System Maintenance	22,907,634	Number of customers	100	3/year	11/year	722-2,646
Total AMI Costs	22,907,634	**Number of customers**	**100**			**3,365-8,850**

Table 10-36. Cost of LEN Controllers

Technology	Total Units	Units	% Sat	Cost/Unit Low $	Cost/Unit High $	Total Cost Low–High $M
LEN Controllers for Local Area Networks on Existing System	464,216	Number of feeders	10	50,000	100,000	2,321-4,642
LEN Controllers for Local Area Networks on New Feeders	67,384	Number of feeders	25	50,000	100,000	842-1,685

Table 10-37.
Smart Grid Costs for Upgrading the Existing Distribution Systems

Technology Group	Total Cost $M	
	Low	High
Distribution Automation	124,134	177,008
Intelligent Universal Transformers	25,000	25,625
Advanced Metering Infrastructure	15,513	41,915
LEN Controllers	2,321	4,642
Total	**166,968**	**249,190**

CUSTOMER SYSTEMS

The cost components of the Smart Grid related to the electricity customer are primarily those related to grid integration. They exclude costs borne by the consumer to purchase appliances and devices. It includes the engineering development costs of Demand Response—Ready (DR-ready) appliances, for example, but not the consumer's purchase air conditioners, washing machines, refrigerators, energy efficient devices, and low-value distributed storage. This distinction is shown in greater detail in the Approach section, "What's In and What's Out" The total cost of the consumer portion of the Smart Grid, as shown in this section, is estimated at $32-56 billion by 2030.

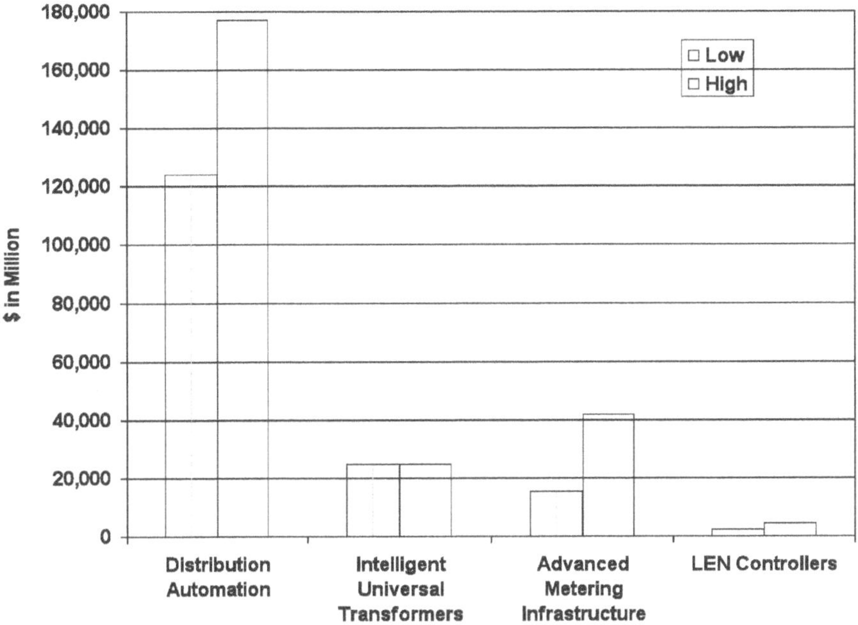

Figure 10-10.
Smart Grid Costs for Upgrading the Existing Distribution System

Table 10-38. Smart Grid Costs for Distribution System to Meet Load Growth

Technology Group	Total Cost $M	
	Low	High
Distribution Automation	38,948	54,059
Intelligent Universal Transformers	25,000	25,625
Advanced Metering Infrastructure	3,365	8,850
LEN Controllers	842	1,685
Total	**68,155**	**90,219**

Consumer/Customer Technologies Cost Components

The key components for the customer portion of Smart Grid costs are listed below:

- Integrated inverter for PV adoption
- Consumer EMS portal and panel

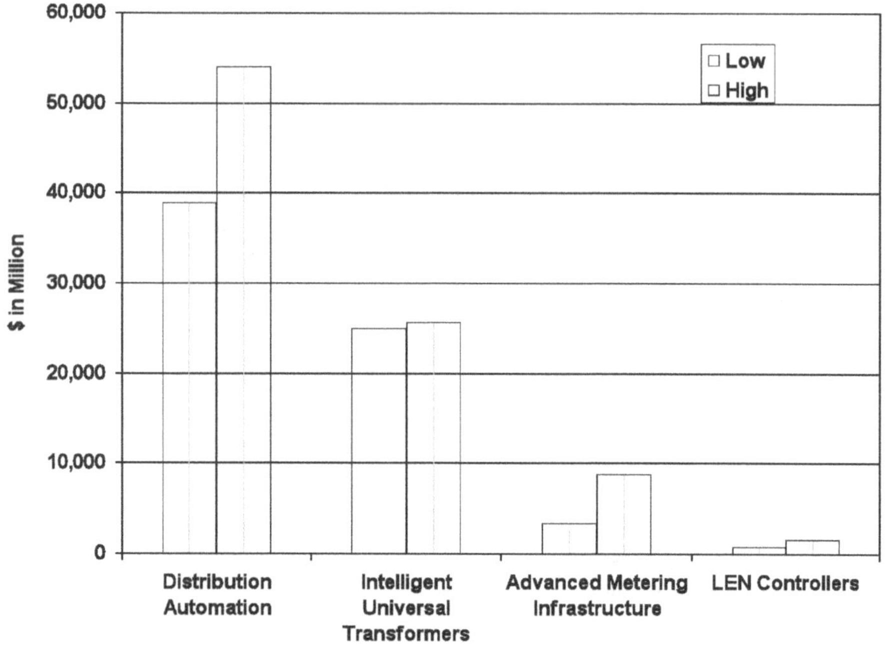

Figure 10-11.
Smart Grid Costs for Distribution System to Meet Load Growth

Table 10-39.
Total Smart Grid Distribution Costs

Costs to Upgrade the Existing System ($M)		
	Low	**High**
Distribution	164,647	249,190
Costs to Embed Smart Grid Functionality While Accommodating Load Growth ($M)		
	Low	**High**
Distribution	67,313	90,219
Total	**231,960**	**339,409**

Figure 10-12. Total Smart Grid Distribution Costs

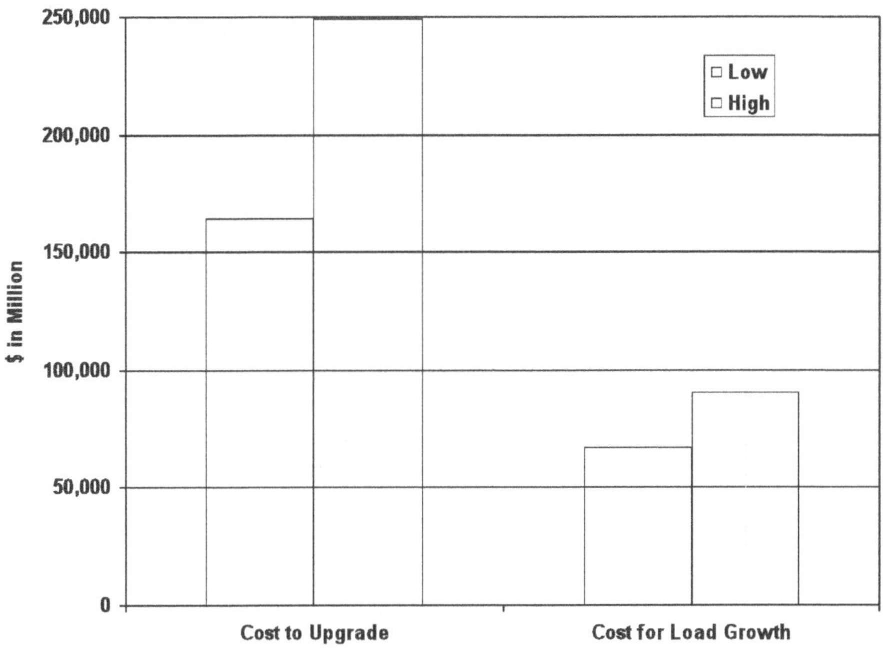

- In-home displays
- Grid-ready appliances and devices
- Vehicle-to-grid two-way power converters
- Residential storage for back-up
- Industrial and commercial storage for power quality
- Commercial building automation

Who Will Bear These Costs?

Costs in this section are labeled as "customer" costs. However, it is not the intent of the author to imply that the consumer directly bears these costs. Nor are they necessarily costs which will be borne by the utility to integrate them, or a combination of customer and utility costs. These are costs which must be borne by society and paid directly, bundled with other goods and services or otherwise included by the utility in its cost of service.

Just like the other technologies, these are critical to achieving the vision of a Smart Grid.

PV Inverters

The aggregate cost of inverter integration with 10MW of PV capacity by 2030 at a unit price of $800-1000/kW at $8-10 billion.

Table 10-40. Cost of PV Inverters

Technology	Total Units	Units	% Sat	Cost/Unit Low $	Cost/Unit High $	Total Cost Low–High $M
Integrated PV Inverter	10,000	kW of distributed PV	100	800	1000	800–1,000

Residential Energy Management System (EMS)

Based on the plurality of the residential EMS architecture paths, the study assumes the residential EMS estimate to 10% of the customer base by 2030 at an average cost $150-300, yielding a total cost of $2.2-4.3 billion.

Table 10-41. Cost of EMS Portals

Technology	Total Units	Units	% Sat	Cost/Unit Low $	Cost/Unit High $	Total Cost Low–High $M
Customer EMS Portal	143,928,676	Number of	10	150	300	2,159–4,318

In-Home Displays and Access to Energy Information

Standards resulting from the NIST PAP 10 work may allow product manufacturers the option to include energy information on multi-purpose devices and "other-purposed" devices. This will tend to eliminate or hide the cost making it difficult to identify the cost of access to the energy consumer energy information. This could be accommodated in the cost estimates by indicating that the cost per customer becomes lower over time. This could also be accommodated by indicating that the penetration rate of the stand-alone single-purpose IHD will not ramp up over time due to the alternative methods of information access. With this understanding in mind, the study team estimated 20% of utility residential customers would have an in-home display by 2030. The average cost was estimated at $20 to $50 per unit leaving the Total Smart Grid cost estimate at $1.4 to 2.9 billion for this item.

Table 10-42. Cost of In-Home Displays

Technology	Total Units	Units	% Sat	Cost/Unit Low $	Cost/Unit High $	Total Cost Low–High $M
In-Home Displays	143,928,676	28,785,735	20	20	50	575-1,439

Grid-Ready Appliances and Devices

At present, the total consumer cost per appliance is estimated to be $40. This appears as higher costs during the first several years. As shown in Figure 10-12 both the engineering cost and the component cost are assumed to reduce over time as is the norm in a product development cycle. In the future, the grid-ready design is expected to become part of a standard appliance design. Furthermore, the team assumed that the appliances will have communication technology built in that is justified for other non-energy usages and additional consumer benefits. This should make this cost become negligible after 10 years.

Penetration by 2030 may be limited by consumer model selection and the fact that certain appliance products (such as cooking and food preservation) may not be nearly as appropriate as others for grid messaging or demand management. Therefore, the study shows the penetration starting to level off as it approaches 40%. The assumptions used by the study are summarized in Table 10-43 and indicate the expected non-linear penetration and costs as smart-grid appliances enter the market.

Variables used to select the cost range are largely dependent on two factors. The first is the cost of the engineering and components added to the appliances. The second significant factor is the rate of market penetration of the Smart-Grid enabled appliances. Note that as diagrammed in Figure 10-13, since the component cost is projected as being higher in the early years, a faster deployment can push the total cost of Smart Grid appliances upward. However, an earlier drop in components cost due to higher volume could counteract this to some degree bringing the cost of common appliance Smart Grid components down earlier in the cycle. The range of cost of grid-read appliances was estimated between $230 to $412 million.

Table 10-43. Assumptions of Grid Ready Appliance Costs

Grid-Ready Appliance Costs	Value	Notations
Total U.S. Households	143,928,676	Includes projected growth to 2030 (1)
Total appliances per household	4.67	Including electric HW and AC (2)
Penetration by 2030	38%	See chart for ramp up (3)
Appliance Life Expectancy	13.91 yrs	Averaged from cooking, cleaning, food preservation (4)
Rate of replacements	1.44	In 20 years each appliance will be purchased 1.44 times (5)
Estimated Total Appliances Purchases	1,173,442,029	58,672,101 per year average (6)
Total Cost	\$230,531,663- \$412,354,482	Assumes penetration ramp up starting between 0.2% and .5% in 2011 (7)

(1) U.S. Census information
(2) Extrapolated from U.S. Census information
(3) Penetration estimated. Note that Smart Grid appliances and other products will ramp up. Estimated from averaging a number of informal sources (that may be changing daily).
(4) By using data from AHAM (Association of Home Appliance Manufacturers) for the life expectancy of major appliances, the team used the average life expectancy of 13.91 years for each appliance.
(5) Simple application of the average life expectancy of 13.91 years for each appliance and dividing this into the 20-year span of the study, the average appliance would be replaced 1.44 times during this time.
(6) Simply the application of the other numbers and averages to determine the number of appliances purchased (conversation with AHAM staff).
(7) The ramp-up rate of products containing Smart Grid enablements. As in (3) above, the ramp rate is arguable and could shift widely over the time period estimated. Currently, the ramp-up happens at an initial pace that is defensible at this point in time. Most manufacturers are not willing to share a lot of sales and projected sales of new and unannounced product models.

Table 10-44. Cost of Grid Ready Appliances

Technology	Total Units	Units	% Sat	Cost/Unit Low \$	Cost/Unit High \$	Total Cost Low–High \$M
Grid-Ready Appliances	143,928,676	Number of	38%	230	412	230-412

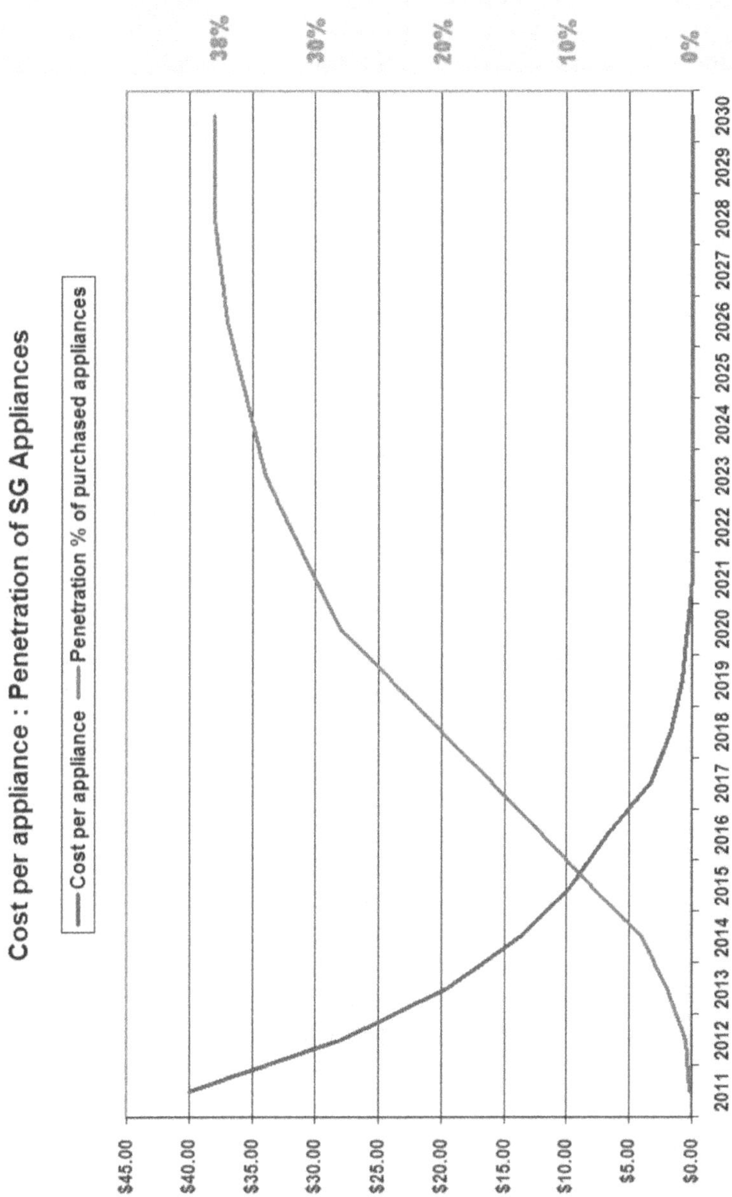

Figure 10-13. Appliance Cost and Presentation

Electric Vehicle Charging Infrastructure

The per-vehicle cost overhead for PLC transceiver is about $20 per vehicle in the near term, reducing to $10 per vehicle longer term, as PLC is already a very widely deployed technology. On the PLC to X bridge aspect, the X in most cases is ZigBee, but WiFi (802.11x-based) is also rapidly emerging as the HAN contender. The per-unit PLC/ZigBee or PLC/WiFi chipset prices vary between $10 to $20 per unit as well. Given that there are likely to be 1.2 charging stations long term for every PEV sold, the per-PEV PLC to X bridge costs will run to $12 to $24. Therefore, the per-PEV infrastructure costs will run to approximately between $25 and $50 for long-term and short-term volumes respectively. Assuming 2030 PEV installed base volume to be about 10 million vehicles, the cost of deploying Smart Grid infrastructure will approach $250 million ($25 per unit times 10 million vehicles) in 2030.

Communication Upgrades for Building Automation

By 2030 some 5% of the 20,178,151 commercial buildings would be upgraded to the level of complete energy automation at a cost of $5,000-20,000 per building. The total Smart Grid cost is estimated between $5-20 billion.

Electric Energy Storage for Customers

By 2030 roughly 1.8 GW of on-site back-up storage will be installed in commercial and industrial facilities at a unit cost of $2300 to 2400/kW. An additional 2.8 GW of battery storage for residential backup applications will be installed at an average unit cost of $2200-2400 kW.

Summary of Customer Costs

The cost to bring the customer interface of the electric infrastructure up to Smart Grid performance levels so that it can support a broad array of customer services—ranging from DR-ready appliances to V2G charging—is estimated at $24-44 billion, as shown in Table 10-46. This cost does not include the sizeable investment that will be made by customer in appliances, PHEVs, HVAC equipment, and the like.

Summary of Results

Over and above the investment to meet electric load growth, Table 10-50 shows that the estimated net investment needed to realize the envisioned power delivery system of the future is between $338 and

Table 10-45. Cost of Vehicle to Grid Converter

Technology	Total Units	Units	% Sat	Cost/Unit Low $	Cost/Unit High $	Total Cost Low–High $M
Vehicle to Grid Power Converter	30,000,000	Number of vehicles	50	300	500	4,500-7,500

Table 10-46. Cost of Communication Upgrades for Building Automation

Technology	Total Units	Units	% Sat	Cost/Unit Low $	Cost/Unit High $	Total Cost Low–High $M
Communication Upgrades for Building Automation	20,178,151	Number of buildings	5	5,000	20,000	5,045-20,180

Table 10-47. Electric Energy Storage Options for Customers

Application	Technology option	Capacity (MWh)	Duration (hours)	Efficiency %	Total cycles	Cost $/kW
Residential	Advanced lead-acid	0.8	8	75%	5000	2300-2400
Commercial & Industrial	Advanced lead-acid	10	2	75%	5000	2200-2400

Table 10-48. Customer Storage Costs

Technology	Total Units	Units	% Sat	Cost/Unit Low $	Cost/Unit High $	Total Cost Low–High $M
Industrial & Commercial Storage for Backup	1,800,000	kW	100	2,300	2,400	4,140-4,534
Residential Storage for Backup	2,800,000	kW	100	2,200	2,400	6,160-6,720

$476 billion. The total value estimate range of between $1,294 and $2,028 billion; and when compared to the Future PDS cost estimate results in a benefit-to-cost ratio range of 2.8 to 6.0. Thus, based on the underlying assumptions, this comparison shows that the benefits of the envisioned Future PDS significantly outweigh the costs.

Table 10-49. Smart Grid Costs for Customers

Technology	Total Units	Units	% Sat	Cost/Unit Low $	Cost/Unit High $	Total Cost Low–High $M
Integrated PV Inverter	10,000	kW of distributed PV	100	800	1000	8.0-10.0
Consumer Energy Management System	143,928,676	Number of	10	150	300	2,159–4,318
In Home Displays	143,928,676	Number of	20	50	100	1,439–2,878
Grid-Ready Appliances	143,928,676	Number of	40%	10	20	222–443
Vehicle to Grid Power Converter	30,000,000	Number of vehicles	50	300	500	4,500–7,500
Communication Upgrades for Building Automation	20, 178,151	Number of buildings	5	5,000	20,000	5,045–20,180
Industrial & Commercial Storage for Backup	1,800,000	kW	100	2,300	2,400	4,140–4,534
Residential Storage for Backup	2,800,000	kW	100	2,200	2,400	6,160–6,720
Ongoing System Maintenance						
Total Cost Customer						**23,672-46,368**
Allocated to Existing Customers						**20,386-39,932**
Allocated to New Customers						**3,286-6,436**

Table 10-50. Summary of Estimated Cost and Benefits of the Smart Grid

	20-Year Total ($billion)
Net Investment Required	338 – 476
Net Benefit	1,294 – 2,028
Benefit-to-Cost Ratio	2.8 – 6.0

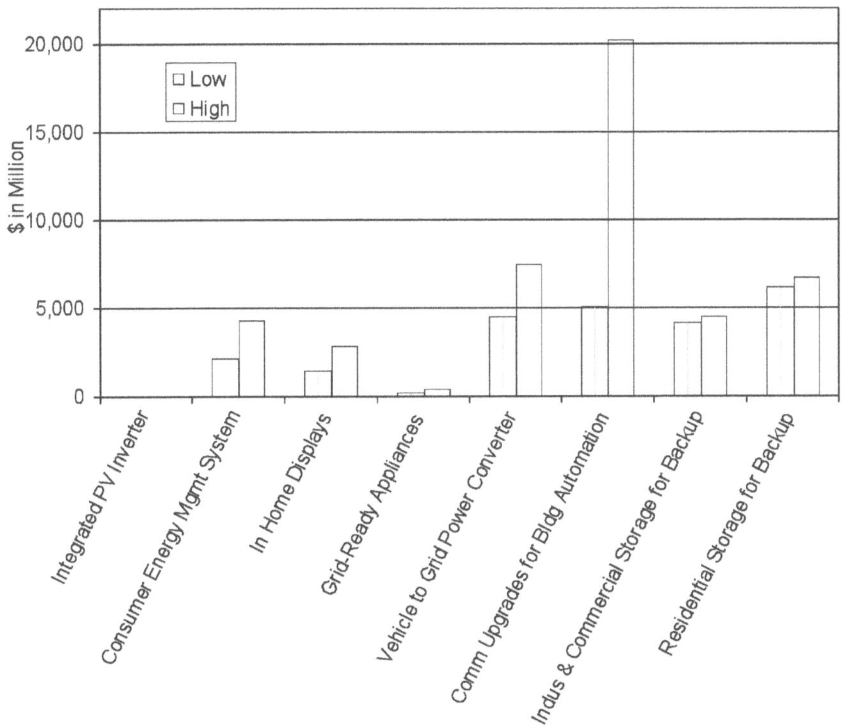

Figure 10-14. Customer Costs for a Smart Grid ($ Billion)

This indicates an investment level of between $17 and $24 billion per year will be required over the next 20 years. The costs cover a wide variety of enhancements to bring the power delivery system to the performance levels required for a Smart Grid. The costs include the infrastructure to integrate distributed energy resources (DER) and to achieve full customer connectivity, but exclude the cost of generation, the cost of transmission expansion to add renewables and to meet load growth, and a category of customer costs for smart-grid ready appliances and devices. Table 10-51 elucidates the total cost.

Included in the estimates of the investment needed to realize the Smart Grid, there are estimated expenditures needed to meet load growth and to enable large-scale renewable power production. As part of these expenditures, the components of the expanded power system will need to be compatible with the Smart Grid.

Table 10-51. Total Smart Grid Costs

Costs to Enable a Fully Functioning Smart Grid ($M)		
	Low	**High**
Transmission and substations	82,046	90,413
Distribution	231,960	339,409
Consumer	23,672	46,368
Total	**337,678**	**476,190**

References

- "Power Delivery System of the Future: A Preliminary Estimate of Costs and Benefits," EPRI, Palo Alto, CA: 2004. 1011001.
- "Characterizing and Quantifying the Societal Benefits Attributable to Smart Metering Investments," EPRI, Palo Alto, CA: 2008. 1017006.
- "The Green Grid: Energy Savings and Carbon Emissions Reductions Enabled by a Smart Grid," EPRI, Palo Alto, CA: 2008. 1016905.
- "Transforming America's Power Industry: The Investment Challenge 2010-2030," prepared by The Brattle Group for The Edison Foundation, November 2008.
- "Methodological Approach for Estimating the Benefits and Costs of Smart Grid Demonstration Projects," EPRI, Palo Alto, CA: January 2010. 1020342.
- "Power Delivery System of the Future: A Preliminary Estimate of Costs and Benefits," EPRI, Palo Alto, CA: 2004. 1011001.
- "Transforming America's Power Industry: The Investment Challenge 2010-2030," prepared by The Brattle Group for The Edison Foundation: November 2008.
- "Energy Storage Market Opportunities: Application Value Analysis and Technology Gap Assessment," EPRI, Palo Alto, CA: 2009. 1017813.
- "Sensor Technologies for a Smart Transmission System," EPRI, Palo Alto, CA: 2009. 1020619
- "Technical Architecture for Transmission Operations and Protection: Envisioning the Transmission Smart Grid," EPRI, Palo Alto, CA: 2008. 1016055
- "Assessment of Commercial Building Automation and Energy Management Systems for Demand Response Applications," EPRI, Palo Alto, CA: 2009. 1017883.
- "Automated Demand Response Tests: An Open ADR Demonstration Project," EPRI, Palo Alto, CA: 2008. 1016082
- "Energy Storage Market Opportunities: Application Value Analysis and Technology Gap Assessment," D. Rastler, EPRI Technical Update 1017813, December 2009.
- "Development of a 15 kV Class Solid-State Current Limiter," EPRI, Palo Alto, CA: 2009, 066335.
- "Assessment of National EHV Transmission Grid overlay Proposals: Cost-Benefit Methodologies and Claims," Christensen Associates Energy Consulting, LLC, February 22, 2010.
- "FirstEnergy Smart Grid Modernization Initiative," their report to the U.S. Department of Energy, 2009.
- "National Electric Transmission Congestion Report," U.S. Department of Energy (US DOE), October 2007.

- "The ElectriNet," C. Gellings and P. Zhang, ELECTRA, CIGRÉ, June 2010.
- "Energy Independence and Security Act of 2007, http://www.eia.doe.gov/oiaf/aeo/otheranalysis/aeo_2008analysispapers/eisa.html.
- "Energy Outlook 2010," Energy Information Agency, www.eia.doe.gov, December 2009.
- "Final Report on the August 14, 2003 Blackout in the United States and Canada: Causes and Recommendations," North American Electric Reliability Corporation for the U.S.-Canada Power System Outage Task Force, April 2004.
- "The Smart Grid: An Estimation of the Energy and CO2 Benefits, Rev. 1," Pacific Northwest National Laboratory (PNNL) for the U.S. Department of Energy, January 2010.
- "The Aging Workforce: Electricity Industry Challenges and Solutions," L.B. Lave, M. Ashworth and C.W. Gellings, Electricity Journal, January 2007.
- "8th FACTS Users Group Meeting: Presented Material at the 8th FACTS Users Group Meeting," EPRI, Palo Alto, CA: 2005. 1010633.
- "The Power to Reduce CO2 Emissions: The Full Portfolio: 2009 Technical Report," EPRI, Palo Alto, CA: 2009. 1020389.
- "Assessment of Achievable Potential from Energy Efficiency and Demand Response Programs in the U.S.: (2010-2030)," EPRI, Palo Alto, CA: 2009. 1016987.
- EPRI's IntelliGridSM Initiative, http://intelligrid.epri.com.
- GridWise Architecture Council, http://www.gridwiseeac.org.
- "The Modern Grid Initiative Version 2.0," conducted by the National Energy Technology Laboratory for the U.S. Department of Energy Office of Electricity Delivery and Energy Reliability, January 2007, http:www.netl.doe.gov/moderngrid/resources.html.
- "Report to NIST on the Smart Grid Interoperability Standards Roadmap," (Contract No. SB1341-09-CN-0031), D. Von Dollen, EPRI, June 17, 2009.
- LogRhythm and NERC CIP Compliance, www.logrhythm.com, 2009.
- "The Cost of Power Disturbances to Industrial and Digital Economy Companies," EPRI, Palo Alto, CA: 2009. 1006274.
- "A National Assessment of Demand Response Potential," Federal Energy Regulatory Commission, June 2009.
- "National Action Plan on Demand Response," Federal Energy Regulatory Commission, June 2010.
- "Smart Grid Strategy and Roadmap," Southern California Edison, 2010.
- "Electric Energy Storage Technology Options: A Primer on Applications, Costs and Benefits," EPRI, Palo Alto, CA: 2010, 1020676.
- "Methodological Approach for Estimating the Benefits and Costs of Smart Grid Demonstration Projects," EPRI, Palo Alto, CA: 2010. 1020342.
- "Sizing Up the Smart Grid," A. Faruqui, R. Heldik and C. Davis, Elster Energy Axis User Conference, February 24, 2009.
- "Characterizing and Quantifying the Societal Benefits Attributable to Smart Metering Investments," EPRI, Palo Alto, CA: 2008. 1017006
- "The U.S. Smart Grid Revolution: KEMA's Perspectives for Job Creation," GridWise Alliance, January 13, 2008.
- "Smart Grid: Enabling the 21st Century Economy," S. Pullins, Governor's Energy Summit, West Virginia, 2008.
- "Estimating the Benefits of the GridWise Initiative," Phase I Report, Pacific Northwest National Laboratory, TR-160-PNNL, Rand Corporation, 2004.
- "GridWise: The Benefits of a Transformed Energy System," Pacific Northwest National Laboratory, PNNL-14396, September 2003.

- "Energy Storage for the Electricity Grid: Benefits and Market Potential Assessment Guide," prepared by Sandia National Laboratories for the U.S. Department of Energy, SAND2010 0815, February 2010.
- "Energy Storage Market Opportunities: Application Value Analysis and Technology Gap Assessment," EPRI, Palo Alto, CA: 2009. 1017813.
- "TAGTM Technical Assessment Guide, Vol. 4: Fundamentals and Methods, End Use," C.W. Gellings and P. Hanser, EPRI, Palo Alto, CA: 1987. P-4463-SR.
- "The Green Grid: Energy Savings and Carbon Emissions Reductions Enabled by a Smart Grid," EPRI, Palo Alto, CA: 2008. 1016905.
- "The Smart Grid: An Estimation of the Energy and CO2 Benefits," Pacific Northwest National Laboratory for the U.S. Department of Energy, PNNL-19112, January 2010.\
- "Smart Meters and Power Planning," C. King, Utility Automation and Engineering T&D, October 2006.
- "Grid Supporting Technologies," J. Wellinghoff presentation to National Electrical Manufacturers Association (NEMA) Energy Storage Council Technical Committee Meeting, October 6, 2008.
- "Illinois Statewide Smart Grid Collaborative: Collaborative Report," Illinois Statewide Smart Grid Collaboration (ISSGC), Illinois Commerce Commission, September 30, 2010.
- "The Modern Grid Initiative Version 2.0," conducted by the National Energy Technology Laboratory (NETL) for the U.S. Department of Energy Office of Electricity Delivery and Energy Reliability, January 2007, http://www.netl.doe.gov/moderngrid/resources.html.
- "Smart Grid VC Investment Surge," greentechgrid, January 21, 2010, www.greentechmedia.com.
- "A Framework for Assessing the Net Benefits of Home Area Networks to Enable Demand Response," S. Mullen and B. Neenan, Electric Power Research Institute (EPRI), Working Paper, 2010.
- "An Updated Annual Energy Outlook (AEO) 2009 Reference Case Reflecting Provisions of the American Recovery and Reinvestment Act and Recent Changes in the Economic Outlook," EIA, 2009.
- "2010 UDI Directory of Electric Power Producers and Distributors," 118th Edition of the Electrical World Director, E. Giles & K.L. Brown, Platts Division of the McGraw-Hill Companies, Inc., New York, NY, 2009.

Chapter 11

Factors Effecting the Demand for Electricity From the Smart Grid

THE DEMAND FOR ELECTRICITY

Electricity demand is driven by a complex combination of the desire for the comfort, convenience, and productivity that energy provides, but is tempered by the price, availability, and functionality of competing energy forms (e.g., natural gas) and the technology that converts electricity into services. Fundamental to this demand is the presence of income (in households) and/or economic activity (in business and industry). In each energy application, the technology that converts electricity or other energy forms into lighting, heating, cooling, motive power, or other energy services plays a key role in the effectiveness of the desired energy service, its economy, and the resultant environmental footprint.

Until recently, the relationship of electricity demand to these factors was straightforward. Now with the advent of photovoltaic power generation, the expanding use of combined heat and power (CHP), increasing use of Plug-In Electric Vehicles and consumer storage and dramatic changes in the end use of electricity, the equation changes. Consumer demand for electricity is now potentially supplied by a combination of grid-supplied energy services and power generated and possibly stored on site.

The advent of new and improved technology also has a substantial impact on the technology landscape and then the overall demand for energy. Recent adoption of plug-in electric vehicles (PEVs) is a prime example of new technologies that can significantly increase the use of electricity. Other new technologies such as tablets, PCs, home entertainment systems, and heat pumps are additional examples.

The Smart Grid will impact the future of energy which could be supplied by the Smart Grid directly and indirectly in several ways. However, there are five outstanding elements where the Smart Grid's capabilities could be critical as follows:

- Energy Efficiency of End-Use Technologies
- Consumer Electronics Technology
- Energy-Efficiency Investments
- New Uses of Electricity
- Industrial Trends
- Customer Generation

US DOE Projections

Historically, the demand for electricity has been increasing steadily in the western world. As shown in Figure 11-1, the United States Department of Energy (DOE)'s Energy Information Agency (EIA) publishes Annual Energy Outlook (AEO) which is the official U.S. Government projection of electricity consumption. It contains an expectation that electricity consumption will increase modestly between now and 2035. A number of factors which cause some practitioners to challenge that assumption.

Figure 11-1 represents electricity consumption, rather than electricity sales by utilities. While consumption overall may increase, it is unclear whether electricity sales by traditional suppliers will increase or be offset by on-site generation. In addition, it is highly likely that these estimates of consumption are far too conservative regarding the predictions of end-use efficiency improvement. A combination of utility-funded, demand-side management programs, State and Federal mandates and market forces will push efficiency beyond the EIA estimates inferred by this figure.

Figure 11-2 illustrates the progressive decline of the AEO forecast since 2008.

The advent of new and improved technology also has a substantial impact on the technology landscape and the overall demand for energy.

The Baseline: Annual Energy Outlook (AEO)

The Annual Energy Outlook 2101 (AEO2012), prepared by the U.S. Energy Information Administration (EIA), presents long-term pro-

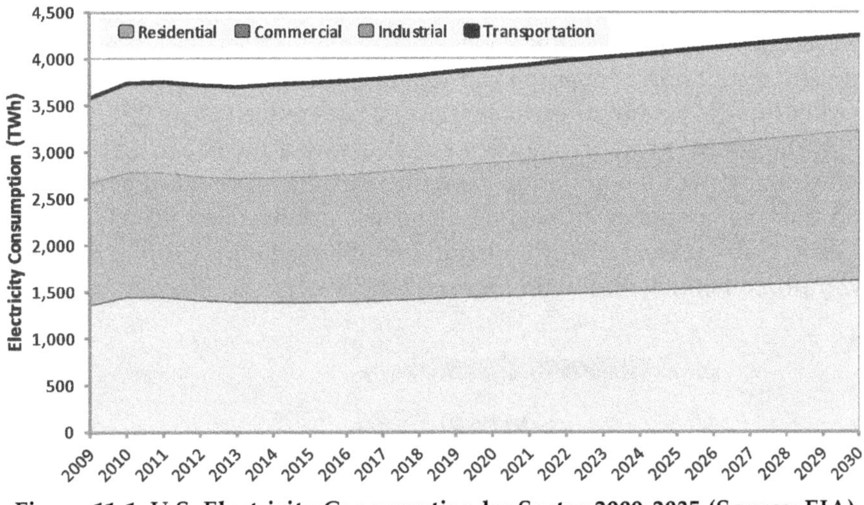

Figure 11-1. U.S. Electricity Consumption by Sector 2009-2035 (Source: EIA)

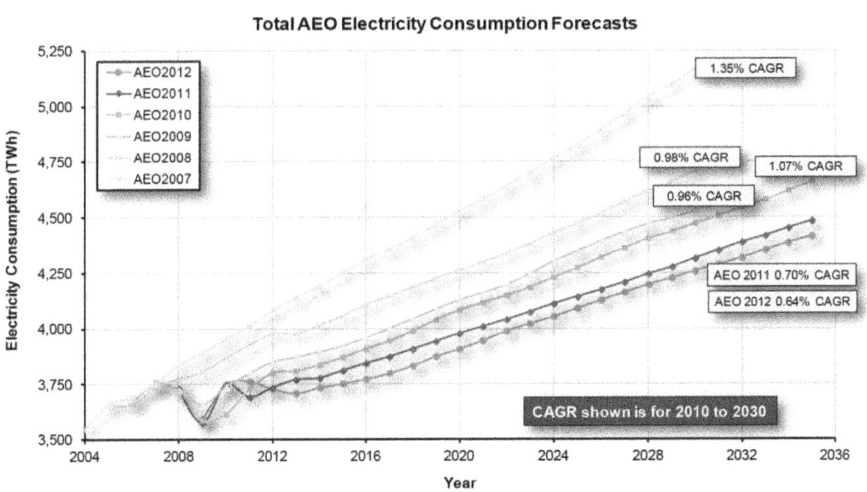

Figure 11-2. A Continuing Decline in the Projection for Future Electricity Demand (AEO)

jections of energy supply, demand, and prices through 2035, based on results from EIA's National Energy Modeling System (NEMS). AEO2012 is used as the baseline in this analysis.

"The projections in the U.S. Energy Information Administration's (EIA's) Annual Energy Outlook 2012 (AEO2012) focus on the factors that shape the U.S. energy system over the long term. Under the assumption that current laws and regulations remain unchanged throughout the projections, the AEO2012 Reference Case provides the basis for examination and discussion of energy production, consumption, technology, and market trends and the direction they may take in the future."

The rate of growth in energy use slows over the projection period, reflecting moderate population growth, an extended economic recovery, and increasing energy efficiency in end-use applications. In AEO 2012, electricity use increased with the number of households despite efficiency improvement, as shown in Figure 11-3.

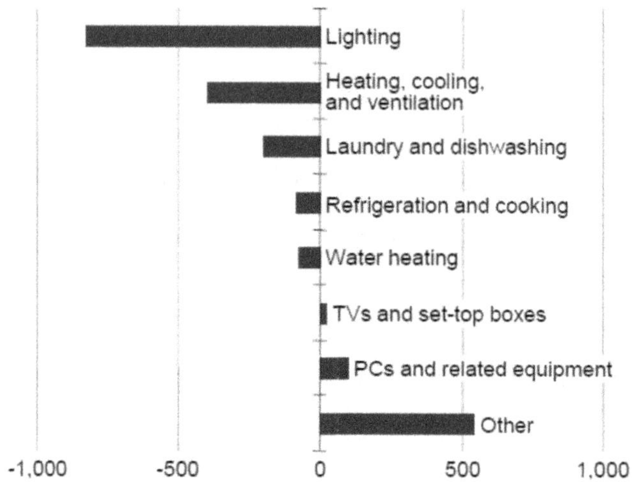

Figure 11-3. Change in Residential Electricity Consumption for Selected End Uses in the Reference Case, 2010-2035 (kWh per household) (Source: AEO2012)

Despite a decrease in electricity consumption per household, total delivered electricity use in the residential sector grows at an average rate of 0.7% per year in the *AEO2012* Reference Case.

> "The increase in efficiency, driven by new standards and improved technology, is not high enough to offset the growth in the number of households and electricity consumption in 'other' uses." (Source: *AEO2012*)

Portions of the federal lighting standards outlined in *EISA2007* took effect on January 1, 2012. Over the next two years, general-service lamps that provide 310 to 2,600 lumens of light are required to consume 30% less energy than typical incandescent bulbs. High-performance incandescent, compact fluorescent, and light-emitting diode (LED) lamps continue to replace incandescent lamps. In 2035, delivered energy for lighting per household in the Reference Case is 827 kWh lower, or 47% below the 2010 level.

AEO2012 indicates that "Electricity use for televisions and set-top boxes grows by an average of 1.1% per year, accounting for 7.3% of total delivered electricity consumption in 2035. Personal computers (PCs) and related equipment account for 4.6% of residential electricity consumption in 2035, averaging 1.8% annual growth from their 2010 level. Electricity use by other household electrical devices, for which market penetration increases with little coverage by efficiency standards, increases by 1.8% annually and accounts for nearly one-fourth of total residential electricity consumption in 2035."

AEO also believes that efficiency standards reduce electric energy intensity in commercial buildings. According to *AEO2012*, electricity, which accounted for 52% of total commercial delivered energy use in 2010, increases to 56% in 2035 in the *AEO2012* Reference Case, as commercial floor space grows at an average annual rate of 1% and new electric end uses become more prevalent. Despite growth, improved efficiency of commercial equipment slows the growth of purchased electricity over the projection period.

Commercial energy intensity in Figure 2-8, defined as the ratio of energy consumption in these appliances to floor space, decreases for most electric end uses from 2010 to 2035 in the Reference Case. Electricity intensity decreases by 1.3% annually for both cooking and refrigeration, by 0.5% annually for lighting, and by 0.7% annually for space conditioning.

End uses such as space heating and cooling, water heating, refrigeration, and lighting are covered by federal efficiency standards that act to limit growth in energy consumption to less than the growth in floor space.

A decline in the electricity share of industrial energy consumption reflects modest growth in combined heat and power (CHP), which offsets purchased electricity requirements, as well as efficiency improvements across industries, primarily as a result of rising standards for motor efficiency. With growth in lumber, paper, and other industries that consume biomass-based byproducts, the renewable share of industrial energy use expands.

Overall, *AEO2012* estimates that residential and commercial sectors dominate electric demand growth. AEO estimates that electricity

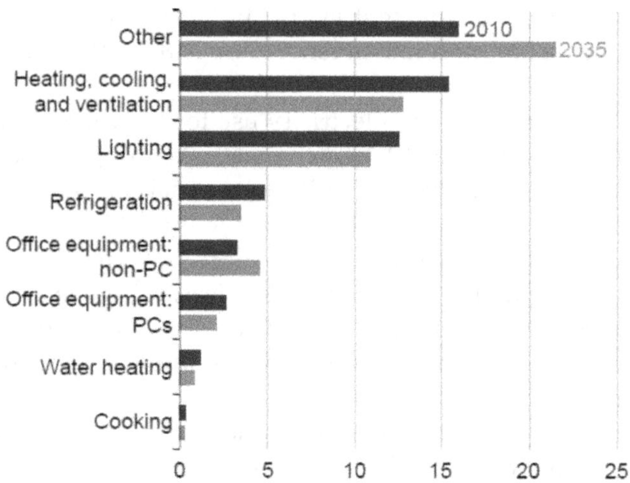

Figure 11-4. Energy Intensity of Selected Commercial Electric End Uses, 2010 and 2035 (thousand Btu per square foot) (Source: AEO2012)

demand (including retail sales and direct use) growth has slowed in each decade since the 1950s, from a 9.8% annual rate of growth from 1949 to 1959 to only 0.7% per year in the first decade of the 21st century. In the *AEO2012* Reference Case, electricity demand growth rebounds a little from those low levels but remains relatively slow, as growing demand for electricity services is offset by efficiency gains from new appliance standards and investments in energy-efficient equipment.

END-USE TECHNOLOGY

Megatrends in energy efficiency have emerged in the last several years. Facilitated by the Smart Grid and driven by combinations of appliance efficiency standards, building energy codes, utility demand-side management (DSM) programs, and market-induced changes in devices and appliances, the efficiency of electric energy utilization has increased dramatically. For example, the expected change in U.S. electricity intensity from 2010 to 2030 ranges from 18% to 47% decreases for major residential end uses and 7% to 25% decreases for commercial building technologies (see Figure 11-5).

Figure 11-5 illustrates the major improvements in major end uses.

The distinction between changes in electricity consumption and

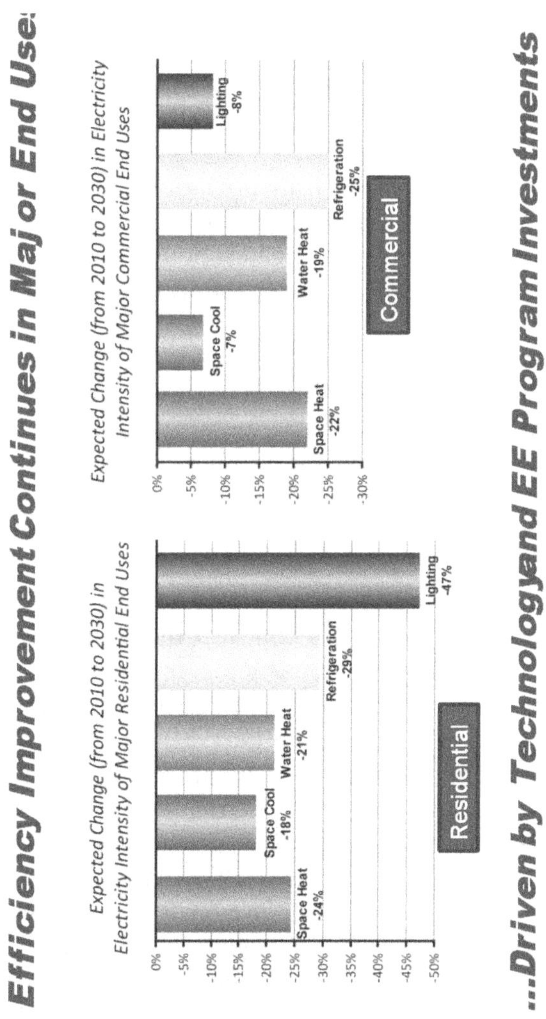

Figure 11-5. 2011 Annual Energy Outlook (AEO) 2030 Prediction (Source: EIA)

changes in peak demand is important. Even in a scenario of low growth or decline in overall electricity sales, air conditioning usage may increase summer peak demand. In addition, as self-generation increases and consumers increasingly rely on the grid to serve excess and off-peak demand, the load factor will deteriorate further. These particular combinations represent a challenging case for utilities. They need to ramp up peak capacity resources, demand response, and additional transmission and distribution infrastructure to meet the growing peak. However, they may not have a sufficient revenue base from traditional electricity sales to support such investments.

Recent advances in electric transportation, particularly plug-in electric vehicles (PEVs) and increasing interest in electric energy-driven heat pumps and heat pump water heaters will offset efficiency gains somewhat, but possibly not enough in the short term to make a substantial impact.

Electricity prices and rates are obviously a variable in the equation that determines the bill consumers pay. The other major determinant is the quantity of energy consumed. The quantity of electric energy consumed is directly proportional to the number, efficiency, size, and type of end-use energy-consuming devices and appliances a consumer owns that are electrically connected, as well as how they are operated.

End-use technologies vary considerably in the efficiency by which they convert electricity into useful energy services. A simple example is replacing incandescent lamps with compact fluorescent lamps (CFLs) Light Emitting Diodes (LEDs) in households. It has been estimated that replacing all incandescent bulbs with CFLs in a typical U.S. household would save approximately 1200 kWh per year. If all U.S. households switched, total residential electricity consumption would be reduced by approximately 10% and total U.S. electricity consumption would be reduced by approximately 3.7% (EPRI).

These consequences are reflected in DOE's Annual Energy Outlook (*AEO2012*) in which the expected change from 2010 to 2035 in electricity consumption is expected to be 31% for lighting. However, AEO projects a move toward more electric appliances than would have normally been applied, showing an increase in 16% for electric space heating and 42% for household electronics.

A 2009 EPRI report (EPRI 1016987) estimated the potential for energy efficiency to reduce electricity demand relative to the *AEO2008* Reference Case. The following refers heavily to EPRI's study.

The *AEO2008* Reference Case already accounted for market-driven efficiency improvements, the impacts of all legislated federal appliance standards at that time and building codes (including the Energy Independence and Security Act of 2007) and rulemaking procedures. The *AEO2008* Reference Case was predicated on a relatively flat electricity price forecast in real dollars between 2008 and 2030. It also assumed continued contributions of existing utility- and government-sponsored energy efficiency and demand response programs established prior to 2008.

EPRI's study determined that *energy-efficiency programs have the potential to reduce electricity consumption in 2030 by 398 to 544 billion kWh.* This represents a range of achievable potential reduction in electricity consumption in 2030—from a "moderate case" or *realistic achievable potential* of 8% to a "high case" or *maximum achievable potential* of 11%.*†

Relative to the *AEO2008* Reference Case, which assumed a level of energy-efficiency program impact, this study identifies between 236 and 382 billion kWh of *additional savings potential* from energy efficiency programs.

Therefore, energy-efficiency programs have the potential to reduce the annual growth rate in electricity consumption forecasted in *AEO2008* between 2008 and 2030 of 1.07% by 22% to 36%, to an annual growth rate of 0.83% to 0.68%.

These estimated levels of electricity savings are achievable through voluntary energy-efficiency programs implemented by utilities or similar entities. The EPRI analysis does not assume the enactment of new energy codes and efficiency standards beyond what is already in law. More progressive codes and standards would yield even greater levels of electricity savings.

In the EPRI study, a reference case was created. In the reference case, by 2030 electricity use is expected to increase to 4,696 TWh, a 25% increase over use in 2008. This Reference Case forecast already includes expected savings from several efficiency drivers including:

---

*The values for realistic- and maximum-achievable potentials in 2030 measured with respect to the baseline forecast are 398 and 544 billion kWh, respectively, or 8 to 11%. These values represent the total savings impact of energy-efficiency programs in 2030 inclusive of savings embedded in the *AEO2008* Reference Case.

†Realistic Achievable Potential (RAP) can be thought of as a "moderate case" for the savings impact of energy-efficiency programs; Maximum Achievable Potential (MAP) can be thought of as a "high case" for the savings impact of energy-efficiency programs. Through the terms may be used interchangeably, the nomenclature of RAP and MAP are used throughout this report.

- Codes and Standards
- Federal, state, and local building efficiency codes already enacted.
- Appliance and equipment standards already enacted; this includes the Energy Independence and Security Act of 2007, which, among its features, mandates higher lighting efficiency standards.
- Other possible related effects, including structural changes in the economy that impact overall electric energy intensity.
- Market-Driven Efficiency
- Trends in customer purchases of energy-efficient equipment attributable to market-driven effects outside of utility programs.
- Implicit Programs
- An estimate of the utility-based energy-efficiency programs adopted prior to 2008, and an estimate of the impact of these existing programs.

The study estimated the aggregate impact of these drivers by developing a "frozen efficiency" case that represents what consumption would be if the electricity energy intensity of the economy (expressed in terms of kWh per dollar of real U.S. GDP) were held fixed at 2008 levels (0.33 kWh/$GDP). This case, depicted in Figure 4-3, maintains the 2.5% growth rate of the previous three decades. The difference between the frozen efficiency forecast and the AEO 2008 Reference Case can be considered to be the cumulative impact of energy-efficiency programs included in AEO2008, market-driven efficiency, efficiency codes and standards, and other effects. Figure 11-6 illustrates these components.

The estimated impact of energy-efficiency programs "embedded" in the *AEO2008* Reference Case was "added back" to construct an adjusted "baseline" forecast, in accordance with standard industry practice. This baseline represents a projection of electricity consumption absent of any assumed impact of energy-efficiency programs.

The baseline forecast does *not* assume any expected savings from future federal or state appliance and equipment standards or building codes not currently enacted. Finally, the baseline embodies the *AEO2008* price forecast, which is relatively flat in real terms over the forecast horizon.

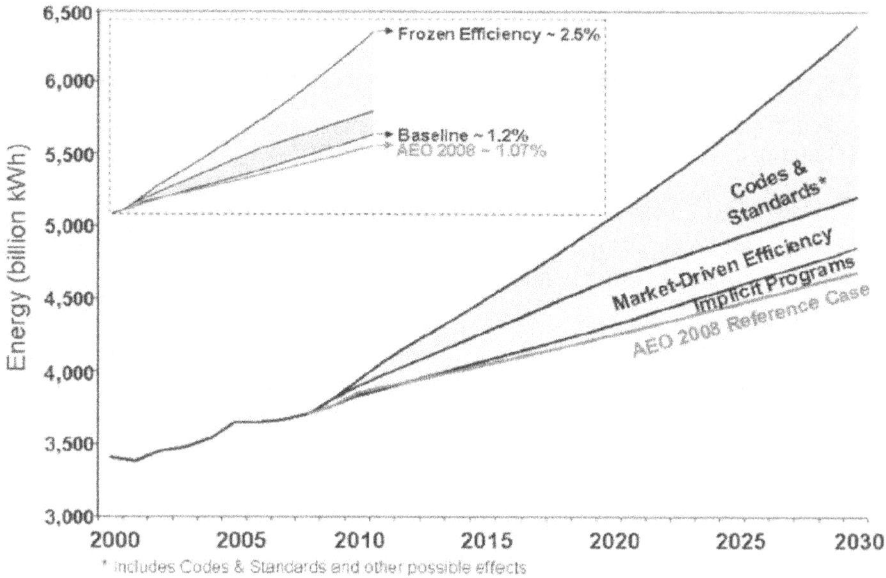

Figure 11-6. Estimated Impact of Energy-Efficiency Drivers Inherent in AEO2008 Reference Case

THE POTENTIAL FOR ELECTRICITY SAVINGS FROM UTILITY PROGRAMS

The analysis of potential savings from utility programs began with a list of energy-efficiency measures. This list includes high-efficiency appliances and equipment for most end uses, many of which have numerous efficiency levels, devices, controls, maintenance actions, and enabling technologies such as programmable thermostats. Table 11-1 summarizes the energy-efficiency measures included in the analysis.

As described above, the full set of measures is included in the estimation of technical potential, while only the subset that passes the economic screen is included in economic and achievable potential.

Table 11-2 presents energy-efficiency potential estimates for the U.S. in 2020 and 2030. Relative to the baseline forecast, in 2030:

- Realistic Achievable Potential is 398 TWh, or an 8% reduction in projected consumption
- Maximum Achievable Potential is 544 TWh, or an 11% reduction in projected consumption

Table 11-1. Summary of Energy Efficiency Measures

Residential Sector Measures	Commercial Sector Measures
Efficient air conditioning (central, room, heat pump)	Efficient cooling equipment (chillers, central AC)
Efficient space heating (heat pumps)	Efficient space heating equipment
Efficient water heating (e.g. heat pump water heaters & solar water heating)	Efficient water heating equipment (heat pumps)
Efficient appliances (refrigerators, freezers, dishwashers, clothes washers, clothes dryers)	Efficient refrigeration equipment & controls (e.g. efficient compressors, floating head pressure controls, anti-sweat heater controls, etc.)
Efficient lighting (CFL, LED, linear fluorescent)	Efficient lighting (interior and exterior; LED exit signs, task lighting)
Efficient power supplies for Information Technology and consumer electronic appliances	Lighting controls (occupancy sensors, daylighting, etc.)
Air conditioning maintenance	Efficient power supplies for Information Technology and electronic office equipment
Heat pump maintenance	Water temperature reset
Duct repair and insulation	Efficient ventilation (air handling and pumps; variable air volume)
Infiltration control	Economizers and energy management systems (EMS)
Whole-house and ceiling fans	Programmable thermostats
Reflective roof, storm doors, external	Duct insulation
Roof, wall and foundation insulation	Retro-commissioning
High-efficiency windows	**Industrial Sector Measures**
Faucet aerators and low-flow	Efficient process heating
Pipe insulation	High-efficiency motors and drives
Programmable thermostats	High-efficiency Heating, Ventilation and Air Conditioning (HVAC)
In-home energy displays	Efficient lighting

Relative to the AEO2008 Reference Case, in 2030:

- Realistic Achievable Potential represents 236 TWh of additional energy-efficiency savings, or a 5% reduction in projected consumption.
- Maximum Achievable Potential represents 382 TWh of additional energy-efficiency savings, or an 8% reduction in projected consumption.

These estimates suggest that energy efficiency-programs can realistically reduce the annual growth rate of U.S. electricity consumption from 2008 to 2030 projected by the AEO2008 Reference Case by 22%, from 1.07% to 0.83%.

From Technical Potential to Realistic Achievable Potential

Figure 11-7 is an example of the residential air conditioner to illustrate the transition from technical potential to realistic achievable potential.

- **Technical Potential**: Central air conditioning (CAC) systems in existing homes are replaced, upon reaching the end of their useful lives, with the highest SEER level equipment available regardless of cost; in new homes, the highest SEER level available in each year is installed. In 2010, this is the SEER 20 air conditioner or the ductless (mini-split) heat pump with variable speed operation.

Table 11-2. Energy Efficiency Potential for the U.S.

	AEO2008 Reference Case	Baseline Forecast	Realistic Achievable Potential	Maximum Achievable Potential
Forecasts (billion kWh)				
2020	4,253	4,319	4,112	3,881
2030	4,696	4,858	4,460	4,314
Savings Relative to *AEO2008* Reference Case (billion kWh)				
2020			141	372
2030			236	382
Savings Relative to Baseline Forecast (billion kWh)				
2020			207	438
2030			398	544

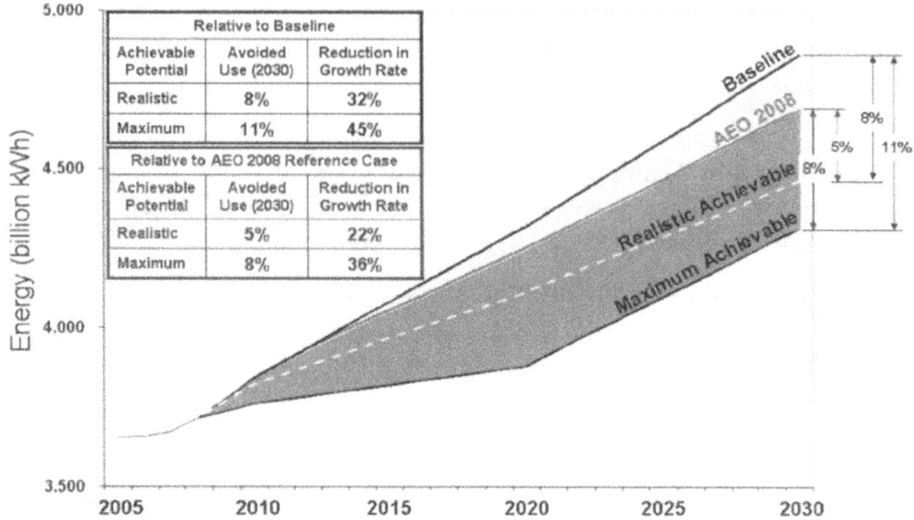

Figure 11-7. U.S. Energy Efficiency Achievable Potential

- **Economic Potential**: CAC systems in existing homes are replaced, upon reaching the end of their useful lives, with the highest SEER level CAC that passes the economic screen; in new homes, the highest SEER level CAC passing the economic screen is installed. The results of the economic screening vary by region. In the Southern region in 2010, for example, the highest-efficiency CAC that passes the economic screen is SEER 15.

- **Maximum Achievable Potential (MAP)**: MAP applies a market-acceptance rate to the economic potential results, based on the best experiences of energy efficiency programs per technology or end-use category, as well as the considered judgment of industry experts. The market acceptance rate for the high-efficiency CAC unit is estimated to be 25% by 2010, and is projected to increase to 75% in 2020 and remain at that level through 2030.

Realistic Achievable Potential (RAP): RAP applies a program implementation factor to MAP. The program implementation factor for the high-efficiency CAC unit is assumed to be 15% in 2010, and is projected to increase to 42% in 2020 and 70% in 2030. The combined effect of the market acceptance rate and program implementation factor for residential central air conditioning gives a realistic achievable potential that is

4% of economic potential in 2010, 32% in 2020 and 53% in 2030. Program implementation factors vary by technology category.

Overview of End Use Opportunities

Figure 11-8 identifies realistically achievable savings by sector and end use. Two broad categories of opportunity include the following:

- First, there continues to be a large opportunity for savings in end uses that already have a long history in energy efficiency, suggesting that there is potentially more "low-hanging fruit" to harvest. Commercial lighting, industrial motors, and residential cooling fall into this category.
- Second, the recent growth in consumer electronics and computing equipment has not only added to the baseline forecast, it creates a sizeable opportunity for efficiency improvements that will result in electricity savings. The industry is only beginning to understand what is possible for these end uses and to exploit the potential for savings.

LIGHTING

Lighting is one key end use that is undergoing dramatic changes. Figures 11-8 and 11-9 illustrate one projection of the likely change in lighting. As of 2011, over 70% of traditional Edison sockets are still inefficient lighting.

Four lighting technologies figure prominently in projections for the reduction in lighting energy use: CFLs, LEDs, MicroHID, and Induction.

CFL (compact fluorescent lighting)

Compact and fluorescent lighting offers reduced power consumption over traditional incandescent, fluorescent, and high intensity discharge (HID) lighting. They are offered in Edison, mogul, and pin bases in the wattage range of 4 W to 250 W. CFLs are sensitive to ambient temperature, vibration, and enclosed can installation, and should be installed by the base not by the tube to reduce risk of premature failure. Despite these sensitivities, they have an estimated life of 3,000 to 8,000 hours. While still highly rebated, many consumers still hold a negative view of the technology due to lamp performance, perceived

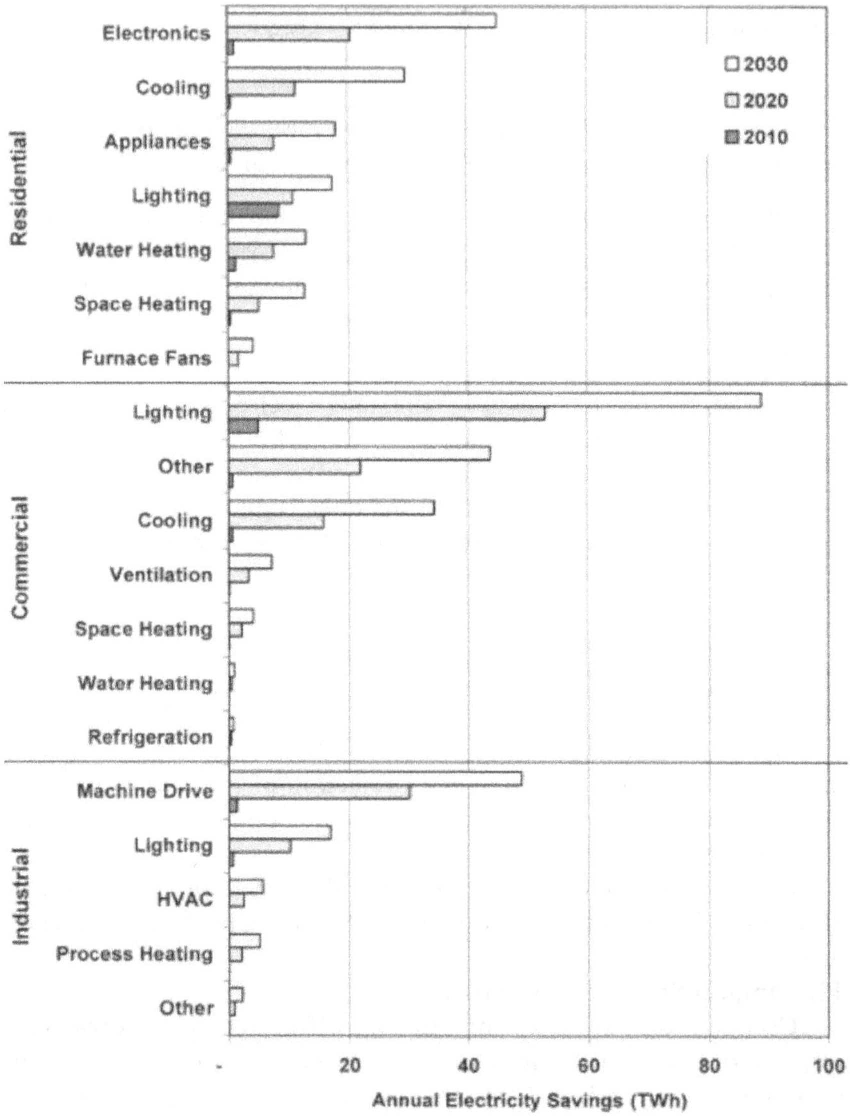

Figure 11-8. Realistic Achievable Potential by End Use (Relative to Baseline)

inferior light output, and noticeable warm-up time. CFLs also contain trace amounts of mercury. (http://www.energystar.gov/ia/partners/promotions/change_light/downloads/Fact_Sheet_Mercury.pdf) CFLs have improved in recent years and are developing some deeper market penetration. Currently, they represent the largest portion of the *efficient screw-based* lighting marketplace.

LED (light-emitting diode)

Light-Emitting Diode offers a replacement to a range of traditional lighting technologies and offers 10% to 70% power savings depending on applications and comparative technology. They are offered in a range of color temperatures, advertise 20,000 to 50,000 hours of life, are typically dimmable, and contain no mercury. LEDs are currently available for use in a range of screw and pin-based applications, a variety of commercial/manufacturing applications, flood/signage lighting, roadway lighting, area lighting, high bay/high mast lighting, architectural lighting, concert/theater lighting, and agricultural lighting. They have a higher cost than competitive technologies (CFL, HID, induction, etc.), but are marketed to offset the cost of acquisition via a combination of power savings and reduced maintenance. LED replacement lamp revenue grew from $2.0 billion in 2011 to $2.58 billion in 2012; LED-based luminaries expanded from $7.2 billion to $9.2 billion in the same timeframe. Packaged LEDs represented $13.7 billion in sales in 2012, but LED-based luminaries are expected to exceed $20 billion by 2017.

Micro HID lighting

Micro HHID Lighting offers a replacement to traditional HID fixtures and offers 30% power savings over traditional HID lamps with 50,000 hours of life. Micro HID delivers full-spectrum lighting, integrated dimming capability within the technology, and compatibility with a range of controls. Today, they are currently utilized in roadway lighting, area lighting, high bay/high mast lighting, architectural lighting, concert/theater lighting, and agricultural lighting. The core technology is primarily made by two companies, but that technology is utilized in a range of fixtures manufactured by other companies. Micro HID currently represents a small portion of the lighting industry, but this share should expand as the technology becomes paired with a range of fixtures and higher wattage fixtures are developed.

Induction lighting

Induction lighting is a promising replacement for office and high bay lighting. These light sources produce light by use of a magnetic field. It is based on work by Nicola Tesla in the late 1800s. Induction lighting wattages range from 40 W to 400 W. They offer approximately 50% power savings over HID technology and claim 100,000 hours of life. Step dimming is available in some configurations. Induction lighting has a growing market share but is still developing. For example, the state of New Jersey Highway Department standardized on induction, and it is utilized by the military and along coasts.

Photonics

Photonics is a growing area of technology that could substantially impact the use of electricity in lighting and consumer electronics. The science of photonics includes the generation, emission, transmission, modulation, signal processing, switching, amplification, and detection/sensing of light.

Daylighting

Improvements in photonics could fundamentally change the ability to use daylighting by enabling natural light to be economically routed to interior spaces. Daylighting is an area within lighting that has growth potential both as a replacement of, and a supplemental light source to, electric lighting. Daylighting offers a large potential for reducing power consumption and for more efficient utilization of produced light. The cost of daylighting systems varies based on the daylighting technology selected and specific application requirements. The lifecycle cost is typically declining due to increasing market share, wider selection of technology options, and the fact that light transfer capacity and the distance that light can be delivered is increasing. Daylighting can be utilized in a range of applications, but the site must be conducive based on geographic location and building orientation, as well as offer a largely unobstructed line of sight to the horizon.

Optics

Improvements in optics, including lenses and reflectors, will potentially improve all light sources, including daylighting. Developments within lens and reflectors will play a critical role in numerous lighting technologies, developing improved output and efficiency numbers.

Lens

Glass, polymers, and plastics that reduce/eliminate yellowing, resist cracking due to heat, maintain their shape (which assures consistent light placement), and that allow high amounts of light throughout will enable all lighting technologies to provide consistent and dependable light delivery.

Reflectors

Though not utilized extensively with point sources like LEDs, organic LEDs (OLEDs), and field-induced polymer luminance (FIPEL), improvements in reflective/mirroring surface will play a critical role in improved efficiencies for HID, micro HID, induction, CFL, and other sources where generated light is lost within the fixture.

FIPEL (Field-Induced Polymer Luminance)

FIPEL lighting is currently at the bench scale and uses minimal power. Current claimed lumen output is far below competitive technologies. Proponents claim power savings over LED lighting without the buzz of fluorescent lighting.

OLED (Organic Light-Emitting Diode)

OLED lighting will eventually become utilized in commercial (and potentially by 2030 residential) lighting, and promises reduced consumption of power of up to 50% over traditional LEDs.

ESL (Electro Stimulated Luminance)

ESL is a niche technology (one lamp shape from one manufacturer currently available) could grow to a larger market share by 2030 due to no mercury content, relatively low cost, even light distribution, and CFL-like power numbers.

Fiber Lighting

Fiber Lighting is a niche technology is primarily in entertainment or hospitality venues. It typically contains metal halide, tungsten halogen or LED lamps. This technology has the ability to gain some market share due to utilizing distributed light from a single source, but may remain a niche market.

Nanowire/Quantum Wires

Nanowire/Quantum wires technology includes development of architectures with a diameter or thickness of sub-tens of nanometers or less. These small devices can be utilized within a variety of applications to deliver improved performance, reduced power consumption and speed.

- Due to the size of nanowire devices, quantum mechanical characteristics are critical to proper operation.
- They were developed to meet the demand for devices that overcome the limits of traditional integrated circuit (IC) chip and metal wire topologies.
- They can be metallic (Ni, Pt, Au), semiconducting (Si, InP,GaN), or insulating (SiO_2, TiO_2), and be either organic or inorganic.
- They are highly scalable and highly efficient.

Optical Interconnects and Semiconductors

Silicon (Si)-based photonic connections and components within chips and circuits allow for the potential development of faster, lower-power, and higher-efficiency components for use in a range of computing and consumer electronic applications.

- Research has determined the force required to move atoms of various elements. This allows for atom-up construction and the development of technologies to meet the specific requirements of a variety of applications.
- Development should help dramatically reduce the cost and energy consumption in data centers.
- This technology can provide intra- and inter-chip solutions.
- This lighting will also assist with the growth of global communication due to increased speed.
- The technology is currently in deep development, but mainstream utilization of actual products are likely ten years away.
- Current Market Barriers
  - One barrier is cost/performance level compared to current technologies by way of high volume fabrication tools paired with low-cost technologies and end-use applications.
  - The technology needs complete supply chain support and integration of optical interconnect technologies, including clear roadmaps for end user and OEM applications.

Optical Photonic Switch

Optical Photonic Switch technology uses development of on-chip light-based nanowire switches (currently using cadmium sulfide) to replace traditional electronic integrated circuit switches.

- This switch allows for optical data processing.
- This technology could lead to consumer photonics.
- The technology needs to be paired with nanowire-based emitters, detectors, modulators, and waveguides to spur development.
- Top-down-based fabrication of nanowire-based optical system components offers improved surface properties and strong optical confinement.
- The technology enables development of components with a footprint the fraction of comparable silicon- based devices.

GaN (gallium nitride)

Gallium nitride is a wide bandgap semiconductor historically used in LEDs has potential advantages for use in optoelectronic, high-power, and high-frequency devices. GaN-based violet diodes are utilized in Blu-ray players.

- Advantages
    - Higher frequency switching and smaller transistors
    - Smaller and fewer periphery components (capacitors and inductors)
    - More efficient than MOSFETs (a type of transistor used for amplifying or switching electronic signals)
    - Lower operational temperature then MOSFETs
    - Operates in higher temperature then traditional IC technologies
- Laboratory tests have shown that GaN transistors are more efficient in simple DC-to-DC conversions.
- As MOSFETs are hitting their theoretical limits, GaN has already surpassed their maximum capabilities.

DEMAND RESPONSE

The Smart Grid enables the connectivity necessary to deploy demand response.

Demand response (DR) is a concept that grew out of demand-side management (DSM). DSM defined six overall load shape objectives that could be attained via changes in the way consumers use electricity. DSM's purpose was to influence the pattern and/or amount of electricity consumption to optimize supply (generation) and demand, creating a portfolio of integrated supply and demand options.

The original six load shapes were: peak clipping, valley filling, energy efficiency, load shifting, flexible reliability, and load growth. Over time peak clipping, load shifting, and flexible reliability have become known as demand response.

These demand response load shape changes have the reduction of peak demand as their primary objective. Reduction of peak demand is aimed at reducing the need for peaking capacity and/or the need to purchase wholesale electricity at peak prices.

While the primary result of effective DR programs is to reduce peak demand, some overall reduction in consumption or efficiency improvements have been observed with DR programs and activities. These may result from the fact that DR programs have the accompanying effect of raising consumer awareness about their consumption habits. Another theory involves the operating efficiency of end-use equipment on peak versus off peak. In DR programs, reductions occur during peak times and high ambient temperatures, when end-use equipment may not operate as efficiently. During recovery, such as when a building is cooled from a DR intervention, the lower ambient temperature improves end-use equipment efficiency.

Consumer Electronics

Consumer electronics is a broad category of one of the fastest growing residential and commercial end uses. It includes handheld devices (cell phones, smart phones, tablets); personal computers (desktop PCs and laptops); televisions, set-top boxes; video games; DVDs; and home audio devices. Table 11-3 lists the principal factors that affect the energy consumption of consumer electronics.

In a 2013 interview (January 30, 2013) on NBC's Today Show, Bill Gates, the former Microsoft CEO, was asked about the future of technology. His response suggested that 20 years in the future, computers will be integrated into the surroundings. In Mr. Gates' opinion—a mouse or keyboard will no longer be needed, walls will be display screens, everything will be integrated, and users need only issue spoken commands.

Again, the Smart Grid will enable many of the features of consumer electronics connectivity.

Early signs of this future are already emerging. For example, based on a visit to the 2012 Consumer Electronics Show held in Las Vegas, the following key themes and trends are already apparent:

- Devices incorporating wireless connectivity between all types of consumer electronics (predominantly Wi-Fi) are now available. According to exhibitors, there is an average of five connected (networked) devices per U.S. home today.
- Tablet sales have been dramatically increasing in the past few years, now reaching 34% penetration and growing. And there is currently a 45% penetration of smart phones; and the use of these devices is accelerating with the average person viewing their mobile devices (phone or tablet) 150 times per day! These smart phones and tablets

Table 11-3. Factors Affecting the Energy consumption of Consumer Electronics

Indicators of Increases in Energy	Indicators of Decreases in Energy
Increase in availability of networking features in electronics	Increased efficiency of power supplies
Increased sales of consumer electronics	Increased efficiency of display technologies
Increased usage of consumer electronics	Increased efficiency of processors
	Consumer migration from more demanding devices (such as computer or set-top box) to less demanding devices (phones/tablets)
	Increased use of "thin clients" that depend on central IT served by data centers
	Increasing number of energy standards (EnergyStar, CEC, DOE and 80PLUS)

have the potential to raise energy awareness as the mobile energy usage display and control interface of the future.

- There is a shift in home area network (HAN) architecture from a piece of hardware (i.e., gateway) as the center of the ecosystem to the use of a Smartphone or tablet that delivers a seamless user experience with content stored in the cloud. There is also a trend to low-cost, "do-it-yourself" (DIY) connected home systems from large industry players, sold as home automation, but which can also enable efficiency and demand response.
- Televisions are increasing in size (85- to 100-inch screens), brighter and sharper (4 to 16x resolution of HD), thinner, 3D-enabled, and Internet-connected to support third-party apps tailored to specific makes and models. The energy information hub of the home may become the television after all. While televisions increase in size, they are also becoming more energy-efficient. Consider, for example, that Samsung's current 46-inch LED TV draws 30 W, whereas a comparably sized flat panel TV produced eight years ago draws 60 W.
- Fifteen of the leading multi-channel video providers and device manufacturers are launching set-top box energy conservation agreements that will result in annual residential electricity savings of $1.5 billion.

Utility Implications of Connected Homes

- Connected home devices have the potential to enable ubiquitous demand response (DR) capability. Currently, most ecosystems connect a single manufacturer's products across different categories on a proprietary basis (e.g., Samsung, LG). The next step is for different manufacturers' products to communicate using a single user interface, with Lowe's appearing closest to this goal.
- From a utility perspective, the maturation of the connected device market provides an opportunity to partner with established retail players such as Lowe's, Ingersoll Rand, and GE in their demand-response and energy-efficiency programs. All large vendors are developing application programming interfaces (APIs) for third parties to access in-home devices through the cloud. By developing custom APIs, utilities can deliver services to their customers as well as obtain data on consumer response that can improve the ro-

bustness of their consumer offerings and programs. This approach bypasses current difficulties such as bandwidth limitation for two-way communications through traditional radio frequency (RF) or through advanced metering infrastructure (AMI).

- Connectivity is being added to products to fulfill a variety of value propositions, including security, safety, enhanced user experience, and energy management.
- Smartphone's and tablets have the potential to raise energy awareness as the mobile in-home display of the future. Their penetration is nearing 50%, and the average person is utilizing them more. This improves the opportunity for setting changes that can improve efficiency and provide DR (e.g., power alerts, energy and demand usage). Because of the battery operation of these wireless devices, the components used in their design are extremely energy-efficient. EPRI testing of an iPad showed that fully charging it every other day consumes 12 kWh per year or about $1.36 per year.
- With connectivity embedded in a greater variety of devices, the potential of social media to influence customer behavior (e.g., purchase decisions, usage habits, and energy consumption profiles) through targeted messaging or viral campaigns could be transformative.
- Cloud-based computing services have the potential to provide a personalized user experience and make DR adoption and participation more viable for individual customers based on their preferences. Personalized apps and services can be tailored to individual customer preferences and learned lifestyle choices. This has the potential to automate DR participation and thereby enable ubiquitous DR that reduces customer perception of inconvenience.
- The communication methods supported by the connected appliances on display appeared consistent with the Association of Home Appliance Manufacturers' (AHAM's) expressed preference for Wi-Fi, ZigBee, and HomePlug Green PHY. None of the appliance manufacturers displayed support for the CEA-2045 Modular Communications Interface option. Moreover, energy hubs/gateways on display typically supported an integrated communication protocol (e.g., ZigBee, Z-Wave) inside the device; and many hubs/gateways also contained a physical USB port, which were associated with

enabling optional support of additional protocols (via plug-in of a USB dongle).

The Digital Economy and Information Technology

The impact of mobile devices and social networking will transform the electricity industry. They are beginning to impart a transformational change in how consumers access information and use that information. This presents an opportunity for the electricity industry to reconnect with the consumers in new ways and enhance operational efficiency.

The tools people use to access information have changed dramatically. For example, in June 2007, the iPhone was introduced and 500 million were shipped in 2011. In April 2010, the iPad was introduced and approximately 60 million were shipped in 2011. And how people interact and process information has changed dramatically as well. In 2007, there were approximately 340 tweets per day. In 2012, there are approximately 340 million tweets per day and 900 million active users on Facebook.

Social networking itself provides numerous opportunities, including consumer engagement, energy-gaming, energy awareness, and behavioral intervention. The industry is beginning to use social media to influence consumption behavior. Social media can complement behavioral interventions through normative comparisons, interaction with trusted peers, individual and group level competitions, and leveraging of huge popular networks.

Social gaming can also influence behavior. "Participants placed an irrational high value on their own creations—and that value was proportional to how long they had worked on it." (Wired Magazine, June 2011) Worldwide revenue generated from the sale of virtual goods increased from \$2 billion in 2007 to \$7 billion in 2010 and is forecasted to increase to \$15 billion by 2014.

Smartphone's and tablets alone provide numerous energy-related opportunities for consumer engagement, energy storage, workforce productivity, and simple load control. These devices are connecting and enabling consumers while eliminating small frictions that can radically alter one's decision. Some utilities are already offering billing, outage, and account information at customer fingertips on handheld devices. Examples such as the Nest (www.nest.com) thermostat offer set-and-forget thermostat simplicity in load control.

Consumers increasingly rely on digital communications, which has become a pervasive part of everyday life and is integral to modern lifestyles. From Smartphone to tablet to laptop, consumers are seemingly always tethered to the on-line work—a trend even more pronounced among younger generations who have grown up in a world of on-line ubiquity and an expectation of information and entertainment on demand.

Demographic changes influence this trend. By 2025, 50% or more of the world's population will have been born and raised in the so-called "digital age" (after 1990). As quoted in a recent USA Today newspaper article on "The End of Online Dating," "this Valentine's Day, fewer singles are tethered to laptops or those other technological relics: desktops."

Similarly, enterprises rely on digital communications and information analytics to run their operations and drive business decisions.

Trends such as consumer entertainment (i.e., streaming HD videos on demand over the Internet, on line HD gaming through gaming consoles, set-top boxes to record television programming, etc.), social networking, and data communications (i.e., Skype video-conferencing) and enterprise digitization (i.e., digitization of medical records in healthcare industry, electronic commerce, etc.) have a profound impact on electricity. There are two electricity-impacting dimensions to consider: the customer-facing front-end and the information-serving back-end.

The residential and commercial sector impact is reflected in the ever-increasing number and variety of electronic devices at consumers' disposal. Each of these devices has to be plugged in to charge, and while the individual electricity consumption of a single Smartphone, PDA, tablet, or laptop computer may be relatively modest (note that gaming consoles and personal computers themselves can be highly electricity-intensive), in aggregate this represents geometric growth in electricity consumption for the previously unspecified "miscellaneous" or "other" category of electricity use.

Electric technologies are evolving dramatically through several different paths. Following are several examples:

The Kickstarter light bulb

The LIFX Kickstarter Project—is the development of a Wi-Fi-enabled, multi-color, energy-efficient LED light bulb that is designed to integrate technology that allows control with an IOS or Android device. The project was launched on the Internet by enabling funders to purchase units in advance for prices depending on bundle size. Once screwed into a light socket, the user simply downloads the app and it is ready to go. Brightness can be selected for a single bulb, a room, or the entire house. It can change color, send push notifications via Twitter, Facebook, and SMS, and even be animated to match the beats of music. (USA Today, September 17, 2012)

The NEST Thermostat

The NEST thermostat integrates with an existing home heating and cooling system. NEST is referred to by its developers as a "learning" thermostat. It is a combination of hardware and software that offers many features at the thermostat and wirelessly through tablets and phones, and is compatible with 95% of home heating and cooling systems. Several key features of the NEST include:

- System Match—Automatically activates various features such as true radiant (auto-activation); heat pump balance (minimize use of auxiliary heat); early-on (warm the house for early risers); or filter reminders.
- Auto-Away—If the system does not detect activity, it adjusts.
- Auto-Schedule—Learns consumer preferences and adjusts accordingly.
- Android, iPhone, and iPad applications.

Additional indicators of the digital economy and IT could include:

- Internet bandwidth consumption (number of people on line, number of people in social networks, amount of streaming content).
- Cloud and data center electric energy consumption.

According to a report by Franhofer USA, residential consumer electronics consumed approximately 193 TWh of electricity in 2010 (see Table 5-2), an amount equal to 13.2% of residential electricity consumption and 5.1% of total U.S. electricity consumption in 2010. This translates into 9.3% and 2.1% of residential and U.S. total primary energy consumption, respectively. (DOE 2011)

Televisions

The focal point of consumer electronics (CE) technology in the home is the television. Of all residential CE devices, the changes in television technology and the continued increase in its use is one area where the demand for electricity will evolve. Table 11-5 lists the factors that

Table 11-4. Residential CE Energy Consumption Summary (Franhofer) Device UEC

Device	UEC (kWh/yr)	Installed Base (million)	AEC (TWh)
Audio Video Receivers	65	99	6.4
Blu-ray Players	14	12	0.2
Computers			
Desktop	220	101	22
Portable	36	132	8.3
Computer Speakers	37	74	2.8
DVD Devices	28	223	6.3
Monitors	97	131	13
Network Devices			
Integrated Access Device	53	42	2.2
Modem	44	46	2.0
Router	44	49	2.1
Set-top Boxes			
Cble	150	87	13
Satellite	112	76	8.5
Stand-alone	54	45	2.4
Telco	115	16	1.8
Televisions	183	353	65
Video Game Consoles	135	109	15
Other CE Devices	18	1,160	21
Total	-	**2,870**	**193**

could impact the demand for electricity from the adoption of television technology.

Overall cost of televisions is decreasing, which makes it more affordable for consumers to upgrade current televisions. However, consumers appear to be purchasing more expensive TVs than before due to the decrease in cost to purchase a model above. This may include greater

Table 11-5. Factor Impacting the Demand for Electricity from Television

Indicators of Increases in Energy	Indicators of Decreases in Energy
Increase in overall television stock	Increased penetration of efficient technologies
Increase in hours of usage per day	Introduction of new standards
Increases in number of televisions per home	Introduction of new and even more efficient television technologies
Increases in average screen size	Stricter regulations and standards

size or more feature-rich models.

Newer TVs have slimmer profiles, reduced energy consumption, better resolution/color quality, and increased screen size. As a result, television energy consumption could increase as consumers have more TVs per household. A countervailing factor is that larger screen sizes consume more energy, but market studies show increase of average screen size purchased but with decreasing slope.

In addition, as Nielsen ratings show, consumers are watching more TV than ever; this is evidenced by the increase in usage of DVRs and other instant video options. However, appliance efficiency standards will place downward pressure on television energy consumption.

Computers and Tablets

As indicated previously, computers, tablets and other hand-held devices are one of the essential consumer electronics, and they are undergoing a substantial evolution. This is an evolution that includes a strong trend to embrace more use of information technology.

But the evolution that is also noticeable is the migration from desktop computers to laptops to tablets to smart phones and ultimately to "thin clients" that will rely on the cloud or data center servers for computational horsepower and data storage.

As shown in Figure 11-9, sales of personal computers (PCs) are stagnant and will fall over time, and the sales of laptops are stagnant and will fall over time. At the same time, sales of tablets and smart phones are growing rapidly. And those new devices are benefitting from improved battery life that forces energy-efficient designs. These new devices are replacing PCs and laptops that are usually recharged every night. As the evolution continues, sales of servers and storage arrays are growing rapidly to fill data centers that operate 24/7 to deliver apps for the tablets and smart phones. Many of these data centers require tens of MWs of power to provide information and services to tablets and smart phones. This trend is illustrated in Figure 11-9.

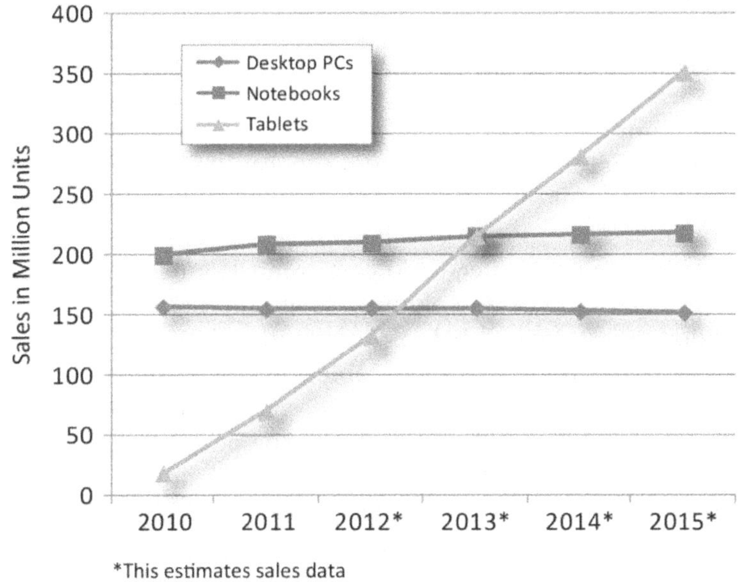

Figure 11-9. Forecast for Global Sales of Tablets, Notebooks and Desktop PCs from 2010 to 2015 (in million units) (Source: IDC, Morgan Stanley, Tablet Landscape Evolution)

A number of trends are making data center loads grow. For example, Facebook had 1 billion users in August 2012, all sharing photos and details on what they're eating for lunch. Over 1.9 billion emailers sent 107 trillion emails in 2010. Apple now has 300,000 iPad apps and 800,000 iPhone apps available. They have a $1 billion, 500,000 square foot data center just to sell apps and iTunes—and many of those apps need "other" data centers to run. Data centers are also supporting cloud computing. This reflects data center services that are being offered without having to build and operate their own data center.

A number of other trends are making data center loads grow. These include the trends toward digitizing electronic medical records at a time when more people are living longer and taking more tests. The Smart Grid will require management of data from tens of millions of smart meters, with utilities' data centers collecting petabytes of data. There is an explosion in interest of gaming. For example, Sony PlayStation Network has approximately 77 million users. Global data traffic is growing by 5% every week.

And the data computation and management will grow very fast. In the 1990s, emails were just short text messages… a few kilobytes. Then in the 2000s, people began attaching things like photos and PowerPoint slides… a few megabytes. Now in the 2010s, users are streaming two-hour high-definition movies from Netflix… at the rate of a few gigabytes. Each of these is an increase of a thousand every ten years. As a result, data centers in the U.S. are using approximately 93 billion kWh per year.

According to Computerworld.com (Matt Hamblen, 2012), Smartphone and tablet growth is exploding, especially in business, resulting in 1.2 billion smart devices being purchased in 2013. IT managers already know that smart phones and tablets are being purchased in vast numbers by workers, and Gartner added credence to that trend today, reporting that 1.2 billion of the devices will be purchased in 2013. "The consumerization trend has hit IT as an unstoppable force," Gartner said in a statement. In all, smart phones and tablets combined make up 70% of all devices sold in 2012. The total is expected to reach 821 million by the end of the year.

Smartphone's and tablets will not entirely replace PCs for most businesses, "but the ubiquity of smart phones and the increasing popularity of tablets are changing the way businesses look at their device strategies," said Carolina Milanesi, one of Gartner's smart device analysts. Gartner also predicted that 66% of mobile workers will own a Smartphone in 2016. More than half of those smart phones—56%—will be Google's Android devices.

Android and Apple's iOS devices will be used in businesses side-by-side, and often instead of BlackBerry devices. Milanesi noted that businesses have moved in the last 12 months from resisting Apple products to accepting them for use in their organizations. According to Gartner, tablet purchases by businesses will reach 13 million in 2012 and triple to 53 million in 2016.

According to an article in Forbes Magazine by Jason Evangehlo on January 30, 2013, Lenovo is experiencing "hyper growth" in their Smartphone and tablet business. They had shipped 9 million smart phones and 800,000 tablets during the final quarter of 2012. That equates to nearly an 80% year-to-year growth. Lenovo has been responsible for some inspired form factors based on Windows 8. Their IdeaPad Yoga and ThinkPad Twist are both multipurpose tablet/ultrabook hybrids with competitive price points and enterprise-friendly features.

Power Supplies

A critical component related to each consumer electronic device is the power supply. Figure 11-10 illustrates the continued improvement in efficiency related to power supplies for consumer electronics. Most of this improvement is driven by codes and standards directed toward power supplies, including:

- EnergyStar
- 80 Plus®
- CEC Title 20, Appliance Efficiency
- EISA 2007, Section 301

In summary, efficiency continues to increase in the current stock of power supplies. Energy consumption may decrease if CEC passes regulations on electronics power consumption. DOE is also proposing new efficiency standards that would decrease energy consumption of battery chargers and external power supplies.

Energy Consumption Outlook in Consumer Electronics

The following standards related to consumer electronics are likely to reduce energy consumption of consumer electronics:

— EISA 2007

— EnergyStar

— Tier II Standard for "State Regulated" EPSs—California

— Battery Charger Standard—California

— Potential decrease if CEC passes regulations on electronics power consumption

— DOE is also proposing new efficiency standards that would decrease energy consumption of battery chargers and external power supplies

Reduced energy consumption of consumer electronics includes set-top boxes, DVD players, home audio, rechargeables, video games, and

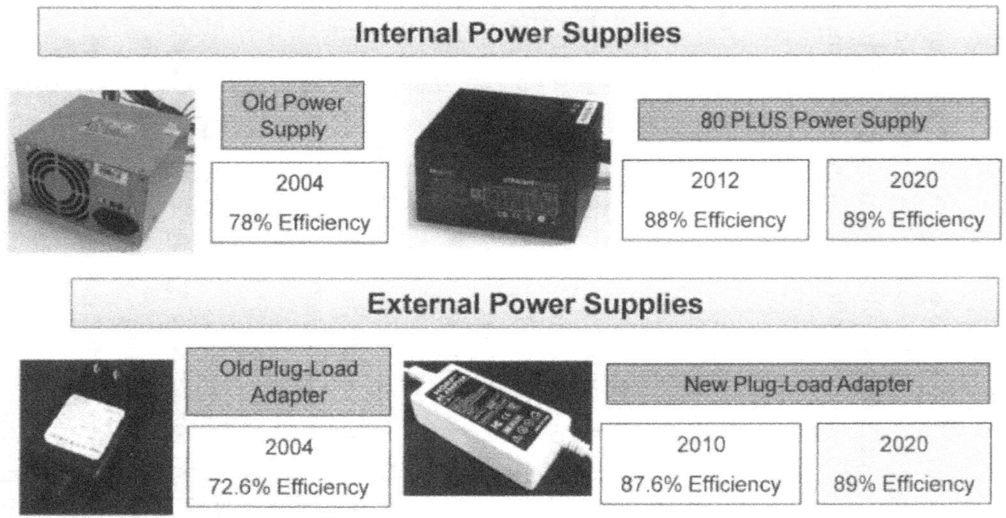

Figure 11-10. Efficiency Catches Up with "Power Supplies" (Source: EPRI Laboratory Testing (2004-20110, see notes for 2020 predictions)

external power supplies, as indicated in Figures 5-15 through Figure 5 21. AEO2012 projections ignore the efficiency potential in these devices. The "Adjusted AEO Energy" reflects estimates of these efficiency improvements.

A number of changes indicate the need to adjust AEO2012 predictions resulting from power supply efficiency improvements. This decreases kWh consumption in 2030 by 23 x 109 kWh.

NEW USES OF ELECTRICITY

In 2009, EPRI conducted a study to address the potential for expanding end-use applications of electricity. The study focused on converting residential, commercial, and industrial equipment and processes—existing or anticipated—from traditional fossil-fueled end-use technologies to more efficient electric technologies.

The study began with development of baseline forecasts of energy use and CO_2 emissions. The forecasts were consistent with the U.S. Department of Energy (DOE) Energy Information Administration's (EIA's) "Reference Case" as presented in its 2008 Annual Energy Outlook (EIA AEO 2008). The study estimated the potential for energy savings and CO_2 emissions reductions during the years 2009 through 2030 for the residential, commercial, and industrial sectors as a function of end-use technology, fuel displaced, and census region. The analysis yielded forecasts of changes in primary energy use and energy-related CO_2 emissions for the U.S.

This study did not investigate replacing fossil-fueled vehicles with plug-in electric vehicles (PEVs). The potential for PEVs has been analyzed in a separate study covered later in this chapter.

Replacing fossil-fueled energy end uses with electric end-use technologies can deliver meaningful net reductions in CO_2 emissions. The most significant reductions can be achieved when implemented in conjunction with low-carbon generation, other efficiency programs, and when specific high-value replacement targets in energy end uses can be identified and acted upon.

Energy Savings

According to EIA AEO 2008, total annual energy consumption for the U.S. in the residential, commercial, industrial, and transportation sectors is estimated at 102 3 quadrillion Btus in 2008, including delivered energy and energy-related losses. The Reference Case forecasts this consumption to increase by 153% to 1180 quadrillion Btus in 2030, an annualized growth rate from 2008 to 2030 of 0.65%.

The Reference Case already accounts for market-driven efficiency improvements, and the impacts of all currently legislated federal appliance standards and building codes (including the Energy Independence and Security Act of 2007) and rulemaking procedures. It is predicated on a relatively flat electricity price forecast in real dollars between 2008 and 2030. It also assumes continued contributions of existing utility- and government-sponsored end-use energy programs established prior to 2008.

Relative to the EIA AEO 2008 Reference Case, this study identifies between 1.71 and 5. 32 quadrillion Btus per year of energy savings in 2030 due to expanded end-use applications of electricity. The lower bound represents the Realistic Potential, while the upper bound represents the Technical Potential. Therefore, expanded end-use applications of electricity have the potential to reduce the annual growth rate in energy consumption forecasted in EIA AEO 2008 between 2008 and 2030 of 0.65% by 10% to 32%, to an annual growth rate of 0.58% to 0.44%

The lower bound of these estimated levels of energy savings are potentially achievable through voluntary fuel conversion programs implemented by utilities or similar entities. The EPRI analysis does not assume the enactment of new codes and standards beyond what is already in law. More progressive codes and standards would yield even greater levels of energy savings.

Reductions in CO_2 Emissions

According to EIA AEO 2008, annual energy-related CO_2 emissions for the U.S. in the residential, commercial, industrial, and transportation sectors is estimated at 5,983 million metric tons in 2008. The Reference Case forecasts emissions to increase by 14.5% to 6,850 million metric tons in 2030, an annualized growth rate from 2008 to 2030 of 0.61%. The Reference Case is predicated on a relatively flat CO_2 intensity of the electricity generation mix between 2008 and 2030.

Relative to the EIA AEO 2008 Reference Case, this study identifies between 1 14 and 320 million metric tons per year of CO_2 emissions reductions in 2030 due to expanded end-use applications of electricity. The lower bound represents the Realistic Potential, while the upper bound represents the Technical Potential. Therefore, expanded end-use applications of electricity have the potential to reduce the annual growth rate in CO_2 emissions forecasted in EIA AEO 2008 between 2008 and 2030 of 061% by 12% to 35%, to an annual growth rate of 054% to 040%.

The lower bound of these estimated levels of CO_2 emissions reductions are potentially achievable through voluntary fuel conversion programs implemented by utilities or similar entities. The EPRI analysis does not assume the enactment of new codes and standards beyond what is already in law, nor does it assume the CO_2 intensity of the electricity generation mix will decrease significantly in the future. More progressive codes and standards and a less carbon-intensive generation mix would yield even greater levels of CO_2 emissions reductions.

Annual Technical and Realistic Potentials by Sector

Table 11-6 summarizes the Technical and Realistic Potential results by end-use sector for the favorable* electric end-use technologies. The results are presented as annual values, and thus, represent the impacts for the given year—they are essentially a snapshot in time. For the Technical Potential, the table shows that the residential sector has the greatest promise for beneficial impacts. The commercial and industrial sectors follow with values that are roughly comparable to each other. The Technical Potential impacts of all three sectors combined are energy savings of 5.32 quadrillion Btus per year and CO_2 emissions reductions of 320 million metric tons per year in 2030 relative to the baseline forecast.

In the Realistic Potential case, the industrial sector has the highest potential for energy savings, followed by the residential sector and then commercial sector. In regards to the Realistic Potential for CO_2 reductions, the residential sector holds the greatest promise, followed by the industrial sector and then the commercial sector. The Realistic Potential impacts of all three sectors combined are energy savings of 1.71 quadrillion Btus per year and CO_2 emissions reductions of 114 million metric tons per year in 2030 relative to the baseline forecast.

These estimates suggest that expanded end-use applications of electricity have the potential to reduce the annual growth rate in energy consumption forecasted in EIA AEO 2008 between 2008 and 2030 of 0.65% by 10% to 32%, to an annual growth rate of 0.58% to 0.44%. In addition, they have the potential to reduce the annual growth rate in CO_2

*Favorable technologies reduce overall energy use and CO_2 emissions despite increasing kWh sales.

Table 11-6. Technical and Realistic Potential: Annual Impacts on Primary Energy Use and CO_2 Emissions by Sector

Baseline	EIA AEO 2008 Baseline Forecast of Primary Energy Use (Quadrillion Btus per Year)			EIA AEO 2008 Baseline Forecast of Energy-Related CO_2 Emissions (Million Metric Tons per Year)		
Sector	2010	2020	2030	2010	2020	2030
Residential	22.2	23.4	25.0	1,259	1,323	1,450
Commercial	18.7	22.0	25.0	1,080	1,265	1,474
Industrial	33.3	34.3	35.0	1,693	1,718	1,733
Transportation	29.0	31.2	33.0	1,980	2,077	2,193
U.S.	**103.3**	**110.8**	**118.0**	**6,012**	**6,383**	**6,850**
Technical Potential	Decrease in Primary Energy Use(Quadrillion Btus per Year)			Decrease in CO_2 Emissions(Million Metric Tons per Year)		
Sector	2010	2020	2030	2010	2020	2030
Residential	0.352	2.20	2.96	21.7	135	178
Commercial	0.118	0.72	1.09	7.75	47.2	69.1
Industrial	0.123	0.72	1.27	7.30	43.7	73.1
U.S.	**0.593**	**3.64**	**5.32**	**36.7**	**226**	**320**
Realistic Potential	Decrease in Primary Energy Use(Quadrillion Btus per Year)			Decrease in CO_2 Emissions(Million Metric Tons per Year)		
Sector	2010	2020	2030	2010	2020	2030
Residential	0.055	0.417	0.633	4.97	34.3	47.3
Commercial	0.018	0.146	0.277	1.93	13.3	21.4
Industrial	0.078	0.457	0.802	4.57	27.1	45.1
U.S.	**0.152**	**1.02**	**1.71**	**11.5**	**74.7**	**114**

emissions forecasted in EIA AEO 2008 between 2008 and 2030 of 0.61% by 12% to 35%, to an annual growth rate of 0.54% to 0.40%.

Cumulative Technical Potential by Technology

Table 11-7 shows the cumulative technical potential impacts on primary energy use and CO_2 emissions by electric technology for the residential sector. The cumulative numbers represent cumulative impacts between the study's start year of 2009 and the given year. Thus, a value listed for a given year includes all impacts accumulated between 2009 and that year. Space heating represents about 74% of residential fossil fuel consumption annually over the forecast period and presents the biggest potential for energy and CO_2 savings. Displacement of fossil-fueled space heating with heat pumps results in cumulative energy savings of 32.3 quadrillion Btus and cumulative CO_2 reductions of 1,970 million metric tons from 2009 to 2030. Fossil-fueled water heating represents about 18% of fossil fuel consumption in the residential sector over the forecast period. Heat pump water heaters have the potential to save 9.59 quadrillion Btus and reduce CO_2 emissions by 573 million metric tons. Finally, heat pump clothes dryers have the potential to save 0.14 quadrillion Btus and to reduce CO_2 emissions by 14.4 million metric tons. In total, these three electric technologies have the potential to reduce primary energy consumption by 42.1 quadrillion Btus and reduce CO_2 emissions by 2,557 million metric tons from 2009 to 2030.

Table 11-7 shows the cumulative impacts on primary energy use and CO_2 emissions by electric technology for the commercial sector. Space heating represents about 40% of fossil fuel consumption annually over the forecast period and presents the biggest potential for energy and CO_2 savings in the commercial sector. Displacement of fossil-fueled space heating with heat pumps results in a cumulative energy savings of 11.4 quadrillion Btus and a cumulative CO_2 reduction of 740 million metric tons from 2009 to 2030. Fossil-fueled water heating represents about

Table 11-7. Technical Potential: Cumulative Impacts on Primary Energy Use and CO_2 Emissions by Favorable Electric End-Use Technology (Residential Sector)

Favorable Electric End-Use Technology	Decrease in Primary Energy Use(Quadrillion Btu)					Decrease in CO_2 Emissions(Million Metric Tons)				
	2010	2015	2020	2025	2030	2010	2015	2020	2025	2030
Heat Pump Space Heating	0.38	3.64	10.3	20.5	32.3	23.6	224	634	1,252	1,970
Heat Pump Water Heater	0.15	1.39	3.92	6.79	9.59	8.88	83.9	235	407	573
Heat Pump Clothes Dryer	0.00	0.02	0.05	0.10	0.14	0.19	1.89	5.43	9.84	14.4
Total	**0.53**	**5.05**	**14.3**	**27.3**	**42.1**	**32.7**	**310**	**875**	**1,669**	**2,557**

16% of fossil fuel consumption in the commercial sector over the forecast period. Heat pump water heaters have the potential to save 2.73 quadrillion Btus and reduce CO_2 emissions by 181 million metric tons. Finally, heat pump space cooling has the potential to save 0.16 quadrillion Btus and to reduce CO_2 emissions by 8.99 million metric tons. In total, these three electric technologies have the potential to reduce primary energy consumption by 14.3 quadrillion Btus and reduce CO_2 emissions by 930 million metric tons between 2009 and 2030.

Table 11-8 lists the cumulative impacts on primary energy use and CO_2 emissions by electric technology for the industrial sector. Each value in the table represents the nationwide cumulative impacts between the start date of 2009 and the given year. By 2030, heat pumps are projected to potentially save 12.2 quadrillion Btus of energy and reduce CO_2 emissions by 634 million metric tons. They represent the technology with the greatest potential for beneficial impacts. Electric arc furnaces are projected to yield energy savings of 1.70 quadrillion Btus and CO_2 reductions of 198 million metric tons between 2009 and 2030. Electrolytic reduction has the potential to reduce energy use by 0.740 quadrillion Btus and CO_2 emissions by 35.6 million metric tons. Electric induction melting could potentially save 0.320 quadrillion Btus while reducing CO_2 emissions by 22.1 million metric tons. Lastly, plasma melting is projected to reduce energy use by 0.374 quadrillion Btus and CO_2 emissions by 21.4 million

Table 11-8. Technical Potential: Cumulative Impacts on Primary Energy Use and CO_2 Emissions by Favorable Electric End-Use Technology (Commercial Sector)

Favorable Electric End-Use Technology	Decrease in Primary Energy Use(Quadrillion Btu)					Decrease in CO_2 Emissions(Million Metric Tons)				
	2010	2015	2020	2025	2030	2010	2015	2020	2025	2030
Heat Pump Space Heating	0.14	1.29	3.59	7.02	11.4	8.93	83.8	234	457	740
Heat Pump Water Heater	0.04	0.37	1.06	1.88	2.73	2.58	24.9	71.0	126	181
Heat Pump Space Cooling	0.00	0.02	0.05	0.10	0.16	0.12	1.08	2.93	5.63	8.99
Total	**0.18**	**1.67**	**4.70**	**9.00**	**14.3**	**11.6**	**110**	**308**	**589**	**930**

metric tons. The combined results of all industrial technologies implemented across the U.S. are cumulative energy savings of 15.3 quadrillion Btus and cumulative CO_2 emissions reductions of 911 million metric tons between 2009 and 2030.

Figures 11-11-and 11-12 graphically depict the impacts of the electric end-use technologies on primary energy use and energy-related CO_2 emissions, respectively. Once again, the values are expressed in terms of the cumulative Technical Potential impacts between 2009 and 2030. In all three sectors, heat pumps are the technology with the greatest promise for saving energy and reducing CO_2 emissions. In the industrial sector, electric arc furnaces have a significant potential for beneficial impacts as well. In addition, electrolytic reduction, electric induction melting, and plasma melting also show promise. Under a less carbon-intensive future generation mix, more technologies would cross the line and become favorable in regards to saving energy and reducing emissions.

In both the residential and commercial sectors, the end-use areas with the most potential for beneficial impacts are space heating and then water heating. Clothes drying (residential) and space cooling (commercial) also exhibit potential. In the industrial sector, process heating is the predominant end-use area showing potential, followed by space heating.

Table 11-9. Technical Potential: Cumulative Impacts on Primary Energy Use and CO_2 Emissions by Favorable Electric End-use Technology (Industrial Sector)

Favorable Electric End-Use Technology	Decrease in Primary Energy Use(Quadrillion Btu)					Decrease in CO_2 Emissions(Million Metric Tons)				
	2010	2015	2020	2025	2030	2010	2015	2020	2025	2030
Heat Pump	0.150	1.37	3.78	7.42	12.2	7.94	71.6	198	389	634
Electric Arc Furnace	0.017	0.181	0.531	1.03	1.70	1.92	21.2	62.0	120	198
Electrolytic Reduction	0.009	0.080	0.223	0.440	0.740	0.444	3.97	11.1	21.9	35.6
Electric Induction Melting	0.003	0.033	0.092	0.187	0.320	0.288	2.46	6.81	13.5	22.1
Plasma Melting	0.004	0.040	0.110	0.221	0.374	0.270	2.37	6.62	13.1	21.4
Total	**0.183**	**1.70**	**4.74**	**9.30**	**15.3**	**10.9**	**102**	**285**	**557**	**911**

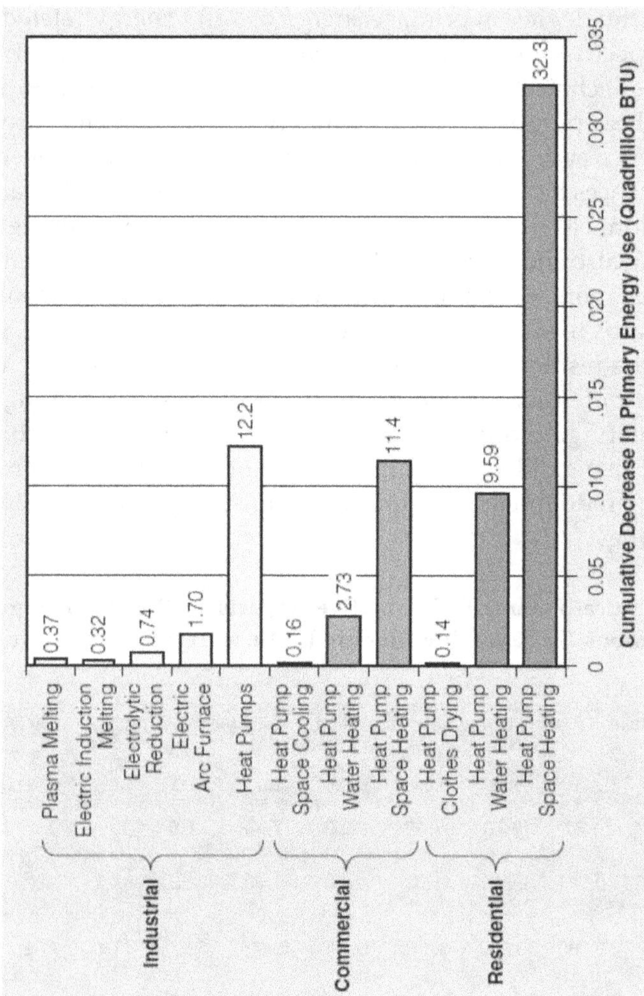

Figure 11-11. Technical Potential: Cumulative Decrease in Primary Energy Use between 2009 and 2030 by Sector and Efficient Electric End-Use Technology

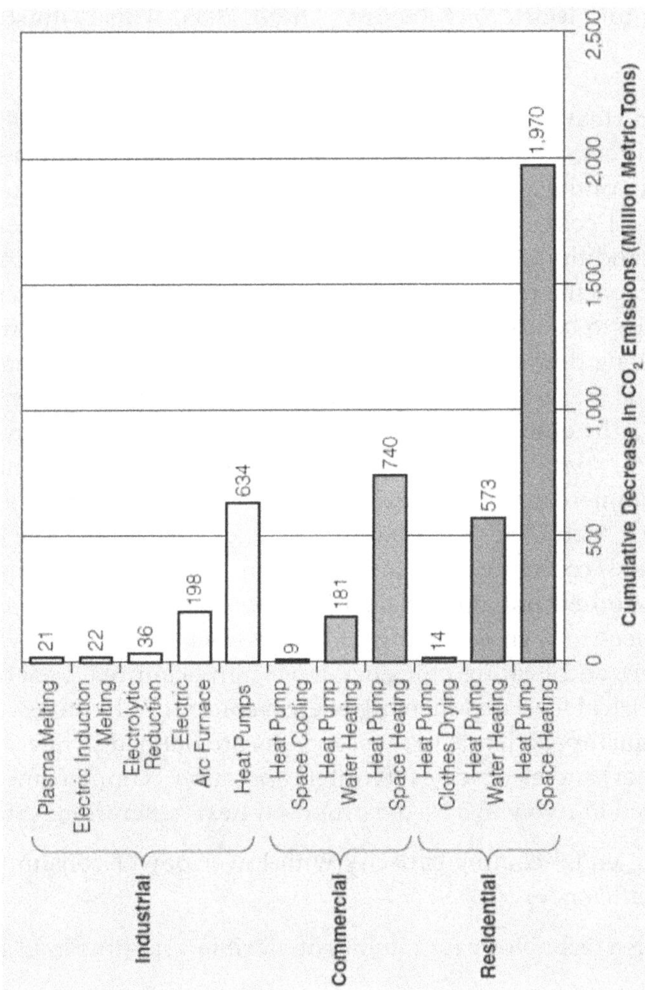

Figure 11-12. Technical Potential: Cumulative Decrease in Energy-Related CO_2 Emissions between 2009 and 2030 by Sector and Efficient Electric End-Use Technology

Not all of the new electric technologies are necessarily fundamentally advanced new uses of electricity. One example is the development of the 4 kV boiler designed to compete with natural gas boilers. The penetration of natural gas boilers in the U.S. is approximately 88% as compared to 5% for electric. However, the efficiency of natural gas boilers is 80 to 85% and for electric, 99% (Source: ORNL 2005). This is illustrated in Figure 11-13.

The Advanced Heat Pump

Electrical technologies for space conditioning, in particular heating, in the form of the air-source heat pump (ASHP), have the potential to offer the most cost-effective and source-energy-efficient method for indoor space conditioning. The inclusion of new, advanced technologies, like variable capacity, increase efficiency substantially, thereby permitting heat pumps to be used in non-traditional colder climates where fossil fuels have long dominated. Heat pumps have the inherent benefit of being a machine that transfers heat rather than produces heat, allowing it to far surpass the efficiency of fossil heating sources in many climates.

In the U.S., space conditioning is the largest load category in residential and commercial sectors, accounting for upwards of one-third of the total electric load [End-Use Energy Efficiency Roadmap]. The combined residential/commercial sector is, in turn, the largest for primary energy use (the others are industrial and transportation) [Gellings, 2008, The Role of Electricity in Reducing CO_2 Emissions]. Improvements in air conditioning and heating efficiency have significant real benefits to managing electrical load and meeting energy-efficiency objectives.

U.S. seasonal peak loads in most regions are defined by air conditioning (summer) and electric heat (winter) operation. Summertime load can be mitigated in two ways by the proposed next-generation ASHP:

- Providing equal cooling capacity with lower power consumption (energy efficiency)
- Providing a technology for intelligent, flexible, real-time load shedding

Wintertime load can be mitigated in four ways:

- The two stated for cooling above
- Flexible capacity capable of providing full heating load at low am-

bient temperatures, eliminating the need for back-up strip heat and allowing heat pump application in northern climates, traditionally served by gas and heating oil

- Advanced frost mitigation, which improves overall system efficiency and eliminates the need for electric resistance make-up heat during defrost

Many advanced technologies in heat transfer, flow distribution, compression, etc., have been developed, but have never been integrated in a comprehensively improved design of the ubiquitous air source heat pump/air conditioner. Strong potential exists for an advanced ASHP design to greatly improve the operating efficiency, thereby expanding geographic applicability into regions traditionally served by gas or heating oil. As an efficient electrotechnology focused on beneficial electrification, this flexible, variable capacity, smart ASHP will serve as a utility resource for energy efficiency, demand response, and renewable integration.

EPRI has already begun interaction with the Air-Conditioning, Heating and Refrigeration Institute (AHRI) Subcommittee on Smart Connected Devices with the aim of bridging the needs of the utility sector with the abilities of future heat pump systems. It is the position of heat pump manufacturers that next-generation systems will have "smart" ability, which must be given proper direction in order to provide a resource (to a utility), while always maximizing the customer experience. The focus of the current EPRI/AHRI interaction is to define approaches for using next-generation heat pumps as resources in non-traditional ways—beyond what they've been used for in the past. For example, this includes methods for varying cooling capacity during peak events to maximize both power reduction and customer comfort.

Heat Pump Water Heaters

Heat pump water heaters (HPWHs) operate on the same principle as heat pumps used for heating, ventilating, and air conditioning (HVAC) applications. They extract heat from a low-temperature source and transfer the heat to a high-temperature sink with the input of electricity. In the case of the HPWH, the low-temperature source is typically the air surrounding the water heater, and the high-temperature sink is the hot water storage tank.

A recent EPRI study estimated the technical potential of replacing fossil-fueled technologies with electric technologies in existing residential buildings and new construction. This study included evaluating the conversion from gas water heaters to electric HPWHs in residential buildings. A high scenario used in the EPRI study assumed an energy factor (EF) of 0.59 for gas water heaters and a COP of 3.4 for electric HPWHs. This scenario assumed that the most efficient HPWHs would replace standard gas water heaters.

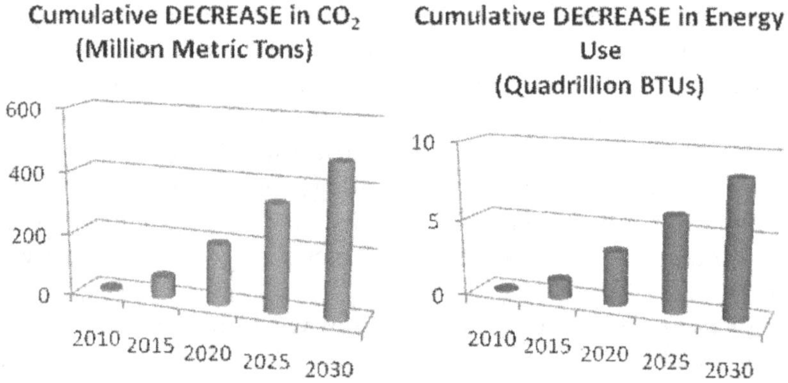

Figure 11-13. High Estimate of Cumulative Potential CO_2 and Fossil Fuel Reductions using HPWH Technology for Selected Years. (Gas water heater EF = 0.59; HPWH COP = 3.4)

There are a number of new technologies on the horizon that are not included in the EPRI analysis (ORNL 2012). These include the following:

- Electron beam melting, which enables precision melting of powder materials and processing of complex geometries not possible through machining.
- Ultrasonic additive manufacturing, which enables a simultaneous additive and subtractive process for manufacturing complex geometries. In addition, a solid-state process allows embedding of optical fibers and sensors.
- Laser metal deposition, which allows site-specific material addition; application of advanced coating materials for corrosion and wear resistance; and the repair of dies, punches, turbines, etc.
- Fused deposition modeling, which is the precision deposition of thermoplastic materials and the development of high-strength

composite materials for industrial applications. It also enables transformation of rapid prototyping to rapid manufacturing.

- High-velocity laser accelerated deposition (HVLAD), which is a new photonic method for producing protective coatings with ultra-high-strength, explosively bonded interfaces. These coatings prevent corrosion, wear, and other modes of degradation in extreme environments. In addition to being crucial to future fusion reactors, HVLAD materials and coatings have other important uses including bearing and shafts for wind turbines, corrosion-resistant structures of offshore platforms, better pipelines for oil and gas transmission, and ultra-hard corrosion-resistant surfaces for naval ships (R&D 100 2011).
- Ultraviolet LED (UV LED) can be applied to industrial curing. This technology offers energy savings compared to mercury lamps as well as to the use of gas to air dry conventional coatings. Energy savings of more than 60% have been reported. The technology radiates ultraviolet light and uses it to cure ultraviolet-sensitive coatings. UV light impacts the photo-initiators in the coating which become excited, passing energy along to other particles in the coating. This stimulates a bonding process between molecules. Excited components, called free radicals, maintain the reaction.

Full deployment of all favorable electric technologies would result in an increase of delivered electricity use (in quadrillion Btu) of 3.38 by 2030. This is equivalent to 991 x 109 kWh (3.38 x 1012/3412).

Plug-in Electric Vehicles

One major new appliance that has begun to populate the residential landscape is the plug-in electric vehicle (PEV). Other appliances that could have a long-term impact are the increased saturation of heat pumps for both space heating and water heating. In the industrial and commercial sectors, electrification may increase through the use of space and process heat pumps, electric material handling, and electric process heating.

As of late 2012, the PEV market included 13-passenger vehicle models, with five more expected by the end of 2013. More than 58,000 PEVs were sold during the 23 months since vehicles became available from major manufacturers, and cumulative sales are expected to surpass 500,000 vehicles by 2015.

However, the PEV market is still very much in its infancy. Some of the vehicle models are destined for limited production, two (Transit Connect Electric and Think City EV) have halted production, and some of the startup manufacturers are having difficulty with recalls or reliability. However, such fits and starts are to be expected with a burgeoning new industry, and overall PEV sales are showing consistent growth every month.

EPRI has made market adoption forecasts of PEVs using its PEV Load Estimator (PEVLE) tool. PEVLE forecasts the number of PEVs sold, cumulative PEV market penetration, PEV electricity consumption, petroleum savings, and greenhouse gas reduction.

The Adoption Forecasts report described a variety of market forces that affect the adoption of plug-in electric vehicles, including:

Refueling Cost

Gasoline and diesel prices are highly volatile, while electricity prices, particularly for residences, are in most cases very stable. By shifting some or all of the vehicle fuel to electricity, PEVs provide at least some help to increase certainty with fuel costs. Furthermore, in most situations, PEVs provide a significant reduction in vehicle refueling cost. With an all-electric vehicle, the cost may be reduced up to 75% or more.* The extent to which cost is reduced and stability is increased depends on the storage capacity of the PEV battery, other attributes of the vehicle design, the availability and usage of time- of- use electricity prices, and the user's driving patterns. PEVs may also provide a benefit of reduced maintenance costs.†

Initial Vehicle Price or Lease Costs

The vehicle "sticker" price is one of the most influential components of a vehicle purchasing decision for vehicle buyers. The upfront price of PEVs is currently higher than a comparable traditional vehicle, but increasing competition in the PEV market will drive prices down. Leasing options may be attractive to some parties who wish to trade off vehicle ownership in favor of potentially lower monthly costs.

**Transportation Electrification: A Technology Overview*. EPRI, Palo Alto, CA: 2011. 1021334. Page 6-1.

†*Total Cost of Ownership for Current Plug-in Electric Vehicles*. EPRI, Palo Alto, CA: 2012. 1024848.

Incentives

Governments, employers, and retailers offer various enticements to PEV owners including tax credits, discounts on PEV charging equipment, discounted or free charging, access to carpool lanes, and discounted parking.

Regulation

There are federal rules covering fuel economy and greenhouse gases, minimum sales requirements for PEVs in California and some other states, and local regulations including building codes and permits that affect the cost of installing charging equipment. These rules impact the cost and availability of PEV s and associated charging facilities.

Environmental and Political Trends

The relative level of concern over national security and the environment—and the public's opinion of how PEVs affect those concerns—affects interest in and acceptance of PEVs. These trends are sure to be different across various regions of the United States, Canada, and other countries.

Figures 11-14 and 11-15 illustrate the result of EPRI's recent forecast. It depicts three cases (high, medium and low). EPRI's high case suggests that new vehicle market share could approach 30% by 2030, resulting in 9 million vehicles sold by 2030. The medium scenario is approximately 17% (3 million vehicles), and the low scenario is approximately 6.5% (1-2 million vehicles).

The 2012 Annual Energy Outlook (AEO2012) projected the sales of hybrid electric, all-electric vehicles, and light duty vehicles (LDVs) will grow substantially. AEO projects that microhybrids, which use start/stop technology to manage engine operation while at idle, account for 6% of total LDV sales in 2035, which AEO considers the largest share for vehicles that use electric storage. AEO projects that gasoline-electric and diesel-electric hybrid vehicles account for 5% of LDV sales in 2035 and plug-in and all-electric hybrid vehicles account for 3% of LDV sales.

The EPRI study estimates that PEVs will demand between 24,956,452 and 100,977,130 MWh. This translates to a possible increase over the AEO2012 estimate of between 20,011,299 and 96,031,887 MWh or between 20 to 96 x 109 kWh.

Figure 11-16 illustrates the energy sales that would result from these penetration estimates.

Table 11-10. AEO Estimates of Sales of Light Duty PEVs and EPRI Estimated kWh Sales (Approximate)

	2013	2020	2030*
Vehicles/yr	28,000		268,000
Cumulative Vehicles	28,000		2,436,000
MWh	56,842		4,945,243
*Estimated from AEO 2020/2035			

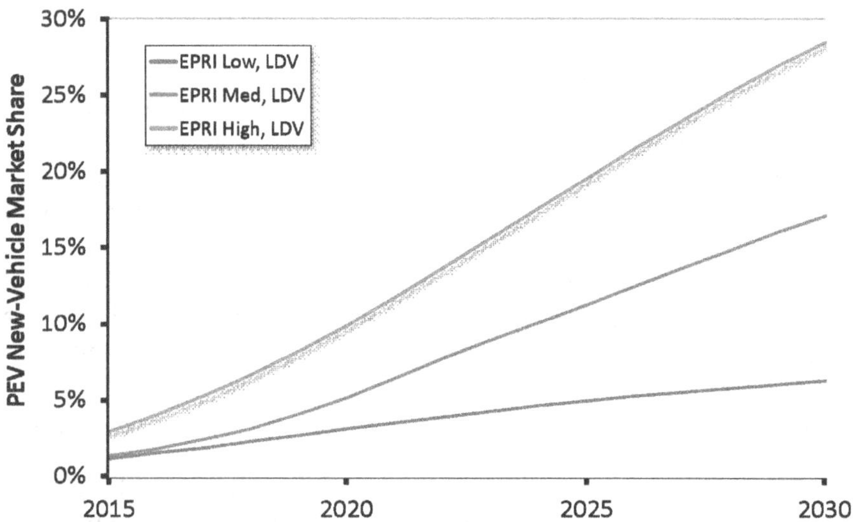

Figure 11-14. PEV New Light-Duty Vehicle Market Share Through 2030

Off-Road Equipment

Estimated trends in increased off-road equipment electricity use presented in Figure 9-8 show significant increases in electricity demand for this sector over the next 20 years. As shown in Table 9-6, electricity use increases are expected to be greatest for industrial equipment, lawn and garden equipment, and shore-side power.

EPRI developed an analysis of future increases in electricity demand from off-road equipment in the lower 48 states. The analysis considers the types of technologies available or expected to be available to electrify off-road equipment now and in the future, and includes

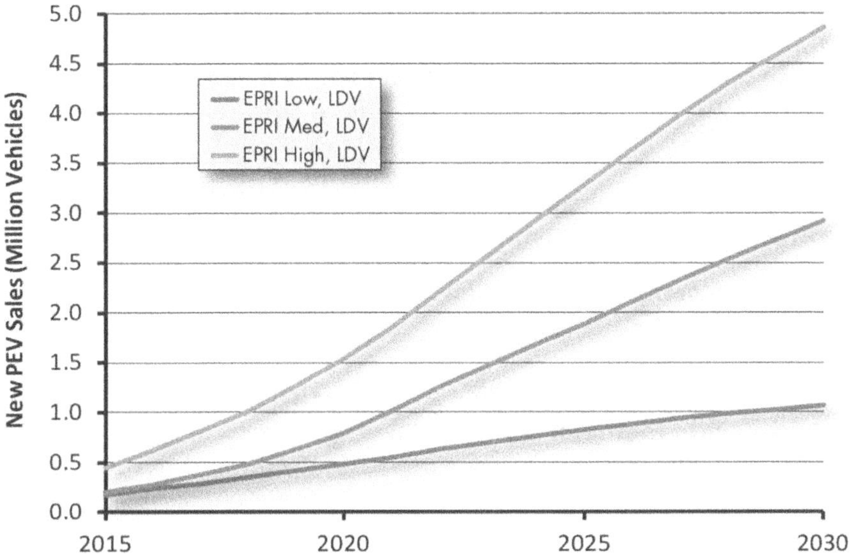

Figure 11-15. U.S. PEV New Light-Duty Vehicle Sales Through 2030

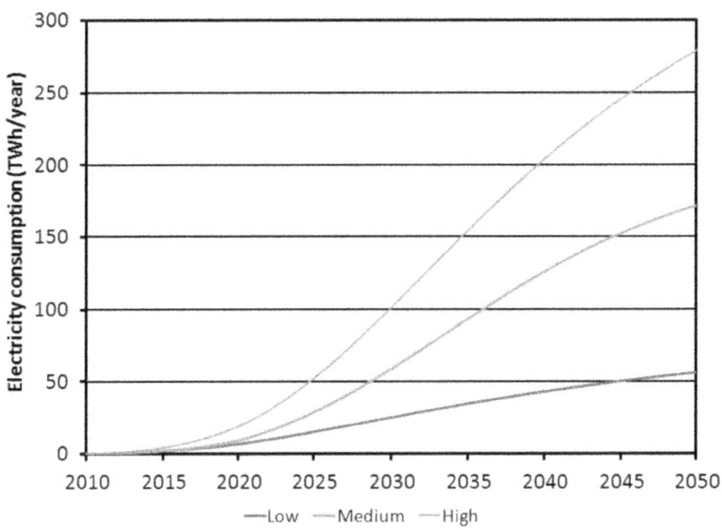

Figure 11-16. Estimates of PEV Growth

estimation of market penetration and fleet penetration of electric units in off-road fleets. Uncertainties in future electrification were taken into account by developing three scenarios:

- High electrification scenario: technology developments and electric equipment deployment occur more quickly and are less constrained.
- Medium or nominal scenario: technology developments and electric equipment deployment proceed as currently anticipated.
- Low scenario: technology developments take longer to deploy and develop in a more limited way.

Electrification technologies such as existing dedicated plug-in, lead-acid, and lithium-ion batteries were considered along with in-development technologies such as zinc-air batteries and hydrogen fuel cells. Estimates of electrification market penetration for each equipment type were used in a fleet turnover model to estimate the number of electrified units in future years.

Off-road equipment types to be electrified were selected based on the potential significance of emissions reductions and electricity demand. Equipment types of note that were not included in this analysis

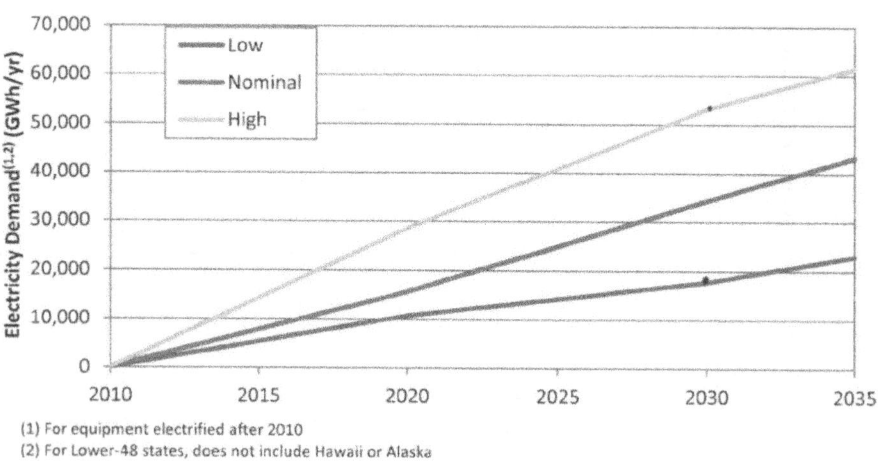

Figure 11-17. Increases in U.S. Electricity Demand from Off-Road Equipment to 2035

Table 11-11. Increases in U.S. Electricity Demand from Off-Road Equipment to 2035

Equipment Type	Electricity Demand in 20 Years (GWh)[1,2,3]
Agricultural Pumps	400 – 600
Aircraft Gate Electrification	80 – 120
Airport Ground Support Equipment	200 – 900
Cargo Handling Equipment (cranes, yard trucks)	1,000 – 2,000
Harbor Craft	900 – 1,700
Industrial Equipment (forklifts, transportation refrigeration units, sweepers)	6,300 – 27,000
Lawn and Garden Equipment	2,900 – 11,000
Ship Shoreside Power	4,400 – 6,600
Switching Locomotives	400 – 900
Recreational Equipment (all-terrain vehicles, off-road motorcycles, golf carts, specialty carts)	1,200 – 2,400
Total	**18,000 – 53,000**

(1) For lower 48 states, does not include Hawaii or Alaska
(2) For equipment electrified after 2010
(3) Year 2030 estimates

are passenger rail, freight rail, and underground mining equipment.

Estimates of off-road equipment populations and annual activity were obtained from emission inventory models (U.S. EPA NONROAD2008a, California ARB OFFROAD2007) or for categories in which emission models are not available by using accepted emission inventory activity datasets.

Data Center Growth

The growth in consumer electronics is reflected in the computing power required to serve, store, and transmit content and applications over the public Internet or corporate intranets. The function is typically associated with data centers, which can range from a server closet in a small business to a room in a mid-sized company, or clusters of buildings over hundreds of thousands of square feet for such technology titans as Google, Facebook, Yahoo, and Amazon. The explosive growth of consumer communications and entertainment technology and overall digitization of the economy has led to geometric growth trends in the electricity consumption of data centers. At present rates, electricity growth of data centers is expected to double every five years, and is projected to reach 30% of U.S. electricity use by 2030. An increase of use from 2% to 10% alone would equal 340 x 109 kWh in 2030.

This slower-than-expected adoption of new data centers as a result of the recent recession leads to lower electricity consumption in the as yet unpublished AEO2013 Reference Case than in AEO2012. In a 2007 EPA Report to Congress, trends at that time called for a doubling of the data center energy consumption every five years. While that projection has been moderated, the recent slow growth will not sustain. However, performance and efficiency improvements are inextricable linked. The rapid rates of growth in data center electricity use that prevailed from 2000 to 2005 slowed significantly from 2005 to 2010, yielding total electricity use by data centers in 2010 of about 1.3% of all electricity use for the world, and 2% of all electricity use for the U.S.

Net Increase in New Uses

The net impact of new uses, plug-in electric vehicles and data centers are listed in Table 11-12.

Table 11-12.
Estimated Impacts of New Uses of Electricity in 2030 (109 kWh)

New Uses	991
Plug-In Electric Vehicles	20 – 96
Off-Road Equipment	18 – 54
Data Centers	340
Total	1369 - 1481

INDUSTRIAL TRENDS

There is some indication that both industrial manufacturing and non-manufacturing sectors are increasing in output. In January 2013, manufacturing output decreased 0.4% following upwardly revised gains of 1.1% in December 2012 and 1.7% in November 2012. For the fourth quarter as a whole, manufacturing production is now estimated to have advanced 1.9% at an annual rate; previously, the increase was reported to have been 0.2% (Source: Federal Reserve).

The PMI™* registered 53.1%, an increase of 2.9 percentage points from December 2012's seasonally adjusted reading of 50.2%, indicating expansion in manufacturing for the second consecutive month. The New Orders Index registered 53.3%, an increase of 3.6% over December 2012's seasonally adjusted reading of 49.7%, indicating growth in new orders. Manufacturing is starting out the year on a positive note, with all five of the PMI™'s component indexes—new orders, production, employment, supplier deliveries and inventories—registering above 50% in January 2013. The January 2013 PMI™ indicates growth for the 44th consecutive month in the overall economy, and indicates expansion in the manufacturing sector for the second consecutive month.

The Manufacturers Alliance for Productivity and Innovation (MAPI) forecasts that manufacturing production will increase 4.5% in 2012, 2.3% in 2013, and 3.3% in 2014. High-tech production is forecast to rise 4.9% in 2012, 5.7% in 2013, and 10.1% in 2014. Non-high-tech or traditional manufacturing, which accounts for 90% of value-added in manufacturing, will grow 4.7% in 2012, 2.4% in 2013, and 3.5% in 2014 (Source: MAPI).

Figure 11-18 illustrates the MAPI assessment of the current evolution of the industrial sector. It highlights areas of accelerating growth, decelerating growth, accelerating decline, and decelerating decline.

Of the 18 manufacturing industries, MAPI reports that 13 are reporting growth in January 2013 in the following order: plastics and rub-

---

*PMI™ (Purchasing Managers Index) is a composite index of five "sub-indicators," which are extracted through surveys to more than 400 purchasing managers from around the country, chosen for their geographic and industry diversification benefits. The five sub-indexes are given a weighting, as follows: Production level (.25), New orders (from customers) (.30), Supplier deliveries (.15), Inventories (.10), Employment level (.20). A reading of 50 or higher generally indicates that the industry is expanding. An index level higher than 42%, over time, is considered the benchmark for economic (GDP) expansion.

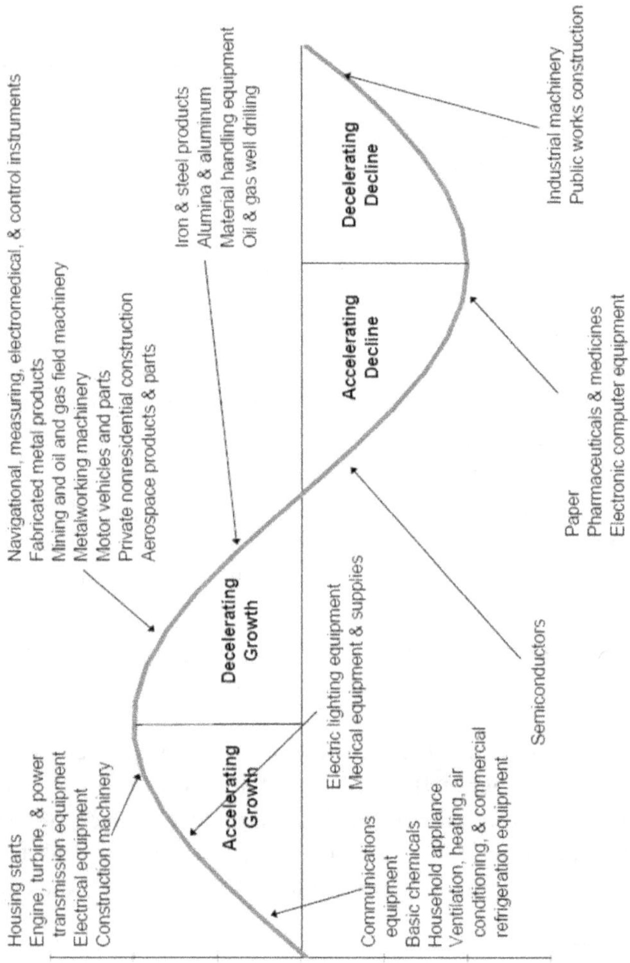

Figure 11-18. Industrial Sector by Phase of Cycle as of September 2012 (MAPI)

ber products; textile mills; furniture and related products; printing and related support activities; apparel, leather and allied products; electrical equipment, appliances and components; miscellaneous manufacturing; fabricated metal products; transportation equipment; petroleum and coal products; machinery; primary metals; and food, beverage and tobacco products. The four industries reporting contraction in January are non-metallic mineral products; computer and electronic products; wood products; and chemical products.

Industries that are showing signs of a comeback include automotive manufacturing. Once a crucial tent pole in America's industrial economy, this industry is poised for a surprising (if slow) comeback. The second and even more surprising growth sector is computer and electronics manufacturing. Apple recently announced plans to invest $100 million in building more of its Mac computer components in the United States. Chinese PC giant Lenovo announced in October 2012 that it would break ground on a manufacturing center in North Carolina. Factors for manufacturing comeback in U.S. include:

- Increased cost of shipping
- Rising foreign labor costs
- Decreasing wages at home
- Growing consumer demand

Collectively, these factors are some of that reasons why cautious optimism about a returning manufacturing base to the U.S. is warranted (Forbes 2012).

Factories in the American South from the Carolinas to Alabama to Kentucky have already experienced such a rebirth. For example, in South Carolina, Boeing builds 787 Dreamliners just north of Charleston, and Starbucks roasts coffee beans in St. Matthews, outside Columbia. General Electric is once again making water heaters and refrigerators at its large Appliance Park plant in Louisville, Kentucky. A Mobile, Alabama, shipyard run by Austal USA is growing so quickly that in the past two years, the Australian company's workforce has swollen from 800 to 3,300. North Carolina's Secretary of Commerce, Keith Crisco, says about 80% of the new companies coming to the state involved some form of manufacturing. "That's a big number for us, and the jobs are better than they were," he said. In Georgia and Tennessee, manufacturing employment grew 3% in the 12 months ending in October 2012, nearly twice the national average of 1.6%, according to the Bureau of Labor Statistics.

However, there are countervailing trends that could affect the growth in manufacturing. According to MAPI, regulations impose costs across the entire manufacturing sector. Major regulations could reduce manufacturing output by 2.3% to 6.0% on average over the next decade. Energy-intensive sectors are most affected. The chemicals and petroleum products sector output could fall by about 9% to 10% per year on average over the next decade.

The Environmental Protection Agency (EPA) imposes the largest number of regulations on the manufacturing sector with respect to number of regulations (972 regulations in total, including 122 major regulations), followed by the Departments of Transportation (880 regulations in total, including 69 major regulations), Labor (214 regulations in total, including 27 major regulations), and Energy (106 regulations in total, including 17 major regulations).

Clearly natural gas prices can have a substantial impact on the movement of industry. This was highlighted in a February 1, 2013 article in the *Wall Street Journal*: "Cheaper Natural Gas Lets Nucor Factory Rise Again on Bayou." As natural gas is now inexpensive in the U.S., several chemical firms have announced plans to build multi-billion-dollar plants in Texas and Louisiana. Figure 10-2 illustrates the range of fuel switching options in industry.

Non-Manufacturing Trends

The (Non-Manufacturing Index) NMI* registered 55.2% in January 2013, 0.5 percentage point lower than the seasonally adjusted 55.7% registered in December 2012. This indicates continued growth at a slightly slower rate in the non-manufacturing sector. The Non-Manufacturing Business Activity Index registered 56.4% in January 2013, which is 4.4 percentage points lower than the seasonally adjusted 60.8% reported in December 2012, reflecting growth for the 42nd consecutive month.

Economic activity in the non-manufacturing sector grew in January 2013 for the 37th consecutive month (Source: Bloomberg).

Continued growth in manufacturing and non-manufacturing out-

*NMI™ (Non-Manufacturing Index) is a composite index based on the diffusion indexes for four of the indicators with equal weights: Business Activity (seasonally adjusted), New Orders (seasonally adjusted), Employment (seasonally adjusted), and Supplier Deliveries. A reading above 50% indicates the non-manufacturing sector economy is generally expanding; below 50% indicates the non-manufacturing sector is generally contracting.

put could cause industrial electricity sales to exceed AEO2012 estimates by 50 x 109 kWh per year in 2030.

DISTRIBUTED GENERATION

There are increasing opportunities for consumers to generate their own electricity. Whether truly cost-effective or influenced by other motivations such as the desire to simply divorce themselves from the "grid," these opportunities result from the decreased cost of generation from photovoltaics and combined heat and power.

Renewable Energy

Customers have self-generated for years. In fact, many industrial facilities generated their own power since World War II, particularly when cogeneration or combined heat and power (CHP) could be economically employed. Interest in customer generation started in earnest in the 1970s in part brought on by energy policy.

The first major energy policy-related regulations in the U.S. related to customer generation were the Solar Energy Research and Development Act of 1974. Federal government incentives were added with the Job Creation and Worker Assistance Act of 2002 that permitted business to take an additional 30% depreciation on solar, wind, and geothermal property in the first year. In 2003, the Job Creation and Tax Relief Reconciliation Act of 2003 increased this additional depreciation to 50%. The U.S. federal government also currently offers a 10% tax credit to businesses that invest in or purchase solar energy property or facilities.

PV Adoption

The overall growth and resultant market penetration of non-utility PV systems is based on a number of factors, as listed in Table 11-13.

Solar Industry Magazine (Solar Industry 2013) recently cited five keys to continued growth for residential solar in 2013:

1. Streamlined permitting
2. Better consumer awareness and understanding
3. Reliable options for financing systems
4. Incentive certainty and a9149P Rev. 30 01/21/15 availability
5. Cooperation from utilities

Table 11-13. Factors Impacting the Adoption of Non-Utility PV Systems

Factor	Implication
Cost	The first cost, operating cost, and financing cost of PV systems based on opportunities from customer-owned and from PV systems installed by third parties on customer premises. This includes access to capital and lease programs.
Retail Price of Electricity	The retail price of electricity is the "competing" energy for customer PV systems. The higher the retail price is – the stronger is the customer's motivation to own/operate and/or lease a system.
Factor	Implication
Incentives	Incentives that result from utility or government rebates, tax credits, and subsidized financing can reduce the consumer's cost and enhance the adoption of PV systems.
Physical Attributes	The rooftops or land area available, the orientation (south, east, west, north) of the site, the structural integrity of the building to support the weight of the system, and other site attributes can limit installations.
Building Codes & Standards	Local and state building codes and standards can inhibit or facilitate PV installations.
Energy Policy	State and federal energy policy can impact the incentives that flow from institutions and governments. These can include renewable portfolio standards that may force utilities to incent renewable power production, including installations on customer premises (see Figure 11-1).
Democratization	Although difficult to quantify, industry practitioners are observing a strong trend toward democratization in electricity. There appears to be a desire for electricity consumers to exercise more control over their electricity purchases. As such, the idea of buying less electricity from energy providers – or even selling electricity to those providers – is highly desirable among consumer segments.
Net Metering & Feed-in Tariffs	Net metering programs allow customers to feed back PV electricity to the grid at times where they have excess beyond their own demand. The feedback is "netted" out of their total use so essentially bought back at retail rates. Feed-in tariffs are where utilities pay consumers a flat rate – typically greater than the retail rate. Both of these schemes can substantially increase PV penetration. According to one estimate, by 2030 there are expected to be 156,000 customers on net metering (Faruqui 2012).

Levelized Cost of PV Systems

According to Bloomberg New Energy Finance (Bloomberg 2013), the levelized cost of electricity (LCOE) from photovoltaic power generation technology continues to decline and now ranges between 8¢/kWh and 35¢/kWh, with most systems averaging approximately 15¢/kWh. Many of these installations already compared favorably with "conventional" (coal, natural gas and nuclear) power generation, ranging between approximately 5¢/kWh and 15¢/kWh. In addition, PV LCOE costs continue to decline dramatically.

Making a direct comparison between PV LCOE and those from conventional systems is difficult because the growth of customer PV systems, which is the subject of this analysis, is based, in part, on retail electricity prices, and conventional LCOE reflect wholesale prices absent any delivery (transmission and distribution) expense. Given the complexity of the incentives for customer PV systems, these PV prices are now being considered in "parity" with the wholesale prices of conventional generation. This is entirely dependent on the regional insolation and individual installation configuration and associated costs.

On-site generation in the U.S. by consumers using increasingly cost-effective photovoltaic systems has increased to nearly 300,000 buildings, growing at a rate of nearly 5% per year (NREL 2011). Photovoltaic system costs recently decreased to less than $1.00 per watt and are projected to decline further.

In addition, as consumers integrate greater amounts of photovoltaic power and other renewable resources, they may leverage solutions to provide flexibility for the inherent variability of wind and solar energy resources. Energy storage technologies offer a viable source of flexibility, and as a result, may significantly increase the amount of solar PV generation consumers can produce. By generating more during the day and storing it for use at night, the consumer's reliance on grid services is further diminished.

As costs of electric energy storage decline, the utilization rate of all forms of distributed generation will likewise increase. On-site generation and storage are potentially the most disruptive technologies to the traditional electric utility's future business model. An open question remains as to whether distribution utilities can turn these disruptive technologies into business opportunities. This trend, coupled with the increase in the use of PVs, is converting increasing numbers of "consumers" into prosumers. Prosumers are both consumers and producers of electricity.

Figure 11-19. Renewable Portfolio Standard Policies (DOE)

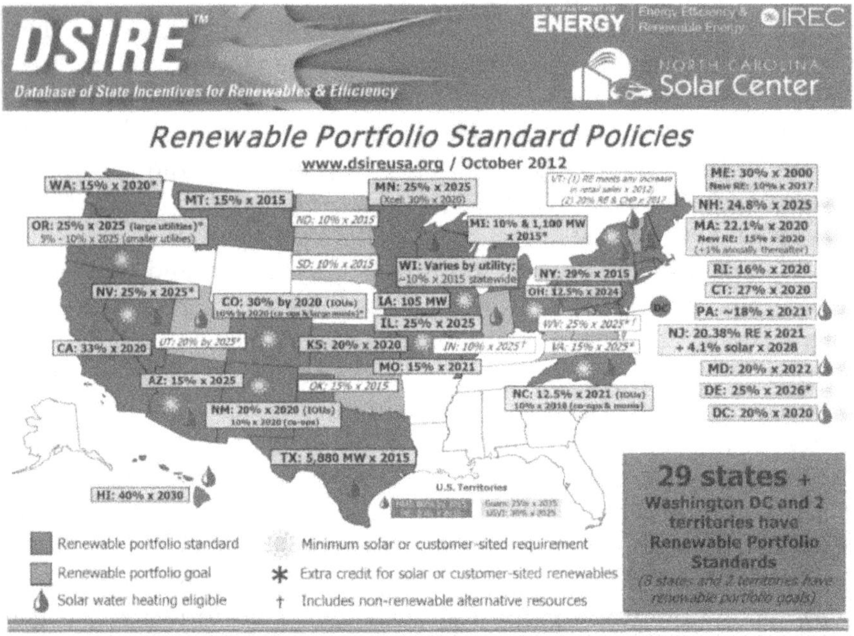

Since 2004, the global solar industry enjoyed extraordinarily strong growth. This growth was stimulated by European incentives. This has led to innovation and also to overheated markets. The current low prices are driving technology manufacturers into bankruptcy. However, the solar industry will emerge from this period of consolidation stronger, more nimble, and more competitive.

Driven primarily by persistent oversupply, PV prices continue to follow a downward trajectory during the latter half of 2012, resulting in rising insolvencies and further supply chain consolidation. The industry now appears to be entering into a phase of development that hinges, more than ever, on strategic marketing efforts to influence consumer sentiment. For one, a number of product manufacturers are emphasizing product and/or service differentiation to combat a growing buyer perception of PV as a commodity good. And more broadly, the PV segment is beginning to tackle product and installation quality issues—including alleged fraud—caused in large measure by the greater integration of product inventories into project pipelines from defunct manufacturers. (EPRI 1025411)

The global PV industry had a 60% compound annual growth rate (CAGR) from 1996 to 2011. From 1996 to 2011, the industry had a 46% CAGR as illustrated in Figure 11-20.

Figures 11-21 and 11-22 illustrate one estimate of U.S. PV demand by market segment from 2005-2010 (SEIA 2010) and the PV installed capacity.

These estimates indicate that the installed PV capacity from non-residential and residential installations reached 636 MW in 2010.

It appears that the residential and non-residential PV installations in 2011 grew by approximately 1050 MW.

In 2012, the Solar Energy Industries Association (SEIA) and GTM Research published their 2011 year-in-review, which indicated that non-residential installations grew to 800 MW in 2011 and residential grew to 758 MW (SEIA 2012).

In more recent estimates, SEIA has reported that approximately 1080 MW of residential and non-residential PV was installed through the third quarter of 2012 and that the fourth quarter of 2012 will see a growth of approximately 44% over 2011, which would assume an added 576 MW and a total of 1656 MW of customer PV installed in 2012 (SEIA 2013).

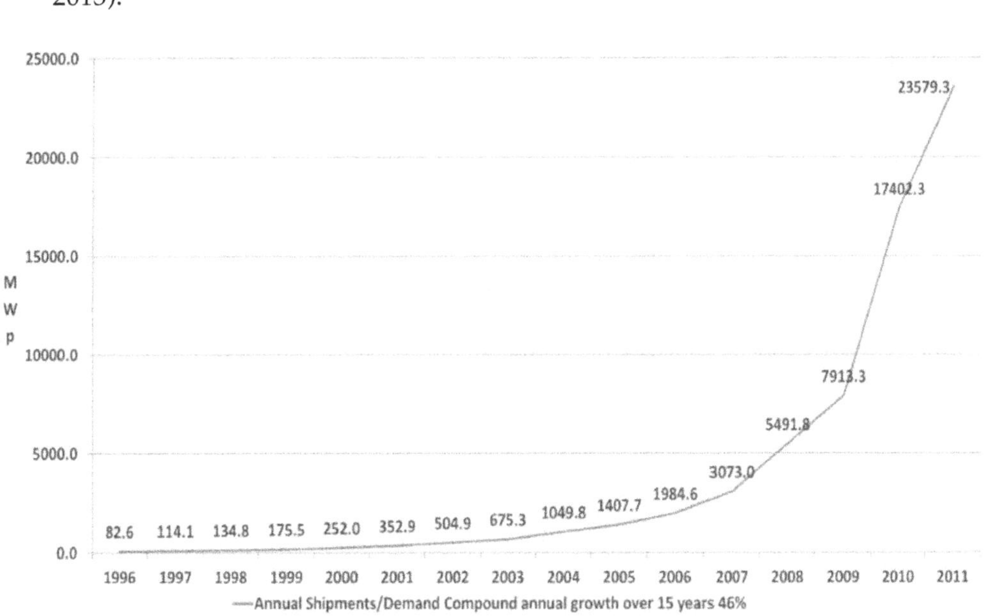

Figure 11-20. Global Demand for PV Systems

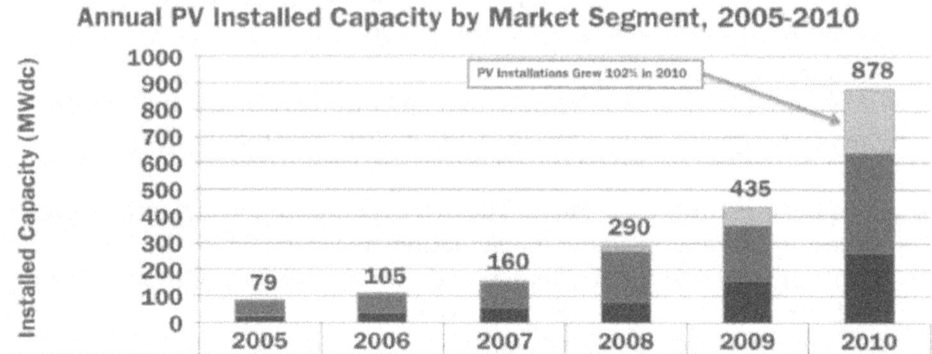

	2005	2006	2007	2008	2009	2010
Total	79	105	160	290	435	878
Utility	1	0	9	22	70	242
Non-Residential	51	67	93	190	208	372
Residential	27	38	58	77	157	264

Figure 11-21. Annual Installed PV Capacity by Market Segment, 2005-2010 (SEIA 2010)

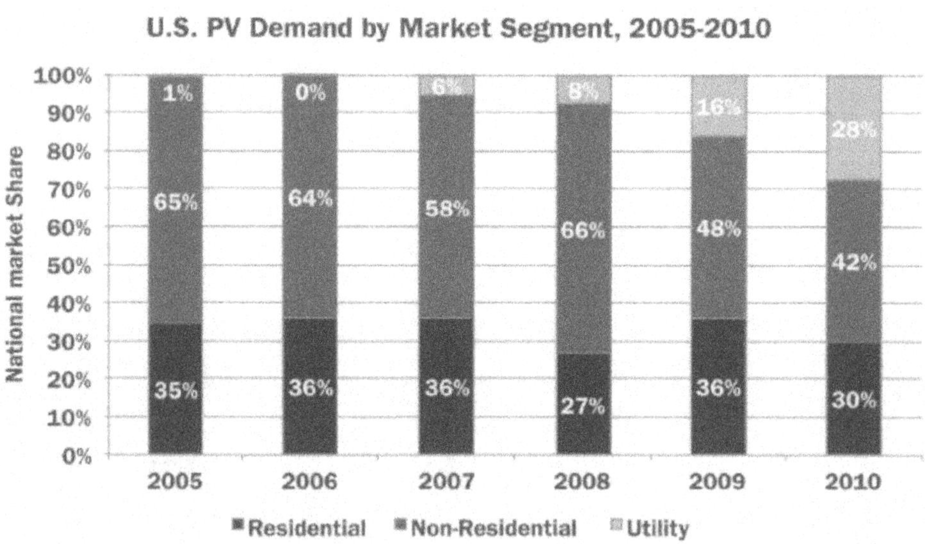

Figure 11-22. U.S. PV Demand by Market Segment, 2005-2010 (SEIA 2010)

Combining these estimates yields a progression of installed residential and non-residential non-utility PVs that has increased to 1656 MW in 2012. This is detailed in Table 11-14.

Table 11-14. Photovoltaic Installations in MW of DC (Source: SEIA and IREC: 2010, 2011 and 2012)

	2005	2006	2007	2008	2009	2010	2011	2012
Residential	51	67	93	190	208	372	840	---
Non-Residential	27	38	58	77	157	264	310	---
Total	78	105	151	167	365	636	1150	1656

Using this sequence of data, EPRI has conducted an analysis using regression to forecast possible penetration through 2030. Several models have been created. However, it was decided to only choose a simple model of straight-line regression. The model and actual data are illustrated in Figure 11-7. The results conclude that over 5000 megawatts of consumer PV could be installed on rooftops in the year 2030 and that total installed PV could total 59.6 GW (see Figure 11-23).

With a total installed capacity of 59.6 MW in 2030, PV production would offset 71.6 x 109 kWh in sales.

These projections are not unlike those made by others. For example, the Institute for Local Self-Reliance (ILSR) published an estimate of the potential installed solar photovoltaic capacity that could result if grid parity were achieved. The ILSR estimate suggests that, by 2022, there is 190,000 megawatts of residential and 122,000 megawatts of commercial solar power possible.

AEO2012 estimates that 11 gigawatts of non-utility solar PV will be added from 2010 to 2035, with approximately 4.6 MW installed by 2030. AEO believes that the growth of solar capacity is concentrated predominantly in the decade 2025 to 2035, when the technology becomes more cost-competitive.

Assuming a 20% capacity factor, 4.6 MW of customer PV would produce 8059 MWh annually.

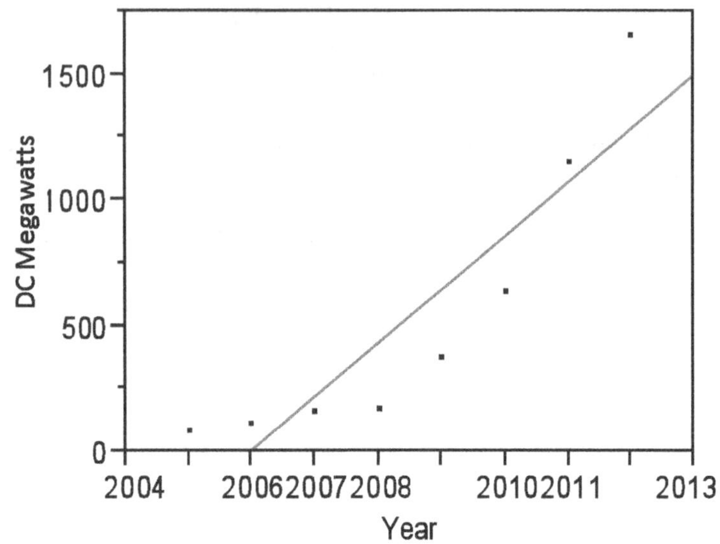

Figure 11-23. PV Adoption Model

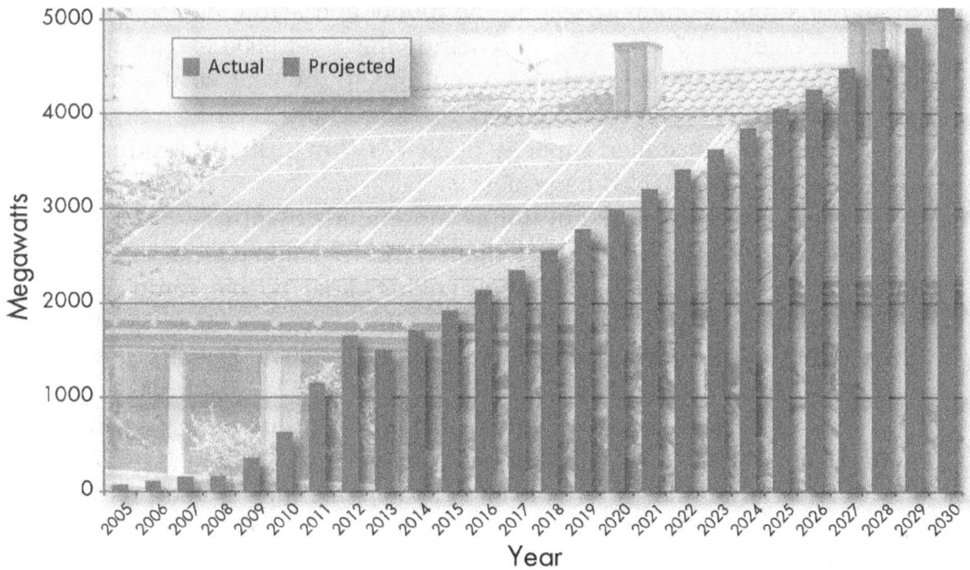

Figure 11-24. Annual and Projected Consumer Photovoltaic Installations—MW DC

Distributors in the Photovoltaic Business

Emerging PV technology brings significant opportunities for many stakeholders, including electric utilities, electric customers, energy-service providers, and PV equipment vendors. The opportunities for utilities range from owning and deploying various PV generation resources and related products, to incentivizing other owners to install PV systems and technology that provide benefits to the power system (EPRI 1018097).

Accelerating PV deployment is paralleled by continued diversification in grid-connected applications. In addition to small rooftop installations on individual homes, distributed PV is in "solar subdivisions" and supplies large commercial office buildings. Solar powered appliances include PV-direct applications such as lights, pumps, and fans. Other PV applications that involve energy storage and controllable power electronic interfaces can deliver services to the electric grid. All of these applications introduce new dimensions to utility business strategies, grid operations, and grid planning.

Going forward, distributed PV likely will bring about changes in consumer behavior, affect markets served by conventional electricity infrastructure, impose new demands on the grid, and create business opportunities. Utilities, distribution companies, and other industry participants can choose one or both of two PV strategies. First, they must respond to interconnection requests, manage grid impacts, and react as the future unfolds. Second, they can also anticipate emerging applications, move into new markets, and work with regulators and other stakeholders to derive economic benefit while helping maximize distributed PV's societal value (EPRI 1018097).

Figure 11-25 illustrates existing and emerging feeder-level PV applications.

Drivers for the Expansion of PV

Today, airlines, railroads, utilities, oil and gas developers, and other industries extensively use equipment leasing as a way to finance expensive equipment. Indeed, equipment leases and similar instruments account for more than $600 billion of business each year. Now equipment financing has become common in the solar energy sector. Such financing systems make solar energy more accessible, allowing customers across a broader range of income levels to buy solar electricity without taking on the upfront cost of purchasing the equipment (Lowman and Medina, 2013).

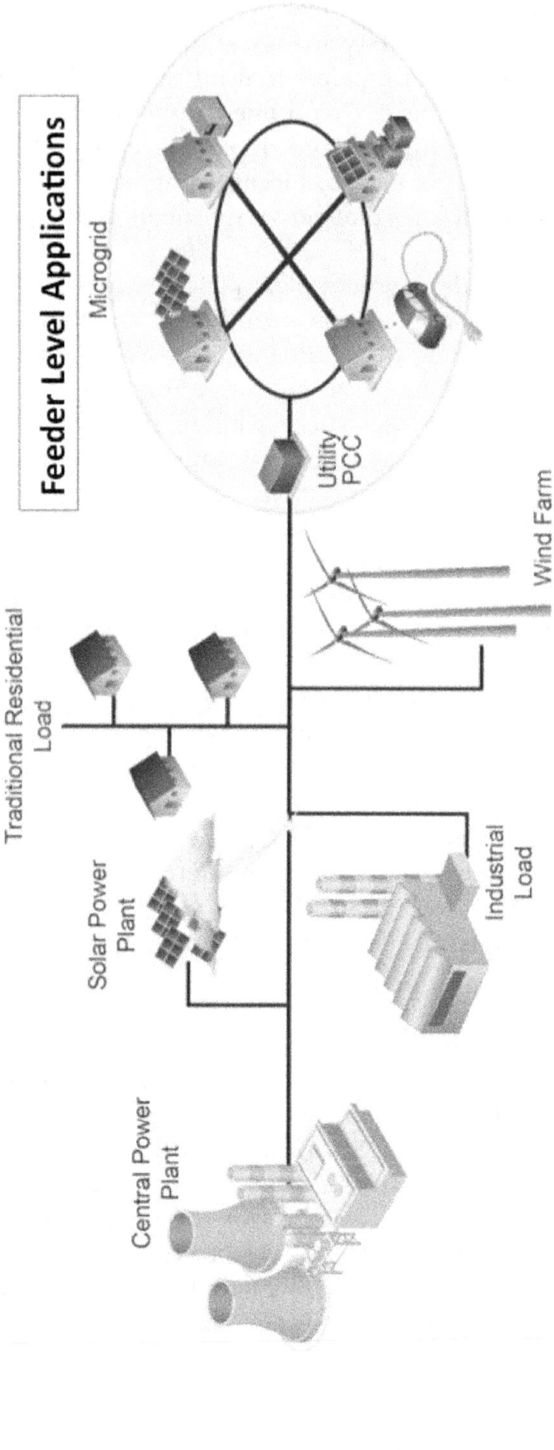

Figure 11-25. Existing and Emerging Feeder-level Applications

Solar project owners are eligible for the Section 48 Investment Tax Credit (ITC), a 30% tax credit expanded in the Energy Act of 2005 to encourage the domestic production of renewable energy. Solar developers are relying on the precedents provided by the equipment leasing industry and the ITC to create financing structures in order to develop new business.

Because solar companies cannot always use all the tax benefits available, solar developers seek out investors who can efficiently use the tax benefits to provide the financing at a lower cost than would be the case with direct borrowing. This low tax method of financing means that more projects can be built at a lower cost to the host customers. This is virtually the same as equipment rental companies financing equipment for end customers through a lease.

This type of equipment financing has made solar power more accessible to customers who otherwise might not have considered it or been able to afford it. A June 2012 report from the California Public Utilities Commission reported that a significant increase in solar participation from customers in lower and medium income areas has coincided with a dramatic rise in third-party-owned systems. The report noted a 364% increase in applications since program inception from households in zip codes with median incomes of less than $50,000, and a 445% increase in applications from households in zip codes with median incomes between $50,000 and $75,000, and attributes this growth to "a sharp increase in third-party-owned systems, such as solar leases and power purchase agreements (PPAs), [which] allow households who cannot afford to own a PV system to go solar."

Combined Heat and Power (CHP)

In this context, CHP includes both the customer-owned and third-party-owned cogeneration installed on customer premises, as well as independent customer generation with heat recovery.

CHP has always been of interest to medium- to large-size industrial concerns that have a significant demand for thermal energy (steam or hot water) and that require significant electric energy. CHP installations were typically sized to run the equipment to meet thermal demand and to utilize the electricity production as a byproduct, while typically relying on grid resources to both balance the residual electrical demand and to provide reliability.

However, as power quality became increasingly important, some industries felt that they could be assured of more consistent digital-grade power by supplementing these power supplies with digital-grade power supplied through various topologies that included self-generation.

Because most CHP installation decisions have always been based, in part, on economics, today's low natural gas prices may well stimulate additional growth in CHP. This coupled with government incentives and technology development could result in further penetration of CHP. In fact, Pike Research, a part of Navigant's Energy Practice, estimates that the market for CHP in commercial buildings alone will grow from $2.2 billion in 2012 to $11.2 billion in 2022 (www.pikeresearch.com/newsroom).

One somewhat recently added incentive is the current interest in assuring resiliency of power supply. Growing out of the storms that crippled the Northeastern U.S. in the end of 2011 and 2012, there is now increased interest in stimulating and enabling more microgrids and self-generation. Proponents suggest that proliferation of these distribution systems could potentially provide greater flexibility and recovery from otherwise wide-scale outages from super storms.

Current estimates indicate that CHP capacity is between 76 and 82 GW. Oak Ridge National Laboratory (ORNL) estimates that CHP installations total 82 GW. This is based on usually favorable regulatory treatment or incentives combined with higher retail electric rates; higher current activity in infrastructure applications such as universities, hospitals, and wastewater treatment; and industrial facilities using coal boilers. Table 11-15 lists the ORNL data.

Existing data sources and forecasts of CHP vary considerably. Some include both customer and utility generation and customer generation without heat recovery. In the cases where there is a forecast, the

Table 11-15. Current CHP Installations (2011) (Source: ORNL/DOE CHP Installation Database)

CHP Market	Market Data
Installed Capacity	81.7 GW (7.5% of current U.S. capacity)
Number of Installations	3,700 industrial and commercial facilities
Locations	Almost half (38 GW) is concentrated in Texas, California, Louisiana and New York

basis for the estimates is unclear. Hence, the estimates of additional CHP contained in AEO are suspect and require further study.

A 2008 ORNL study predicted the potential to increase CHP capacity from 85 GW to 241 GW in 2030 or 20% of generation. ORNL assumed that CHP would attract $234 billion in private sector investment and create nearly 1 million new jobs. The details are shown in Table 11-16.

In 2010, the U.S. Department of Energy (DOE) projected an increase in CHP through 2030 as illustrated in Figure 11-26 (Cogeneration & On-Site Power Production, 2010).

Table 11-16. Current and Forecasted CHP Installations (ORNL, 2008)

National Averages: $9.2/MBtu for commercial buildings, $5.2/MBtu for industrial facilities	2006	2030
CHP Capacity	85 GW	241 GW
Annual Fuel Savings	1.9 Quads	5.3 Quads
Total Annual CO_2 Reduction	248 MMT	848MMT
Cars Taken Off Road (equivalent)	45 million	154 million

Table 11-17. Existing CHP Capacity by Producer Type, 2010 (EIA)

Producer Type	# of Generators	Generator Nameplate Capacity (MW)	Net Summer Capacity (MW)	Net Winter Capacity (MW)
Electric Power Sector				
Electric Utilities	9,520	654,959	602,151	622,326
Independent Power Producers	5,708	407,978	370,887	385,804
Total	**15,228**	**1,062,937**	**973,038**	**1,008,130**
CHP Sector				
Electric Power[1]	628	41,613	36,250	39,129
Commercial[2]	683	2,796	2,490	2,596
Industrial[2]	1,612	31,294	27,359	28,893
Total	**2,923**	**75,702**	**66,099**	**70,618**
Total All Sectors	**18,151**	**1,138,638**	**1,039,137**	**1,078,748**

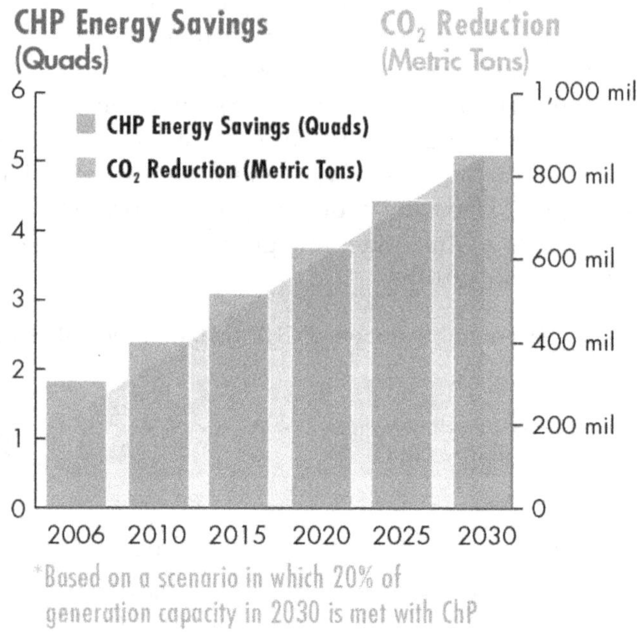

Figure 11-26. Potential Benefits of Increased CHP Capacity (U.S. DOE)

The federal government, through U.S. DOE and policies of the Department of Defense, continues to encourage the growth of combined heat and power. For example, the White House issued an executive order on August 30, 2012 titled "Accelerating Investment in Industrial Energy Efficiency" which calls for the agencies involved to "coordinate and strongly encourage efforts to achieve a national goal of deploying 40 gigawatts of new, cost effective CHP in the United States by the end of 2020."

Table 11-18 summarizes CHP estimates from various sources.

CONCLUSIONS

This section describes an attempt to sort through the "puts and takes" of each of the major markers described in preceding sections. Figure 11-27 summarizes the key "puts and takes" of changing electricity demand relative to the 2012 Annual Energy Outlook.

The results indicate that each of the "puts and takes" indicated in

Table 11-18. CHP Estimates

	2010	2015	2025	2030 (est.)	2035
AEO2012 GW Capacity*	0	13.7	28.2	38.1	47.9
AEO2012 kWh x 10^6	156	197	276	341	406
DOE CO_2 (x 10^6 tons)		450	740	850	---
DO EkWh x 10^9		875	1439	1653	
DOE kWh x 10^9 > *AEO2012*		678	1163	1312	

*Includes combined heat and power plants and electricity-only plants in the commercial and industrial sectors. Includes small on-site generating systems in the residential, commercial, and industrial sectors used primarily for own-use generation, but which may also sell some power to the grid. Excludes off-grid photovoltaics and other generators not connected to the distribution or transmission systems.

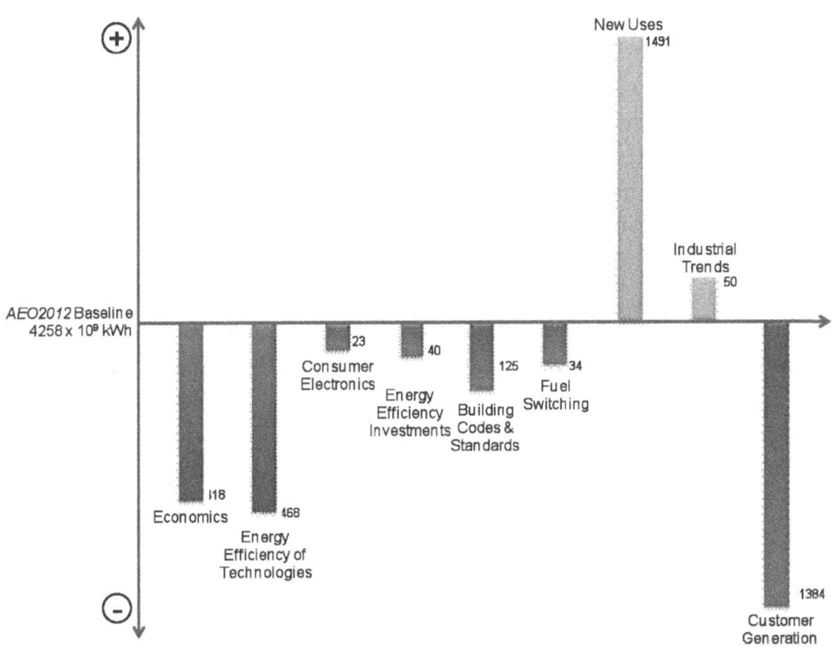

Figure 11-27. Changing Electricity Demand (Year 2030)

the figure can have a substantial impact on the resulting demand for electricity. In each case, there is substantial uncertainty in the estimates provided. EPRI has identified each of these variables as key markers in tracking the demand for electricity from grid-related services. In this report, EPRI uses the Annual Energy Outlook (AEO) 2012 as the base case.

Leveraging the Smart Grid

Each of the factors described in this chapter can be viewed as a "threat" to traditional business of the distribution utility. However, they can also be viewed as an opportunity to the distributor. The distributor, the electric distribution utility or "load-serving entity" as defined by the North American Electric Reliability Corporation (NERC) will potentially face the greatest technology needs and business environment changes of all of the industry stakeholders. The future of the distributor will be impacted by the demand for electricity, as previously outlined as well as a number of other factors, some driven by changes in demand, others by related changes in technology and weather patterns.

In addition, information technology (IT) is becoming an increasingly integral part of both transmission and distribution owners and operators. However, it is in the distribution system that the implications of IT will be most profound. In tomorrow's distribution system, the proliferation of sensors, controls, intelligent electronic devices (IEDs), and the resultant "marriage" of customers to the utility system will lead to an increased complexity in distribution system operations.

As efficiency accelerates, distributed generation proliferates, and demand response and new electric technologies get added to the mix, the distribution utility will need technologies to facilitate an increasingly complex system. As such, the concept of distribution management as we know it now will need to evolve. With the appropriate business models in place, the distribution system can become an enabler allowing seamless integration with consumers and their devices and enabling complex and two-way flow of energy.

Numerous research implications follow including the need to reduce the cost of technologies to help the industry prepare for integration. In addition, these complexities will necessitate a far more complex distribution control system incorporating advanced distribution simulation.

The concept of the virtual power plant, shunned by many distribution utilities, may become the means by which regulated utilities can sustain today's business model as part of a no-regrets strategy. Several utilities involved in EPRI's Smart Grid demonstrations are evaluating the feasibility of viewing segments of their distribution systems as virtual utilities. In the virtual power plant, efficiency, electrification and demand response become part of the no-regrets strategy.

In the distributor's business is social networking and related social gaming programs like Zinga show signs of how consumers may engage with providers, including electric utilities in the future. Active management of energy by residential and commercial consumers using devices like the NEST thermostat can further reduce electricity sales by distributors and make the distribution system more complex to operate. Understanding customer preferences for energy services and their behavior regarding purchase and use of those appliances as well as utility programs and initiatives will be critical for survival.

A key takeaway is that distribution landscape is a large part of where the technology action is in the future. Certain investments will need to continue such as in new distribution facilities including transformers to handle load growth—but, will a deteriorating load factor these investments signal low returns for investor-owned utilities.

State of New York Reforming the Energy Vision ("REV")

In partial response to the substantial changes in retail electricity markets as summarized in this chapter, the New York State Department of Public Service ("DPS") issued an order in April 2014 instituting a proceeding called Reforming the Energy Vision ("REV") aiming to reorient both the electric industry and the ratemaking paradigm surrounding it toward a consumer-centered approach that harnesses technology and markets. In the DPS' view, distributed energy resources (DER) will be integrated into the planning and operation of electric distribution systems, to achieve optimal system efficiencies, secure universal, affordable service, and enable the development of a resilient, climate-friendly energy system. They referred to "DER" as a potentially wide variety of distributed energy resources, including end-use energy efficiency, demand response, distributed storage, and distributed generation. The assumption by DPS is that DER will principally be located on customer premises, but may also be located on distribution system facilities. (DPS 2014)

The goals of REV are expected to occur over a period of years through the mutual efforts of industry, customers, non-governmental advocates, and regulatory partners. The April 2014 Order stated six objectives for the current initiative:

1. Enhanced customer knowledge and tools that will support effective management of the total energy bill;
2. Market animation and leverage of customer contributions;
3. System wide efficiency;
4. Fuel and resource diversity;
5. System reliability and resiliency; and
6. Reduction of carbon emissions.

The Commission's actions are formed around two tracks: one which focuses on developing distributed resource markets; and one focused on reforming utility ratemaking practices.

REV hopes to establish markets so that customers and third parties can be active participants, to achieve dynamic load management on a system wide scale, resulting in a more efficient and secure electric system including better utilization of bulk generation and transmission resources. As a result of this market animation, distributed energy resources will become integral tools in the planning, management and operation of the electric system. The system values of distributed resources will be monetized in a market, placing DER on a competitive par with centralized options. Customers, by exercising choices within an improved electricity pricing structure and vibrant market will create new value opportunities and at the same time drive system efficiencies and help to create a more cost-effective and secure integrated grid. The more efficient system will be designed and operated to make optimal use of cleaner and more efficient generation technologies. Enabling these markets will require modernization of infrastructure and operations, particularly communication and data management capabilities.

In their view, the reformed electric system will be driven by consumers and non-utility providers, and it will be enabled by utilities acting as Distributed System Platform (DSP) providers. Utilities are responsible for reliability, and the functions needed to enable distributed markets are integrally bound to the functions needed to ensure reliability. Technology innovators and third party aggregators will develop products and services that enable full customer engagement. The utilities acting in concert will constitute a statewide platform that will provide uniform market access to customers and DER providers. Each utility will serve as the platform for interface among its customers, aggregators, and the distribution system. Simultaneously the utility will serve as a seamless interface between aggregated customers and the New York Independent System Operator (NYISO.)

Reforming the Commission's ratemaking practices will be critical to the success of the REV vision. The objective of rate reform is to enable utility earnings to become more dependent on creating value for customers and achieving policy objectives. Rather than simply building infrastructure, utilities could find earning opportunities in enhanced performance and in transactional revenues.

Key to REV is embracing utilities as Distribution Service Providers (DSPs) under the PSC regulatory authority and supervision. Having the utility expand its responsibility to include DSP functionality enhances the opportunity for integrated operation of the distribution system and for realizing the economic value of DER investment. Utilities' behavior with respect to DER has been responsive to the regulatory structure under which they have operated. Reforming the regulatory model, and by extension utility behavior, is a critical component of the REV initiative. Unlike the wholesale market, the markets that will be enabled and potentially operated by the DSP will not establish commodity prices. Commodity prices, the prices for capacity, energy and bulk ancillary services will be set by the NYISO. If DSPs are failing to meet the objectives of REV, the Commission will consider options to allow other entities to serve that function.

References

- Annual Household Income: U.S. Statistical Abstract (U.S. Census Bureau), http://www.census.gov/compendia/statab.
- Annual Household Income Trends: Household Income Trends: Sentier Research, http://www.sentierresearch.com.
- Baseline Economic Assumptions for Baseline Electricity Demand Forecast: U.S. Energy Information Administration Annual Energy Outlook, www.eia.gov/forecasts/aeo/assumptions/pdf/introduction/pdf.
- The Price of Electricity: U.S. EIA "Monthly Electricity Sales and Revenue Report with State Distributions Report," http://www.eia.gov/electricity/data.cfm.
- Spot Prices of Natural Gas: Natural Gas Weekly Update—EIA, www.eia.gov/naturalgas/weekly.
- Retail Price of Natural Gas: Natural Gas Monthly (NGM)—Energy Information Administration, www.eia.gov/naturalgas/monthly.
- Ground Source Heat Pumps: Overview of Market Status, http://www:eere.energy.gov/geothermal/pdfs/gshp_overview.pdfHome.
- Appliance Sales—Association of Home Appliance Manufacturers (AHAM), http://

www.aham.org/.

- Consumer Electronics Market Trends—Consumer Electronics Association (CEA), http://www.ce.org/.
- Electric Vehicle Market Penetration, http://greenautoblog.com.
- Air-Conditioning, Heating and Refrigeration Institute, http://www.ahrinet.org/.
- Consumption and Efficiency—U.S. Energy Information Agency (EIA), http://www.eia.gov/consumption/.
- Assessment of Achievable Potential from Energy Efficiency and Demand Response Programs in the U.S.: (2010-2030). EPRI, Palo Alto, CA: 2009. 1016987.
- Energy Consumption of Consumer Electronics in U.S. Homes in 2010—Final Report to the Consumer Electronics Association (CEA), B. Urban, V. Tiefenbeck, and K. Roth, Franhofer USA, December 2011.
- Highlights of the 2013 International Consumers Electronics Show—Technologies Related to Residential Ecosystems, EPRI, Palo Alto, CA: 2013. 3002000202.
- Energy Consumption of Consumer Electronics in U.S. Homes in 2010, Fraunhofer USA, Final Report to the Consumer Electronics Association (CEA), December 2011.
- TV Energy Consumption Trends and Energy-Efficiency Improvement Options, W.Y. Park, et al., Lawrence Berkeley National Laboratory, LBNL-5024E, July 2011.
- Data Base of State Incentives for Renewables & Efficiency, U.S. DOE Energy Efficiency and Renewable Energy.
- U.S. and Canadian Efficiency Program Budgets 2007-2011, Consortium for Energy Efficiency.
- Federal Energy Regulatory Commission (FERC), Demand Response Assessment.
- U.S. DOE Multi-Year Schedule of Energy Efficiency Standards Updates, http://www1.eere.energy.gov/buildings/appliance_standards/pdfs/multiyear_schedule_aug_2011.pdf.
- Alliance to Save Energy (periodic policy commentary and analysis), http://ase.org/.
- ACEEE (periodic policy commentary and analysis), http://aceee.org/.
- Edison Foundation Institute for Electric Efficiency (periodic policy commentary and analysis) http://www.edisonfoundation.net/iee/.
- The Plug-In Electric Vehicle Market: Current Status and Long-Term Outlook. EPRI, Palo Alto, CA: 2012. 1024103.
- The Potential to Reduce CO_2 Emissions by Expanding End-Use Applications of Electricity—Executive Summary. EPRI, Palo Alto, CA: 2009. 1018906.
- Distributed Generation Penetration, http://www.nrel.gov/doc/fy03osti/34715.pdf.
- U.S. Solar Market InsightTM: 2010 Year-in-Review—Executive Summary, and 2011 Year-in-Review, Solar Energy Industries Association (SEIA) and GTM Research, 2010, www.gtmresearch.com/solarinsight.
- Reforming the Energy Vision, New York State Department of Public Service (DPS) Staff Report and Proposal, Case 14-M-0101: Proceeding on Motion of the Commission in Regard to Reforming the Energy Vision, 4/24/2014

Index

P

R

S

For Product Safety Concerns and Information please contact our EU representative GPSR@taylorandfrancis.com
Taylor & Francis Verlag GmbH, Kaufingerstraße 24, 80331 München, Germany

www.ingramcontent.com/pod-product-compliance
Lightning Source LLC
LaVergne TN
LVHW011101300926
849160LV00008B/245
* 9 7 8 8 7 7 0 0 4 4 8 3 7 *